Animal Physiology and Morphology

Animal Physiology and Morphology

Edited by
Quinn Coyle

Larsen & Keller
www.larsen-keller.com

Animal Physiology and Morphology
Edited by Quinn Coyle
ISBN: 978-1-63549-027-5 (Hardback)

Larsen & Keller
Published by Larsen and Keller Education,
5 Penn Plaza,
19th Floor,
New York, NY 10001, USA

Cataloging-in-Publication Data

Animal physiology and morphology / edited by Quinn Coyle.
p. cm.
Includes bibliographical references and index.
ISBN 978-1-63549-027-5
1. Physiology. 2. Morphology (Animals). I. Coyle, Quinn.
QP31.2 .A55 2017
612--dc23

Printed and bound in China.

For more information regarding Larsen and Keller Education and its products, please visit the publisher's website www.larsen-keller.com

Table of Contents

Preface

Animal physiology studies the intricate systems that regulate biological life in animals while morphology studies the physical structure of various animals and forms the basis of taxonomical classification. These sciences shed light into the evolutionary and ecological processes that animals undergo. This text attempts to understand the multiple branches that fall under these disciplines and how such concepts have practical applications. The topics included in this book on animal physiology and morphology are of utmost significance and are bound to provide incredible insights to readers. This textbook, with its detailed analyses and data, will prove immensely beneficial to students involved in this area at various levels.

Given below is the chapter wise description of the book:

Chapter 1- Physiology is the study of the functions of living organisms. This field is very large and can be divided into animal physiology, plant physiology, bacterial physiology and viral physiology. This chapter will provide an integrated understanding of animal physiology and morphology.

Chapter 2- Animal locomotion is the movement of animals from one place to another. Some forms of locomotion are running, jumping, flying and hopping. Role of skin in locomotion, arboreal locomotion, terrestrial locomotion, aquatic locomotion and fin and flipper locomotion are some important topics related to animal locomotion.

Chapter 3- Fish physiology is the study of the physiological functions of a fish and how the parts of fishes function together. There is a difference between fish physiology and fish anatomy. Fish jaw, swim bladder and fish fin are some of the features of fish physiology. The major aspects of fish physiology are discussed in this chapter.

Chapter 4- Insect physiology studies the physiology of an insect's biological system. Insects are usually made up of three main body regions, the head, thorax and abdomen. Insect flight, insect olfaction, pupa and insect reproductive system are also explained in the following section.

Chapter 5- Anatomy studies the physical structure of organisms and their composition. Comparative anatomy studies the similarities and differences that are found in different species. The topics elucidated in the section are homology, convergent evolution, connective tissue, muscle tissue, nervous tissue etc. This section on anatomy offers an insightful focus, keeping in mind the subject matter.

Chapter 6- Vertebrates in fishes include the jawless fish and also the jawed vertebrates. Invertebrates, neuraxis and limbless vertebrates are also explained in the chapter. The chapter discusses the anatomy of vertebrate and invertebrates in a critical manner providing key analysis to the subject matter.

Chapter 7- One of the main modes of locomotion in animals is flight. Flight is used as the primary mode of locomotion by majority of the world's bird species. A keel is a part of the bird's structure; it runs axially along the midline of the sternum and helps in providing an adequate leverage for flight. The aspects explained in the section provide a better understanding of flight in animals.

At the end, I would like to thank all those who dedicated their time and efforts for the successful completion of this book. I also wish to convey my gratitude towards my friends and family who supported me at every step.

Editor

1

Introduction to Animal Physiology and Morphology

Physiology is the study of the functions of living organisms. This field is very large and can be divided into animal physiology, plant physiology, bacterial physiology and viral physiology. This chapter will provide an integrated understanding of animal physiology and morphology.

Physiology

Physiology is the scientific study of the normal function in living systems. A sub-discipline of biology, its focus is in how organisms, organ systems, organs, cells, and biomolecules carry out the chemical or physical functions that exist in a living system. Given the size of the field, it is divided into, among others, animal physiology (including that of humans), plant physiology, cellular physiology, microbial physiology (microbial metabolism), bacterial physiology, and viral physiology. The Nobel Prize in Physiology or Medicine is awarded to those who make significant achievements in this discipline by the Royal Swedish Academy of Sciences. In medicine, a physiologic state is one occurring from normal body function, rather than pathologically, which is centered on the abnormalities that occur in animal diseases, including humans.

Oil painting depicting Claude Bernard, the father of modern physiology, with his pupils

History

Physiological studies date back to ancient civilizations of India and Egypt alongside anatomical studies, but did not utilize dissection or vivisection.

The study of human physiology as a medical field dates back to at least 420 BC to the time of Hippocrates, also known as the "father of medicine." Hippocrates incorporated his belief system called the theory of humours, which consisted of four basic substance: earth, water, air and fire. Each substance is known for having a corresponding humour: black bile, phlegm, blood and yellow bile, respectively. Hippocrates also noted some emotional connections to the four humours, which Claudis Galenus would later expand on. The critical thinking of Aristotle and his emphasis on the relationship between structure and function marked the beginning of physiology in Ancient Greece. Like Hippocrates, Aristotle took to the humoral theory of disease, which also consisted of four primary qualities in life: hot, cold, wet and dry. Claudius Galenus (c. ~130–200 AD), known as Galen of Pergamum, was the first to use experiments to probe the functions of the body. Unlike Hippocrates though, Galen argued that humoral imbalances can be located in specific organs, including the entire body. His modification of this theory better equipped doctors to make more precise diagnoses. Galen also played off of Hippocrates idea that emotions were also tied to the humours, and added the notion of temperaments: sanguine corresponds with blood; phlegmatic is tied to phlegm; yellow bile is connected to choleric; and black bile corresponds with melancholy. Galen also saw the human body consisting of three connected systems: the brain and nerves, which are responsible for thoughts and sensations; the heart and arteries, which give life; and the liver and veins, which can be attributed to nutrition and growth. Galen was also the founder of experimental physiology. And for the next 1,400 years, Galenic physiology was a powerful and influential tool in medicine.

Jean Fernel (1497–1558), a French physician, introduced the term "physiology".

In 1858, Joseph Lister studied the cause of blood coagulation and inflammation that resulted after previous injuries and surgical wounds. He later discovered and implemented antiseptics in the operating room, and as a result decreases death rate from surgery by a substantial amount.

In 1891, Ivan Pavlov performed research on "conditional reflexes" that involved dogs' saliva production in response to a plethora of sounds and visual stimuli.

In the 19th century, physiological knowledge began to accumulate at a rapid rate, in particular with the 1838 appearance of the Cell theory of Matthias Schleiden and Theodor Schwann. It radically stated that organisms are made up of units called cells. Claude Bernard's (1813–1878) further discoveries ultimately led to his concept of *milieu interieur* (internal environment), which would later be taken up and championed as "homeostasis" by American physiologist Walter B. Cannon in 1929. By homeostasis, Cannon meant "the maintenance of steady states in the body and the physiological processes through which they are regulated." In other words, the body's ability to regulate its internal environment. It should be noted that, William Beaumont was the first American to utilize the practical application of physiology.

The Physiological Society was founded in London in 1876 as a dining club The American Physiological Society (APS) is a nonprofit devoted to fostering education, scientific research, and dissemination of information in the physiological sciences. The Society was founded in 1887 with 28 members.

In the 20th century, biologists became interested in how organisms other than human beings function, eventually spawning the fields of comparative physiology and ecophysiology. Major figures in these fields include Knut Schmidt-Nielsen and George Bartholomew. Most recently, evolutionary physiology has become a distinct subdiscipline.

In 1920, August Krogh won the Nobel Prize for discovering how, in capillaries, blood flow is regulated.

In 1954, Andrew Huxley and Hugh Huxley, alongside their research team, discovered the sliding filaments in skeletal muscle, known today as the sliding filament theory.

Women in Physiology

Initially, women were largely excluded from official involvement in any physiological society. The American Physiological Society, for example, was founded in 1887 and included only men in its ranks. In 1902, the American Physiological Society elected Ida Hyde as the first female member of the society. Hyde, a representative of the American Association of University Women and a global advocate for gender equality in education, attempted to promote gender equality in every aspect of science and medicine.

Soon thereafter, in 1913, J.S. Haldane proposed that women be allowed to formally join The Physiological Society, which had been founded in 1876. On 3 July 1915, six women were officially admitted: Florence Buchanan, Winifred Cullis, Ruth C. Skelton, Sarah C. M. Sowton, Constance Leetham Terry, and Enid M. Tribe. The centenary of the election of women was celebrated in 2015 with the publication of a book "Women physiologists: centenary celebrations and beyond for The Physiological Society ISBN 978-0-9933410-0-7.

Prominent women physiologists include:

- Gerty Cori, along with husband Carl Cori, received the Nobel Prize in Physiology or Medicine in 1947 for their discovery of the phosphate-containing form of glucose known as glycogen, as well as its function within eukaryotic metabolic mechanisms for energy production. Moreover, they discovered the Cori cycle, also known as the Lactic acid cycle, which describes how muscle tissue converts glycogen into lactic acid via lactic acid fermentation.
- Barbara McClintock was rewarded the 1983 Nobel Prize in Physiology or Medicine for the discovery of genetic transposition McClintock is the only female recipient who has won an unshared Nobel Prize.
- Gertrude Elion, along with George Hitchings and Sir James Black, received the Nobel Prize for Physiology or Medicine in 1988 for their development of drugs employed in the treatment of several major diseases, such as leukemia, some autoimmune disorders, gout, malaria, and viral herpes.
- Linda B. Buck, along with Richard Axel, received the Nobel Prize in Physiology or Medicine in 2004 for their discovery of odorant receptors and the complex organization of the olfactory system.
- Françoise Barré-Sinoussi, along with Luc Montagnier, received the Nobel Prize in Physiology or Medicine in 2008 for their work on the identification of the Human Immunodeficiency Virus (HIV), the cause of Acquired Immunodeficiency Syndrome (AIDS).
- Elizabeth Blackburn, along with Carol W. Greider and Jack W. Szostak, was awarded the 2009 Nobel Prize for Physiology or Medicine for the discovery of the genetic composition and function of telomeres and the enzyme called telomerase.

Subdisciplines

There are many ways to categorize the subdiscplines of physiology:

- based on the taxa studied: human physiology, animal physiology, plant physiology, microbial physiology, viral physiology
- based on the level of organization: cell physiology, molecular physiology, systems physiology, organismal physiology, ecological physiology, integrative physiology
- based on the process that causes physiological variation: developmental physiology, environmental physiology, evolutionary physiology
- based on the ultimate goals of the research: applied physiology (e.g., medical physiology), non-applied (e.g., comparative physiology)

Human Physiology

Human physiology seeks to understand the mechanisms that work to keep the human body alive and functioning, through scientific enquiry into the nature of mechanical, physical, and biochemical functions of humans, their organs, and the cells of which they are composed. The principal level of focus of physiology is at the level of organs and systems within systems. The endocrine and nervous systems play major roles in the reception and transmission of signals that integrate function in animals. Homeostasis is a major aspect with regard to such interactions within plants as well as animals. The biological basis of the study of physiology, integration refers to the overlap of many functions of the systems of the human body, as well as its accompanied form. It is achieved through communication that occurs in a variety of ways, both electrical and chemical.

Much of the foundation of knowledge in human physiology was provided by animal experimentation. Physiology is the study of function and is closely related to anatomy which is the study of form and structure. Due to the frequent connection between form and function, physiology and anatomy are intrinsically linked and are studied in tandem as part of a medical curriculum.

Morphology (Biology)

Morphology of a male Caprella mutica

Morphology is a branch of biology dealing with the study of the form and structure of organisms and their specific structural features.

This includes aspects of the outward appearance (shape, structure, colour, pattern, size), i.e. **external** morphology (or eidonomy), as well as the form and structure of the internal parts like bones and organs, i.e. internal morphology (or anatomy). This is in contrast to physiology, which deals primarily with function. Morphology is a branch of life science dealing with the study of gross structure of an organism or taxon and its component parts.

Morphology

The word "morphology" is from the Ancient Greek *don, morphé*, meaning "form", and *lógos*, meaning "word, study, research". The biological concept of morphology was developed by Johann Wolfgang von Goethe (1790) and independently by the German anatomist and physiologist Karl Friedrich Burdach (1800).

Among other important theorists of morphology are Lorenz Oken, Georges Cuvier, Étienne Geoffroy Saint-Hilaire, Richard Owen, Karl Gegenbaur and Ernst Haeckel.

In English-speaking countries, the term "molecular morphology" has been used for some time for describing the structure of compound molecules, such as polymers and RNA. The term "gross morphology" refers to the collective structures or an organism as a whole as a general description of the form and structure of an organism, taking into account all of its structures without specifying an individual structure.

Divisions of Morphology

- Comparative Morphology is analysis of the patterns of the locus of structures within the body plan of an organism, and forms the basis of taxonomical categorization.
- Functional Morphology is the study of the relationship between the structure and function of morphological features.
- Experimental Morphology is the study of the effects of external factors upon the morphology of organisms under experimental conditions, such as the effect of genetic mutation.
- "Anatomy" is a "branch of morphology that deals with the structure of organisms".

Morphology and Classification

Most taxa differ morphologically from other taxa. Typically, closely related taxa differ much less than more distantly related ones, but there are exceptions to this. Cryptic species are species which look very similar, or perhaps even outwardly identical, but are reproductively isolated. Conversely, sometimes unrelated taxa acquire a similar appearance as a result of convergent evolution or even mimicry. In addition, there can be morphological differences within a species, such as in *Apoica flavissima* where queens are significantly smaller than workers. A further problem with relying on morphological data is that what may appear, morphologically speaking, to be two distinct species, may in fact be shown by DNA analysis to be a single species. The significance of these differences can be examined through the use of allometric engineering in which one or both species are manipulated to phenocopy the other species.

3D Cell Morphology: Classification

Invention and development of microscopy enable the observation of 3-D cell morphology with both high spatial and temporal resolution. The dynamic processes of these cell morphology which are controlled by a complex system play an important role in varied important biological process, such as immune and invasive responses.

References

- Prosser, C. Ladd (1991). Comparative Animal Physiology, Environmental and Metabolic Animal Physiology (4th ed.). Hoboken, NJ: Wiley-Liss. pp. 1–12. ISBN 0-471-85767-X.
- Hall, John (2011). Guyton and Hall textbook of medical physiology (12th ed.). Philadelphia, Pa.: Saunders/Elsevier. p. 3. ISBN 978-1-4160-4574-8.
- Feder, ME; Bennett, AF; WW, Burggren; Huey, RB (1987). New directions in ecological physiology. New York: Cambridge University Press. ISBN 978-0-521-34938-3.
- Mägdefrau, Karl (1992). Geschichte der Botanik [History of Botany] (2 ed.). Jena: Gustav Fischer Verlag. ISBN 3-437-20489-0.

2

Understanding Animal Locomotion

Animal locomotion is the movement of animals from one place to another. Some forms of locomotion are running, jumping, flying and hopping. Role of skin in locomotion, arboreal locomotion, terrestrial locomotion, aquatic locomotion and fin and flipper locomotion are some important topics related to animal locomotion.

Animal Locomotion

Animal locomotion, in ethology, is any of a variety of movements or methods that animals use to move from one place to another. Some modes of locomotion are (initially) self-propelled, e.g., running, swimming, jumping, flying, hopping, soaring and gliding. There are also many animal species that depend on their environment for transportation, a type of mobility called passive locomotion, e.g., sailing (some jellyfish), kiting (spiders) and rolling (some beetles and spiders).

A cheetah chasing prey. In its brief hunting sprints, it is the fastest of all land animals.

Animals move for a variety of reasons, such as to find food, a mate, a suitable microhabitat, or to escape predators. For many animals, the ability to move is essential for survival and, as a result, natural selection has shaped the locomotion methods and mechanisms used by moving organisms. For example, migratory animals that travel vast distances (such as the Arctic tern) typically have a locomotion mechanism that costs very little energy per unit distance, whereas non-migratory animals that must frequently move quickly to escape predators are likely to have energetically costly, but very fast, locomotion.

A bee in flight

Etymology

The term "locomotion" is formed in English from Latin *loco* "from a place" (ablative of *locus* "place") + *motio* "motion, a moving".

Locomotion in Different Media

Animals move through, or on, four types of environment: aquatic (in or on water), terrestrial (on ground or other surface, including arboreal, or tree-dwelling), fossorial (underground), and aerial (in the air). Many animals—for example semi-aquatic animals, and diving birds—regularly move through more than one type of medium. In some cases, the surface they move on facilitates their method of locomotion.

Aquatic

Swimming

Dolphins surfing at Snapper Rocks, Queensland, Australia

In water, staying afloat is possible using buoyancy. If an animal's body is less dense than water, it can stay afloat. This requires little energy to maintain a vertical position, but requires more energy for locomotion in the horizontal plane compared to less buoyant animals. The drag encountered in water is much greater than in air. Morphology is therefore important for efficient locomotion, which is in most cases essential for basic functions such as catching prey. A fusiform, torpedo-like body form is seen in many aquatic animals, though the mechanisms they use for locomotion are diverse.

The primary means by which fish and marine mammals generate thrust is by oscillating the body, from side to side in fish, up and down in marine mammals, the resulting wave motion ending at a large tail fin or flippers. Finer control, such as for slow movements, is often achieved with thrust from pectoral fins (or front limbs in marine mammals). Some fish propel themselves without a wave motion of the body, as in the slow-moving seahorses and *Gymnotus*. Other animals, such as cephalopods, use jet propulsion to travel fast, taking in water then squirting it back out in an explosive burst. Other swimming animals may rely predominantly on their limbs, much as humans do when swimming. Though life on land originated from the seas, terrestrial animals have returned to an aquatic lifestyle on several occasions, such as the fully aquatic cetaceans, now very distinct from their terrestrial ancestors.

Dolphins sometimes ride on the bow waves created by boats or surf on naturally breaking waves.

Benthic

Benthic locomotion is movement by animals that live on, in, or near the bottom of aquatic environments. In the sea, many animals walk over the seabed. Echinoderms primarily use their tube feet to move about. The tube feet typically have a tip shaped like a suction pad that can create a vacuum through contraction of muscles. This, along with some stickiness from the secretion of mucus, provides adhesion. Waves of tube feet contractions and relaxations move along the adherent surface and the animal moves slowly along. Some sea urchins also use their spines for benthic locomotion.

Crabs typically walk sideways (a behaviour that gives us the word *crabwise*). This is because of the articulation of the legs, which makes a sidelong gait more efficient. However, some crabs walk forwards or backwards, including raninids, *Libinia emarginata* and *Mictyris platycheles*. Some crabs, notably the Portunidae and Matutidae, are also capable of swimming, the Portunidae especially so as their last pair of walking legs are flattened into swimming paddles.

A stomatopod, *Nannosquilla decemspinosa*, can escape by rolling itself into a self-propelled wheel and somersault backwards at a speed of 72 rpm. They can travel more than 2 m using this unusual method of locomotion.

Aquatic Surface

Velella moves by sailing.

Velella, the by-the-wind sailor, is a cnidarian with no means of propulsion other than sailing. A small rigid sail projects into the air and catches the wind. *Velella* sails always align along the direction of the wind where the sail may act as an aerofoil, so that the animals tend to sail downwind at a small angle to the wind.

While larger animals such as ducks can move on water by floating, some small animals move across it without breaking through the surface. This surface locomotion takes advantage of the surface tension of water. Animals that move in such a way include the water strider. Water striders have legs that are hydrophobic, preventing them from interfering with the structure of water. Another form of locomotion (in which the surface layer is broken) is used by the basilisk lizard.

Aerial

Active Flight

A pair of brimstone butterflies in flight. The female, above, is in fast forward flight with a small angle of attack; the male, below, is twisting his wings sharply upward to gain lift and fly up towards the female.

Gravity is the primary obstacle to flight. Because it is impossible for any organism to have a density as low as that of air, flying animals must generate enough lift to ascend and remain airborne. One way to achieve this is with wings, which when moved through the air generate an upward lift force on the animal's body. Flying animals must be very light to achieve flight, the largest living flying animals being birds of around 20 kilograms. Other structural adaptations of flying animals include reduced and redistributed body weight, fusiform shape and powerful flight muscles; there may also be physiological adaptations. Active flight has independently evolved at least four times, in the insects, pterosaurs, birds, and bats. Insects were the first taxon to evolve flight, approximately 400 million years ago (mya), followed by pterosaurs approximately 220 mya, birds approximately 160 mya, then bats about 60 mya.

Gliding

Rather than active flight, some (semi-) arboreal animals reduce their rate of falling by gliding. Gliding is heavier-than-air flight without the use of thrust; the term "volplaning" also refers to this mode of flight in animals. This mode of flight involves flying a greater distance horizontally than vertically and therefore can be distinguished from a simple descent like a parachute. Gliding has evolved on more occasions than active flight. There are examples of gliding animals in several major taxonomic classes such as the invertebrates (e.g., gliding ants), reptiles (e.g., banded flying snake), amphibians (e.g., flying frog), mammals (e.g., sugar glider, squirrel glider).

Flying fish taking off

Some aquatic animals also regularly use gliding, for example, flying fish, octopus and squid. The flights of flying fish are typically around 50 meters (160 ft), though they can use updrafts at the leading edge of waves to cover distances of up to 400 m (1,300 ft). To glide upward out of the water, a flying fish moves its tail up to 70 times per second. Several oceanic squid, such as the Pacific flying squid, leap out of the water to escape predators, an adaptation similar to that of flying fish. Smaller squids fly in shoals, and have been observed to cover distances as long as 50 m. Small fins towards the back of the mantle help stabilize the motion of flight. They exit the water by expelling water out of their funnel, indeed some squid have been observed to continue jetting water while airborne providing thrust even after leaving the water. This may make flying squid the only animals with jet-propelled aerial locomotion. The neon flying squid has been observed to glide for distances over 30 m, at speeds of up to 11.2 m/s.

Soaring

Soaring birds can maintain flight without wing flapping, using rising air currents. Many gliding birds are able to "lock" their extended wings by means of a specialized tendon. Soaring birds may alternate glides with periods of soaring in rising air. Five principal types of lift are used: thermals, ridge lift, lee waves, convergences and dynamic soaring.

Examples of soaring flight by birds are the use of:

- Thermals and convergences by raptors such as vultures
- Ridge lift by gulls near cliffs
- Wave lift by migrating birds
- Dynamic effects near the surface of the sea by albatrosses

Ballooning

Ballooning is a method of locomotion used by spiders. Certain silk-producing arthropods, mostly small or young spiders, secrete a special light-weight gossamer silk for ballooning, sometimes traveling great distances at high altitude.

Terrestrial

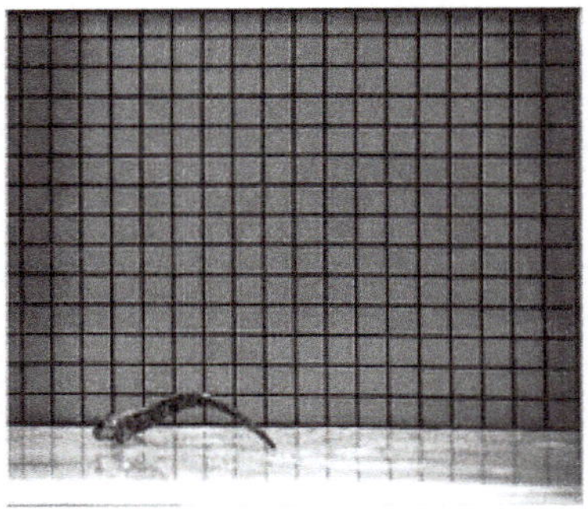

Pacific leaping blenny *Alticus arnoldorum* hopping

Pacific leaping blenny climbing up a vertical piece of Plexiglas

Forms of locomotion on land include walking, running, hopping or jumping, dragging and crawling or slithering. Here friction and buoyancy are no longer an issue, but a strong skeletal and muscular framework are required in most terrestrial animals for structural support. Each step also requires much energy to overcome inertia, and animals can store elastic potential energy in their tendons to help overcome this. Balance is also required for movement on land. Human infants learn to crawl first before they are able to stand on two feet, which requires good coordination as well as physical development. Humans are bipedal animals, standing on two feet and keeping one on the ground at all times while walking. When running, only one foot is on the ground at any one time at most, and both leave the ground briefly. At higher speeds momentum helps keep the body upright, so more energy can be used in movement.

Jumping

Jumping (saltation) can be distinguished from running, galloping, and other gaits where the entire body is temporarily airborne by the relatively long duration of the aerial phase and high angle of initial launch. Many terrestrial animals use jumping (including hopping or leaping) to escape predators or catch prey—however, relatively few animals use this as a primary mode of locomotion. Those that do include the kangaroo and other macropods, rabbit, hare, jerboa, hopping mouse, and kangaroo rat. Kangaroo rats often leap 2 m and reportedly up to 2.75 m at speeds up to almost 3 m/s (6.7 mph). They can quickly change their direction between jumps. The rapid locomotion of the banner-tailed kangaroo rat may minimize energy cost and predation risk. Its use of a "move-freeze" mode may also make it less conspicuous to nocturnal predators. Frogs are, relative to their size, the best jumpers of all vertebrates. The Australian rocket frog, *Litoria nasuta*, can leap over 2 metres (6 ft 7 in), more than fifty times its body length.

Peristalsis

Other animals move in terrestrial habitats without the aid of legs. Earthworms crawl by a peristalsis, the same rhythmic contractions that propel food through the digestive tract.

Sliding

Due to its low coefficient of friction, ice provides the opportunity for other modes of locomotion. Penguins either waddle on their feet or slide on their bellies across the snow, a movement called *tobogganing*, which conserves energy while moving quickly. Some pinnipeds perform a similar behaviour called *sledding*.

Brachiation

Some animals are specialized for moving on non-horizontal surfaces. One common habitat for such climbing animals is in trees, for example the gibbon is specialized for arboreal movement, traveling rapidly by brachiation. Another case is animals like the snow leopard living on steep rock faces such as are found in mountains. Some light animals are able to climb up smooth sheer surfaces or hang upside down by adhesion. Many insects can do this, though much larger animals such as geckos can also perform similar feats.

Walking and Running

Species have different numbers of legs resulting in large differences in locomotion.

Modern birds, though classified as tetrapods, usually have only two functional legs, which some (e.g., ostrich, emu, kiwi) use as their primary, Bipedal, mode of locomotion. Few modern mammalian species are habitual bipeds whose normal method of locomotion is two-legged. These include the macropods, kangaroo rats and mice, springhare, hopping mice, pangolins and hominінan apes. Bipedalism is rarely found outside terrestrial animals—though at least two types of octopus walk bipedally on the sea floor using two of their arms, so they can use the remaining arms to camouflage themselves as a mat of algae or floating coconut.

Bipedal ostrich

There are no three-legged animals—though some macropods, such as kangaroos, that alternate between resting their weight on their muscular tails and their two hind legs, could be looked at as an example of tripedal locomotion in animals.

Many familiar animals are quadrupedal, walking or running on four legs. A few birds use quadrupedal movement in some circumstances. For example, the shoebill sometimes uses its wings to right itself after lunging at prey. The newly hatched hoatzin bird has claws on its thumb and first finger enabling it to dexterously climb tree branches until its wings are strong enough for sustained flight. These claws are gone by the time the bird reaches adulthood.

Hexapedal stick-insect

A relatively few animals use five limbs for locomotion. Prehensile quadrupeds may use their tail to assist in locomotion and when grazing, the kangaroos and other macropods use their tail to propel themselves forward with the four legs used to maintain balance.

Insects generally walk with six legs—though some insects such as nymphalid butterflies do not use the front legs for walking.

Arachnids have eight legs. Most arachnids lack extensor muscles in the distal joints of their appendages. Spiders and whipscorpions extend their limbs hydraulically using the pressure of their hemolymph. Solifuges and some harvestmen extend their knees by the use of highly elastic thickenings in the joint cuticle. Scorpions, pseudoscorpions and some harvestmen have evolved muscles that extend two leg joints (the femur-patella and patella-tibia joints) at once.

Octopedal locomotion by a spider

The scorpion *Hadrurus arizonensis* walks by using two groups of legs (left 1, right 2, Left 3, Right 4 and Right 1, Left 2, Right 3, Left 4) in a reciprocating fashion. This alternating tetrapod coordination is used over all walking speeds.

Centipedes and millipedes have many sets of legs that move in metachronal rhythm. Some echinoderms locomote using the many tube feet on the underside of their arms. Although the tube feet resemble suction cups in appearance, the gripping action is a function of adhesive chemicals rather than suction. Other chemicals and relaxation of the ampullae allow for release from the substrate. The tube feet latch on to surfaces and move in a wave, with one arm section attaching to the surface as another releases. Some multi-armed, fast-moving starfish such as the sunflower seastar

(*Pycnopodia helianthoides*) pull themselves along with some of their arms while letting others trail behind. Other starfish turn up the tips of their arms while moving, which exposes the sensory tube feet and eyespot to external stimuli. Most starfish cannot move quickly, a typical speed being that of the leather star (*Dermasterias imbricata*), which can manage just 15 cm (6 in) in a minute. Some burrowing species from the genera *Astropecten* and *Luidia* have points rather than suckers on their long tube feet and are capable of much more rapid motion, "gliding" across the ocean floor. The sand star (*Luidia foliolata*) can travel at a speed of 2.8 m (9 ft 2 in) per minute. Sunflower starfish are quick, efficient hunters, moving at a speed of 1 m/min (3.3 ft/min) using 15,000 tube feet.

Many animals temporarily change the number of legs they use for locomotion in different circumstances. For example, many quadrupedal animals switch to bipedalism to reach low-level browse on trees. The genus of *Basiliscus* are arboreal lizards that usually use quadrupedalism in the trees. When frightened, they can drop to water below and run across the surface on their hind limbs at about 1.5 m/s for a distance of approximately 4.5 metres (15 ft) before they sink to all fours and swim. They can also sustain themselves on all fours while "water-walking" to increase the distance travelled above the surface by about 1.3 metres. When cockroaches run rapidly, they rear up on their two hind legs like bipedal humans; this allows them to run at speeds up to 50 body lengths/s, equivalent to a "couple hundred miles per hour, if you scale up to the size of humans". When grazing, kangaroos use a form of pentapedalism (four legs plus the tail) but switch to hopping (bipedalism) when they wish to move at a greater speed.

Multi-legged millipede

Powered Cartwheeling

The Moroccan flic-flac spider (*Cebrennus rechenbergi*) uses a series of rapid, acrobatic flic-flac movements of its legs similar to those used by gymnasts, to actively propel itself off the ground, allowing it to move both down and uphill, even at a 40 percent incline. This behaviour is different than other huntsman spiders, such as *Carparachne aureoflava* from the Namib Desert, which uses passive cartwheeling as a form of locomotion. The flic-flac spider can reach speeds of up to 2 m/s using forward or back flips to evade threats.

Subterranean

Some animals move through solids such as soil by burrowing using peristalsis, as in earthworms, or other methods. In loose solids such as sand some animals, such as the golden mole, marsupial mole, and the pink fairy armadillo, are able to move more rapidly, 'swimming' through the loose substrate. Burrowing animals include moles, ground squirrels, naked mole-rats, tilefish, and mole crickets.

Arboreal Locomotion

A brachiating gibbon

Arboreal locomotion is the locomotion of animals in trees. Some animals may only scale trees occasionally, while others are exclusively arboreal. These habitats pose numerous mechanical challenges to animals moving through them, leading to a variety of anatomical, behavioural and ecological consequences as well as variations throughout different species. Furthermore, many of these same principles may be applied to climbing without trees, such as on rock piles or mountains. The earliest known tetrapod with specializations that adapted it for climbing trees, was *Suminia*, a synapsid of the late Permian, about 260 million years ago. Some invertebrate animals are exclusively arboreal in habitat, for example, the tree snail.

Brachiation is a form of arboreal locomotion in which primates swing from tree limb to tree limb using only their arms. During brachiation, the body is alternately supported under each forelimb. This is the primary means of locomotion for the small gibbons and siamangs of southeast Asia. Some New World monkeys such as spider monkeys and muriquis are "semibrachiators" and move through the trees with a combination of leaping and brachiation. Some New World species also practice suspensory behaviors by using their prehensile tail, which acts as a fifth grasping hand.

Energetics

Animal locomotion requires energy to overcome various forces including friction, drag, inertia and gravity, although the influence of these depends on the circumstances. In terrestrial environments, gravity must be overcome whereas the drag of air has little influence. In aqueous environments, friction (or drag) becomes the major energetic challenge with gravity being less of an influence. Remaining in the aqueous environment, animals with natural buoyancy expend little energy to maintain a vertical position in a water column. Others naturally sink, and must spend energy to remain afloat. Drag is also an energetic influence in flight, and the aerodynamically efficient body shapes of flying birds indicate how they have evolved to cope with this. Limbless organisms moving on land must energetically overcome surface friction, however, they do not usually need to expend significant energy to counteract gravity.

Newton's third law of motion is widely used in the study of animal locomotion: if at rest, to move forwards an animal must push something backwards. Terrestrial animals must push the solid ground, swimming and flying animals must push against a fluid (either water or air). The effect of forces during locomotion on the design of the skeletal system is also important, as is the interaction between locomotion and muscle physiology, in determining how the structures and effectors of locomotion enable or limit animal movement. The energetics of locomotion involves the energy expenditure by animals in moving. Energy consumed in locomotion is not available for other efforts, so animals typically have evolved to use the minimum energy possible during movement. However, in the case of certain behaviors, such as locomotion to escape a predator, performance (such as speed or maneuverability) is more crucial, and such movements may be energetically expensive. Furthermore, animals may use energetically expensive methods of locomotion when environmental conditions (such as being within a burrow) preclude other modes.

The most common metric of energy use during locomotion is the net [also termed "incremental"] cost of transport, defined as the amount of energy (e.g., Joules) needed above baseline metabolic rate to move a given distance. For aerobic locomotion, most animals have a nearly constant cost of transport - moving a given distance requires the same caloric expenditure, regardless of speed. This constancy is usually accomplished by changes in gait. The net cost of transport of swimming is lowest, followed by flight, with terrestrial limbed locomotion being the most expensive per unit distance. However, because of the speeds involved, flight requires the most energy per unit time. This does not mean that an animal that normally moves by running would be a more efficient swimmer; however, these comparisons assume an animal is specialized for that form of motion. Another consideration here is body mass—heavier animals, though using more total energy, require less energy *per unit mass* to move. Physiologists generally measure energy use by the amount of oxygen consumed, or the amount of carbon dioxide produced, in an animal's respiration. In terrestrial animals, the cost of transport is typically measured while they walk or run on a motorized treadmill, either wearing a mask to capture gas exchange or with the entire treadmill enclosed in a metabolic chamber. For small rodents, such as deer mice, the cost of transport has also been measured during voluntary wheel running.

Energetics is important for explaining the evolution of foraging economic decisions in organisms; for example, a study of the African honey bee, *A. m. scutellata*, has shown that honey bees may trade-off the high sucrose content of viscous nectar for the energetic benefits of warmer, less concentrated nectar, which also reduces their consumption and flight time.

Passive Locomotion

Passive locomotion in animals is a type of mobility in which the animal depends on their environment for transportation.

Hydrozoans

The Portuguese man o' war (*Physalia physalis*) lives at the surface of the ocean. The gas-filled bladder, or pneumatophore (sometimes called a "sail"), remains at the surface, while the remainder is submerged. Because the Portuguese man o' war has no means of propulsion, it is moved by a combination of winds, currents, and tides. The sail is equipped with a siphon. In the event of a surface attack, the sail can be deflated, allowing the organism to briefly submerge.

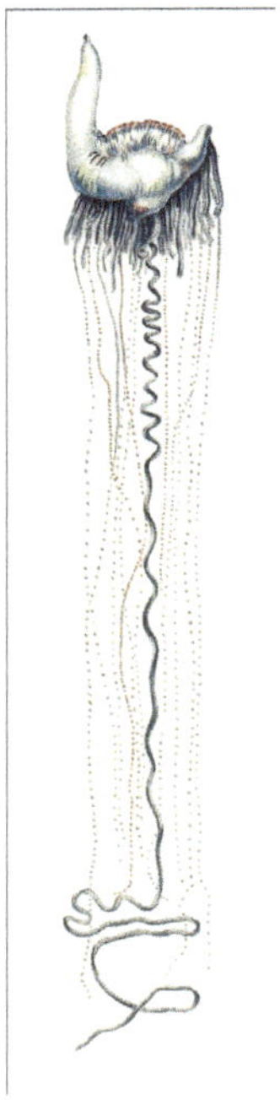

Physalia physalis

Arachnids

The wheel spider (*Carparachne aureoflava*) is a huntsman spider approximately 20 mm in size and native to the Namib Desert of Southern Africa. The spider escapes parasitic pompilid wasps by flipping onto its side and cartwheeling down sand dunes at speeds of up to 44 turns per second. If the spider is on a sloped dune, its rolling speed may be 1 metre per second.

A spider (usually limited to individuals of a small species), or spiderling after hatching, climbs as high as it can, stands on raised legs with its abdomen pointed upwards ("tiptoeing"), and then releases several silk threads from its spinnerets into the air. These form a triangle-shaped parachute that carries the spider on updrafts of winds, where even the slightest breeze transports it. The Earth's static electric field may also provide lift in windless conditions.

Insects

The larva of *Cicindela dorsalis*, the eastern beach tiger beetle, is notable for its ability to leap into the air, loop its body into a rotating wheel and roll along the sand at a high speed using wind to propel itself. If the wind is strong enough, the larva can cover up to 60 metres (200 ft) in this manner. This remarkable ability may have evolved to help the larva escape predators such as the tiphiid wasp *Methocha*.

Members of the largest subfamily of cuckoo wasps, Chrysidinae, are generally kleptoparasites, laying their eggs in host nests, where their larvae consume the host egg or larva while it is still young. Chrysidines are distinguished from the members of other subfamilies in that most have flattened or concave lower abdomens and can curl into a defensive ball when attacked by a potential host, a process known as conglobation. Protected by hard chitin in this position, they are expelled from the nest without injury and can search for a less hostile host.

Fleas can jump vertically up to 18 cm and horizontally up to 33 cm, however, although this form of locomotion is initiated by the flea, it has little control of the jump - they always jump in the same direction, with very little variation in the trajectory between individual jumps.

Crustaceans

Although stomatopods typically display the standard locomotion types as seen in true shrimp and lobsters, one species, *Nannosquilla decemspinosa*, has been observed flipping itself into a crude wheel. The species lives in shallow, sandy areas. At low tides, *N. decemspinosa* is often stranded by its short rear legs, which are sufficient for locomotion when the body is supported by water, but not on dry land. The mantis shrimp then performs a forward flip in an attempt to roll towards the next tide pool. *N. decemspinosa* has been observed to roll repeatedly for 2 metres (6.6 ft), but they typically travel less than 1 m (3.3 ft). Again, the animal initiates the movement but has little control during its locomotion.

Animal Transport

Some animals change location because they are attached to, or reside on, another animal or moving structure. This is arguably more accurately termed "animal transport".

Remoras

Some remoras, such as this Echeneis naucrates, may attach themselves to scuba divers.

Remoras are a family (*Echeneidae*) of ray-finned fish. They grow to 30–90 cm (0.98–2.95 ft) long, and their distinctive first dorsal fins take the form of a modified oval, sucker-like organ with slat-like structures that open and close to create suction and take a firm hold against the skin of larger marine animals. By sliding backward, the remora can increase the suction, or it can release itself by swimming forward. Remoras sometimes attach to small boats. They swim well on their own, with a sinuous, or curved, motion. When the remora reaches about 3 cm (1.2 in), the disc is fully formed and the remora can then attach to other animals. The remora's lower jaw projects beyond the upper, and the animal lacks a swim bladder. Some remoras associate primarily with specific host species. They are commonly found attached to sharks, manta rays, whales, turtles, and dugongs. Smaller remoras also fasten onto fish such as tuna and swordfish, and some small remoras travel in the mouths or gills of large manta rays, ocean sunfish, swordfish, and sailfish. The remora benefits by using the host as transport and protection, and also feeds on materials dropped by the host.

Angler Fish

In some species of anglerfish, when a male finds a female, he bites into her skin, and releases an enzyme that digests the skin of his mouth and her body, fusing the pair down to the blood-vessel level. The male becomes dependent on the female host for survival by receiving nutrients via their shared circulatory system, and provides sperm to the female in return. After fusing, males increase in volume and become much larger relative to free-living males of the species. They live and remain reproductively functional as long as the female lives, and can take part in multiple spawnings. This extreme sexual dimorphism ensures, when the female is ready to spawn, she has a mate immediately available. Multiple males can be incorporated into a single individual female with up to eight males in some species, though some taxa appear to have a one male per female rule.

Parasites

Many endoparasites and ectoparasites, due to their parasitic behaviour, are transported by other animals. For example, tapeworms attach themselves to the inside of the alimentary tracts of other animals and do not locomote within the animal. They do however depend on movement of the host to distribute their eggs.

Other parasites may locomote within, or on, their host, which in turn might be active or stationary. For example, an adult dog flea may crawl about the skin of its sleeping canine host (locomotion), but when the dog awakes and moves, it could be argued the flea is being transported.

Changes between Media

Some animals locomote between different media, e.g., from aquatic to arial. This often requires different modes of locomotion in the different media and may require a distinct transitional locomotor behaviour.

There are a large number of semi-aquatic animals (animals that spend part of their life cycle in water, or generally have part of their anatomy underwater). These represent the major taxons of mammals (e.g., beaver, otter, polar bear), birds (e.g., penguins, ducks), reptiles (e.g., anaconda, bog turtle, marine iguana) and amphibians (e.g., salamanders, frogs, newts).

Fish

Some fish use multiple modes of locomotion. Walking fish may swim freely or at other times "walk" along the ocean or river floor, but not on land (e.g., the flying gurnard —which does not actually fly—and batfishes of the Ogcocephalidae family). Amphibious fish, are fish that are able to leave water for extended periods of time. These fish use a range of terrestrial locomotory modes, such as lateral undulation, tripod-like walking (using paired fins and tail), and jumping. Many of these locomotory modes incorporate multiple combinations of pectoral, pelvic and tail fin movement. Examples include eels, mudskippers and the walking catfish. Flying fish can make powerful, self-propelled leaps out of water into air, where their long, wing-like fins enable gliding flight for considerable distances above the water's surface. This uncommon ability is a natural defense mechanism to evade predators. The flights of flying fish are typically around 50 meters, though they can use updrafts at the leading edge of waves to cover distances of up to 400 m (1,300 ft). They can travel at speeds of more than 70 km/h (43 mph). Maximum altitude is 6 m (20 ft) above the surface of the sea. Some accounts have them landing on ships' decks.

Marine Mammals

Pacific white-sided dolphins porpoising

When swimming, several marine mammals such as dolphins, porpoises and pinnipeds, frequently leap above the water surface whilst maintaining horizontal locomotion. This is done for various reasons. When travelling, jumping can save dolphins and porpoises energy as there is less friction while in the air. This type of travel is known as "porpoising". Other reasons for dolphins and porpoises performing porpoising include orientation, social displays, fighting, non-verbal communication, entertainment and attempting to dislodge parasites. In pinnipeds, two types of porpoising have been identified. "High porpoising" is most often near (within 100 metres) the shore and is often followed by minor course changes; this may help seals get their bearings on beaching or rafting sites. "Low porpoising" is typically observed relatively far (more than 100 metres) from shore and often aborted in favour of anti-predator movements; this may be a way for seals to maximize sub-surface vigilance and thereby reduce their vulnerability to sharks

Some whales raise their (entire) body vertically out of the water in a behaviour known as "breaching".

Birds

Some semi-aquatic birds use terrestrial locomotion, surface swimming, underwater swimming and flying (e.g., ducks, swans). Diving birds also use diving locomotion (e.g., dippers, aulks). Some birds (e.g., ratites) have lost the primary locomotion of flight. The largest of these, ostriches, when being pursued by a predator, have been known to reach speeds over 70 km/h (43 mph), and can maintain a steady speed of 50 km/h (31 mph), which makes the ostrich the world's fastest two-legged animal: Ostriches can also locomote by swimming. Penguins either waddle on their feet or slide on their bellies across the snow, a movement called *tobogganing*, which conserves energy while moving quickly. They also jump with both feet together if they want to move more quickly or cross steep or rocky terrain. To get onto land, penguins sometimes propel themselves upwards at a great speed to leap out the water.

Changes During the Life-cycle

An animal's mode of locomotion may change considerably during its life-cycle. Barnacles are exclusively marine and tend to live in shallow and tidal waters. They have two nektonic (active swimming) larval stages, but as adults, they are sessile (non-motile) suspension feeders. Frequently, adults are found attached to moving objects such as whales and ships, and are thereby transported (passive locomotion) around the oceans.

Function

Animals locomote for a variety of reasons, such as to find food, a mate, a suitable microhabitat, or to escape predators.

Food Procurement

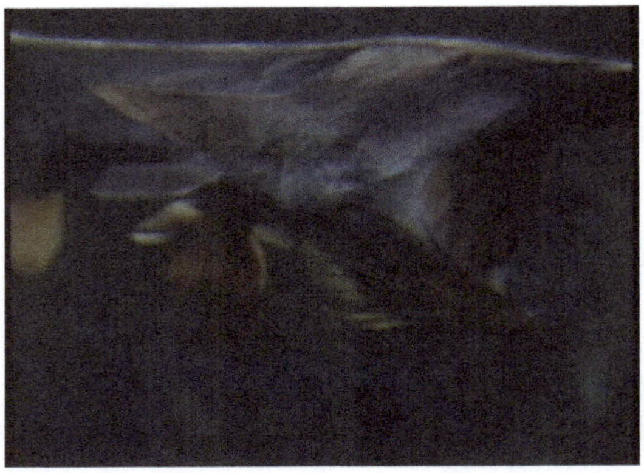

Paddlefish ram suspension-feeding zooplankton in aquarium

Animals use locomotion in a wide variety of ways to procure food. Terrestrial methods include ambush predation, social predation, grazing. Aquatic methods include filterfeeding, grazing, ram feeding, suction feeding, protrusion and pivot feeding. Other methods include parasitism and parasitoidism.

Methods of Study

Documentary film, shot at 1200 fps, used to study the locomoton of a cheetah. The end of the video shows the methods used for filming.

A variety of methods and equipment are used to study animal locomotion:

- Treadmills are used to allow animals to walk or run while remaining stationary with respect to external observers. This technique facilitates filming or recordings of physiological information from the animal (e.g., during studies of energetics). Motorized treadmills are also used to measure the endurance capacity (stamina) of animals.
- Racetracks lined with photocells or filmed while animals run along them are used to measure acceleration and maximal sprint speed.
- Kinematics is the study of the motion of an entire animal or parts of its body. It is typically accomplished by placing visual markers at particular anatomical locations on the animal and then recording video of its movement. The video is often captured from multiple angles, with frame rates exceeding 2000 frames per second when capturing high speed movement. The location of each marker is determined for each video frame, and data from multiple views is integrated to give positions of each point through time. Computers are sometimes used to track the markers, although this task must often be performed manually. The kinematic data can be used to determine fundamental motion attributes such as velocity, acceleration, joint angles, and the sequencing and timing of kinematic events. These fundamental attributes can be used to quantify various higher level attributes, such as the physical abilities of the animal (e.g., its maximum running speed, how steep a slope it can climb), neural control of locomotion, gait, and responses to environmental variation. These, in turn, can aid in formulation of hypotheses about the animal or locomotion in general.
- Force plates are platforms, usually part of a trackway, that can be used to measure the magnitude and direction of forces of an animal's step. When used with kinematics and a sufficiently detailed model of anatomy, inverse dynamics solutions can determine the forces not just at the contact with the ground, but at each joint in the limb.
- Electromyography (EMG) is a method of detecting the electrical activity that occurs when muscles are activated, thus determining which muscles an animal uses for a given movement. This can be accomplished either by surface electrodes (usually in large animals) or implanted electrodes (often wires thinner than a human hair). Furthermore, the intensity of electrical activity can correlate to the level of muscle activity, with greater activity implying (though not definitively showing) greater force.
- Sonomicrometry employs a pair of piezoelectric crystals implanted in a muscle or tendon to continuously measure the length of a muscle or tendon. This is useful because surface kinematics may be inaccurate due to skin movement. Similarly, if an elastic tendon is in series with the muscle, the muscle length may not be accurately reflected by the joint angle.
- Tendon force buckles measure the force produced by a single muscle by measuring the strain of a tendon. After the experiment, the tendon's elastic modulus is determined and used to compute the exact force produced by the muscle. However, this can only be used on muscles with long tendons.
- Particle image velocimetry is used in aquatic and aerial systems to measure the flow of fluid around and past a moving aquatic organism, allowing fluid dynamics calculations to determine pressure gradients, speeds, etc.

- Fluoroscopy allows real-time X-ray video, for precise kinematics of moving bones. Markers opaque to X-rays can allow simultaneous tracking of muscle length.

These methods can be combined. For example, studies frequently combine EMG and kinematics to determine *motor pattern*, the series of electrical and kinematic events that produce a given movement.

Role of Skin in Locomotion

Role of skin in locomotion describes how the integumentary system is involved in locomotion. Typically the integumentary system can be thought of as skin, however the integumentary system also includes the segmented exoskeleton in arthropods and feathers of birds. The primary role of the integumentary system is to provide protection for the body. However, the structure of the skin has evolved to aid animals in their different modes of locomotion. Soft bodied animals such as starfish rely on the arrangement of the fibers in their tube feet for movement. Eels, snakes, and fish use their skin like an external tendon to generate the propulsive forces need for undulatory locomotion. Vertebrates that fly, glide, and parachute also have a characteristic fiber arrangements of their flight membranes that allows for the skin to maintain its structural integrity during the stress and strain experienced during flight.

Soft Bodied Locomotion in Invertebrates

The arms of octopus are muscular hydrostats

The term "Soft Bodied" refers to animals which lack typical systems of skeletal support - included in these are most insect larvae and true worms. Animals that are soft bodied are constrained by the geometry and form of their bodies. However it is the geometry and form of their bodies that generate the forces they need to move. The structure of soft bodied skin can be characterized by a patterned fiber arrangement, which provides the shape and structure for a soft bodied animals. Internal to the patterned fiber layer is typically a liquid filled cavity, which is used to generate hydrostatic pressures for movement. Some animals that exhibit soft bodied locomotion include starfish, octopus, and flatworms.

Hydrostatic Skeleton

A hydrostatic skeleton uses hydrostatic pressure generated from muscle contraction against a liquid filled cavity. The liquid filled cavity is commonly referred to as the hydrostatic body. The liquid within the hydrostatic body acts as an incompressible fluid and the body wall of the hydrostatic body provides a passive elastic antagonist to muscle contraction, which in turn generates a force, which in turn creates movement. This structure plays a role in invertebrate support and locomotor systems and is used for the tube feet in starfish and body of worms . A specialized version of the hydrostatic skeleton is a called a muscular hydrostat, which consists of a tightly packed array of three-dimensional muscle fibers surrounding a hydrostatic body. Examples of muscular hydrostats include the arms of octopus and elephant trunks.

Fiber Arrangement

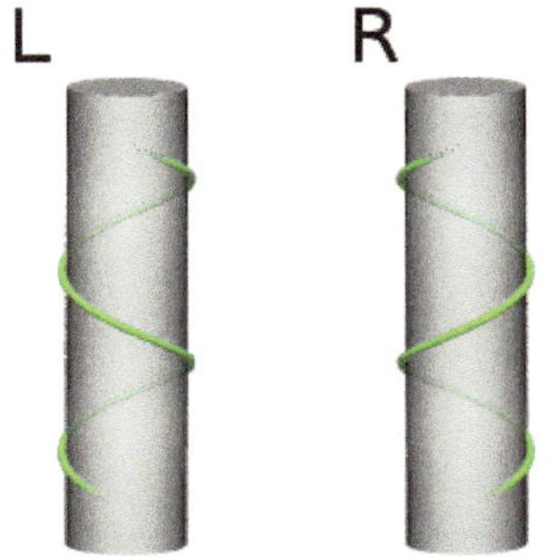

left- and right-handed helices

The arrangement of the connective tissue fibers and muscle fibers create the skeletal support of a soft bodied animal. The arrangement of the fibers around a hydrostatic body limits the range of movement of the hydrostatic body (the "body" of a soft bodied animal) and defines the way the hydrostatic body moves.

Muscle Fibers

Typically muscle fibers surround the hydrostatic body. There are two main types of muscle fibers orientations that are responsible for the movement: the circular orientations and longitudinal orientations. Circular muscles decrease the diameter of a hydrostatic body, resulting in an increase in the length of the body, whereas longitudinal muscles shortens the length of a hydrostatic body, resulting in an increase in the diameter of the body. There are four categories of movements of a hydrostatic skeleton : elongation, shortening, bending and torsion. Elongation, which involves an increase in the length of a hydrostatic body requires either circular muscles, a transverse muscle arrangement, or radial muscle arrangement. For a transverse muscle arrangement, parallel sheets of muscle fibers that extend along the length of a hydrostatic body. For a radial muscle arrangement, radial muscles radiate from a central axis along the axis perpendicular to the long axis. Shortening involves the contraction of the longitudinal muscle. Both shortening and bending involve the contraction of longitudinal muscle, but for bending motion some of the antagonistic muscles work synergistically with longitudinal muscles. The amplitude of movements are based upon the antagonistic muscles forces and the amount of leverage the antagonistic muscle provides for movement. For the torsion motion, muscles are arranged in helical layers around a hydrostatic body. The fiber angle (the angle the fiber makes with the long axis of the body) plays a critical role in torsion, if the angle is greater than 54°44', during muscle contraction, torsion and elongation will occur. If the fiber angle is less than 54°44', torsion and shortening will occur.

Connective Tissue Fibers

The arrangement of connective tissue fibers determines the range of motion of a body, and serves as an antagonist against muscle contraction. The most commonly observed connective tissue arrangement for soft bodied animals consists of layers of alternating right and left-handed helices of connective tissue fibers which surround the hydraulic body. This cross helical arrangement is seen in the tube feet starfish, different types of worms and suckers in octopus. This cross helical arrangement allows for the connective tissue layers to evenly distribute force throughout the hydrostatic body. Another commonly observed connective tissue fiber range is when the connective tissue fibers are embedded within a muscle layer. This arrangement of connective tissue fibers creates a stiffer body wall and more muscle antagonism, which allows for more elastic force to be generated and released during movement. This fiber arrangement is seen in the mantle of squid and the fins in sharks.

Specialized Function in Vertebrates

Swimming and Undulatory Locomotion

The skin of these animal that use undulatory motion to locomote have several distinct characteristics. The skin of these animals consists of **cro**ss-helical arrangement of collagen and elastin fibers embedded in the dermal layer of skin, a two-dimensional stiffness. which permits bending at small curvatures and resists bending at high curvatures and skin is attached directly to the underlying muscles. Fish, shark, and snakes are all examples of animals that locomote using undulatory locomotion.

Eel

Closeup image of the skin of a marbled eel

The cross helical fiber arrangement of the two dermal fibers types collagen and elastin, are responsible for the mechanical properties of the skin such as the two dimensional stiffness seen in the eel skin. In the longitudinal direction, eel skin behaves like a pure fiber system, with a lessertensile strength than skin in the hoop direction. The skin in the hoop direction exhibits a higher elastic modulus than the skin in the longitudinal direction. The two dimensional stiffness allows for the body of the eel to be modeled a pressurized cylinder with the fiber angle of the cross helical arrangement dictating the method by which the eel moves. Eel skin behaves like skin having a fiber angle greater than 45°. In an eel with the cross helical fiber arrangement, muscle contraction in the anterior region bends the fish, and so the skin on the convex side is extended in the longitudinal direction. The extension in the longitudinal direction produces contraction in the hoop direction as the fiber angle decreases until these dimensional changes are resisted by the body of the eel. The skin becomes skin, and additional longitudinal force(applied by skin) results in force being transmitted along the tail. Therefore, changes in fiber angle of the cross helical arrangement in eel skin allows for the transmission of force through the skin during swimming. The skin act like an external tendon allowing for an eel to generate a greater propulsive force per muscle contraction. In addition to the eel skin acting as an external tendon, the skin attaches directly to the underlying muscle, which allow for the eel to generate an even greater force per muscle contraction.

Longnose Gar

Closeup of the skin of the longnose gar

Due to the heavily scaled skin of the Longnose gar, some of the mechanical properties differ from model of describing how eel skin adds movement. The scale row resists longitudinal forces, which unlike eel skin, makes the skin stiffer in the longitudinal direction, providing myomeres with leverage and anchorage for pulling tendons. At low curvatures, it appears that the dermis is slack on both the concave and convex sides of the body. When the dermis is placed in tension, and resistance to bending is developed, which is referred to as flexural stiffness of the fish skin. The flexural stiffness is a result of the two dimensional stiffness of fish with heavily scaled skin, such as the longnose gar. This mechanical property of fish skin is important to the way a fish swims, because this mechanical property passively stiffens the body, which would otherwise would have been done muscularly. The flexural stiffness of fish skin act in a manner similar to the mechanism by which eel skin acts as an external tendon, however in the case of fish skin, the flexural stiffness acts as a mechanism to decelerate body movement rather than to generate a propulsive force.

Snake

Snakes are one of the few vertebrates in which the skin alone is sufficient for locomotion. During Rectilinear locomotion, the skeleton remains fixed, while the skin is alternately lifted and pulled forward, and then allowed to contact the ground and pulled backwards, propelling the body forward.

One of the interesting aspects of snake skin are folds of intersquamous skin between longitudinally oriented scale rows. The function of these folds is to permit the circumference of the snake to increase, allowing prey to pass into the stomach during feeling. Snakes differ from eels in the direction in which the skin is stiffer, the dorsal scale rows are more flexible in snake than in eels because the dorsal scale row associated with stretching. Differences in the local dermal structures, such as variations in the diameters and orientation of collagen fibers within the intersquamous skin create local differences in the mechanical properties of the snake skin, thus allowing it to adapt to the stresses and strains during the feeding process.

Aerial Locomotion

Gliding, Flying and Parachuting are some of the some methods of aerial locomotion used by animals. Vertebrates have altered the structure of the skin to accommodate the stresses and strains of flight. Typically mammalian skin consists of collagen fibers arranged in a felt-work pattern, with no preferential fiber orientation. However,the structures of skin in bats, birds, and gliding lizards are very different from those of typical mammalian skin. The structural arrangement of the fibers within bat wing skin enables the bat to act like a spring during the down-stroke of flapping. The scales of gliding lizards are arranged in a regular rib like pattern to enable to lizard to act as an airfoil. Avain skin must be structurally arranged such that "the coat of feathers" remains smooth and intact during flight.

Bats

Drawing of the bat wing skin showing the fibers of "mesh like scaffolding"

A closeup view of the felt like fiber pattern seen in elephant skin

Bats rely on skin on their wings to generate lift and thrust used in flight. Therefore, the structure of the bat wing skin is different from the skin of the bat body. Bat wing skin consists of two thin layers of epidermis with a thin layer of dermis/hypodermis located between the epidermal layers whereas the skin of the bat body consists of a single layer of epidermis with a thicker layer of dermis internal to the epidermis. Within the dermal and epidermal layer of bat wing skin, the connective tissue and muscle fibers provide the structural support. The connective tissue fibers within bat wing skin consists of collagen and elastin fiber bundles arranged in a "regular mesh like scaffolding", which the nerves, skeletal muscle fibers and blood vessels embed themselves into. Of the muscles that insert themselves into the mesh scaffolding, larger muscles anchor the skin to the bone and control the membrane tension and camber of the bat wing during flight, whereas smaller muscles, which originate from within the mesh scaffolding, attach to collagen fibers within the fiber network and modulate bone loading and allow for precise control of wing shape and tension. As seen in snakes, local structural differences within the arrangement of the fibers change the mechanical properties of local area, but there are general characteristics that describe the mechanical behavior of bat wing skin. Within the mesh scaffolding of bat wing skin, collagen fibers cross bones perpendicular to the long axes of the bones, therefore mechanical properties of bat wing skin oriented perpendicular to the long axes of the bones exhibit a lower stiffness than the skin that is oriented parallel to the long axes of the bodies. Stiffer skin is necessary for bat wing skin oriented in the direction parallel to the long axes of the bones to prevent too much deformation of bat wing skin during flight(with respect to the bone), resulting in the shearing of the bat wing skin off of the bone. Flexible skin is necessary for the direction perpendicular to the long axes of the bones for facilitating the shape changes needed for movement and control during flight. This anisotropy of bat wing skin is also useful as method of storing and releasing elastic energy, particularly during the downstroke. During the downstroke, the bat extends its wing and the wing skin experiences an aerodynamic force. The wing skin expands and counteracts the aerodynamic force. After the wing is fully extended, the orientation of the wing is nearly vertical and the aerodynamic forces on the wing are reduced. As the aerodynamic force is reduced, the wing recoils, drawing the digits of the wings together in preparation for the upstroke.

Gliding Lizards

There are two different mechanisms by which lizards glide through the air. Both mechanisms involve the patagia.

Active parachuting mechanism

In the active mechanism, skeletal supports and muscles run through the patagia of lizards. The skeletal supports and muscle erect the flight membrane and control the gliding using the patagia. Most of the lizards that exhibit this active gliding mechanism are agamine lizards such lizards in the genus Draco. For the passive mechanism of gliding in lizards, the patagia is unfurled by air pressure alone. The patagia of the passive mechanism differs from patagia of the active mechanism; there is the lack of skeleton support and musculature in patagia of the gliding lizards with the passive gliding mechanism . The passive mechanism of gliding is seen in smaller lizards such as the geckos of the genus Ptychozoon. For the passive mechanism of gliding, body movements are believed to control the descent of the gliding lizard.

Passive parachuting mechanism

The surface area to body ratios of lizards with different gliding mechanisms are similar, but how the surface area is distributed is different. The difference in the distribution of surface area indicates the differences in the role of the patagia and accessory areas for the different flight mechanisms. Lizards with passive gliding mechanisms tend to have smaller patagia relative to lizards with active flight mechanisms. However,lizards with passive flight mechanism have,ore surface area located in accessory areas(i.e. webbed toes,tail) than lizards with the active gliding mechanism.

The structure of the skin of the patagia and accessory areas for the patagia of a Ptychozoon kuhli, flying gecko, exhibiting the passive gliding mechanism consists of five layers; a layer of adipose tissue is surrounded by a layer of dermis on each side (ventral and dorsal) and a layer of epidermis is external to the two dermal layers. The distribution of the adipose tissue IS thickest close to the body wall. This thick layer of adipose tissue at close to the body wall is believed to provide a "safety factor" for the structural elements of the skin (i.e. collagen fibers) near the body wall. The thick layer of adipose tissue is more compliant than the structural elements of the body wall (i.e. ribs, muscles), therefore will more readily deform(absorb force) before the structural elements of the skin experience a force. The layer of adipose tissue also aids in the creation of the domed and cambered shape of the patagia. With regards to the structure of the dermal layer of the patagia, there is a thick layer of collagen fibers oriented parallel to the axis of the patagial ribs. These collagen fibers act as the structural support for the shape of the patagia, and provide the stiffness necessary to resist shape chang . The most prominent features of the epidermal layer of the patagia are the scales. The morphology of the dorsal scales of the patagia change as a result of their functional role. A large portion of the dorsal scales of the patagia are arranged in regular rib-like pattern, which guide the flow of air and allow for the lizard to behave as an airfoil. However at the hinge joints (places where patagia folds and attaches to limbs), the regular rib like structure of scales breakdown into a more random distribution of scales. This breakdown of scales is believed to aid in the mechanical loading of the patagia during the unfurling process and also determining the extent the patagia unfurling during flight.

Birds

The wing of a woodpecker

Avian skin is a bit more complicated than the skin of gliding lizards or bats because the presence of feathers. In addition to the counteracting stresses and stains associated with flight, avian skin must provide a means to monitor and anchor a "coat of feathers", thus the structure of avian skin is different from skin of other flying and gliding animals. To better understand the structure of avian skin, avian skin has been broken down into three different functional components:

- hydraulic skeleto-muscular apparatus of the feathers

This functional component consists of the only of the structural features of the feather tracts, namely the cutis, and connective tissue layer fascia superficialis. This functional component was named "hydraulic skeletal" due to the fact that the fat bodies embedded within cutis and fascia act similar to the hydrostatic bodies within a hydrostatic skeleton. However the functional role of the fat bodies within the hydraulic skeleto-muscular apparatus of the feathers is to counteract forces generated by the erector and depressor muscle of the feathers tracts. rather than to facilitate movement within of a body.

- dermo-subcutaneous muscular system of integument

This functional component of avian skin consists of the smooth muscle of the apertia and striated subcutaneous muscles. The smooth muscles of the apertia counteract the horizontal forces experienced by the feather follicles. The striated subcutaneous muscles also adjust the position of the feather follicles in the directions the smooth muscle cannot. Together this system acts as an integrated muscular system that properly positions the feather tracts on the body of the bird.

- subcutaneous hydraulic skeletal system

This functional component of avian skin consists of the fat bodies of the fascia superficialis and Fascia subcutanea. The majority of the fat bodies are located either between fascia superficialis and the Fascia subcutanea. These fat bodies are stratically located at depression within the body of the bird and function to even out depressions so that feather tracts of the skeleto-muscular apparatus function properly.

Arboreal Locomotion

Arboreal locomotion is the locomotion of animals in trees. In habitats in which trees are present, animals have evolved to move in them. Some animals may scale trees only occasionally, but others are exclusively arboreal. The habitats pose numerous mechanical challenges to animals moving through them and lead to a variety of anatomical, behavioral and ecological consequences as well as variations throughout different species. Furthermore, many of these same principles may be applied to climbing without trees, such as on rock piles or mountains.

The earliest known tetrapod with specializations that adapted it for climbing trees was *Suminia*, a synapsid of the late Permian, about 260 million years ago.

Some invertebrate animals are exclusively arboreal in habitat, such as the tree snail.

Leopards are good climbers and can carry their kills up their trees to keep them out of reach from scavengers and other predators.

Biomechanics

Arboreal habitats pose numerous mechanical challenges to animals moving in them, which have been solved in diverse ways. These challenges include moving on narrow branches, moving up and down inclines, balancing, crossing gaps, and dealing with obstructions.

Diameter

Moving along a narrow surface poses special difficulties to animals. During locomotion on the ground, the location of the center of mass may swing from side to side, but during arboreal locomotion, this would result in the center of mass moving beyond the edge of the branch, resulting in a tendency to topple over. Additionally, foot placement is constrained by the need to make contact with the narrow branch. This narrowness severely restricts the range of movements and postures an animal can use to move.

Incline

Branches are frequently oriented at an angle to gravity in arboreal habitats, including being vertical, which poses special problems. As an animal moves up an inclined branch, they must fight the force of gravity to raise their body, making movement more difficult. Conversely, as the animal descends, it must also fight gravity to control its descent and prevent falling. Descent can be particularly problematic for many animals, and highly arboreal species often have specialized methods for controlling their descent.

Balance

Due to the height of many branches and the potentially disastrous consequences of a fall, balance is of primary importance to arboreal animals. On horizontal and gently sloped branches, the primary problem is tipping to the side due to the narrow base of support. The narrower the branch, the greater the difficulty in balancing a given animal faces. On steep and vertical branches, tipping becomes less of an issue, and pitching backwards or slipping downwards becomes the most likely failure. In this case, large-diameter branches pose a greater challenge, since the animal cannot place its forelimbs closer to the center of the branch than its hindlimbs.

Gibbons are very good brachiators because their elongated limbs enable them to easily swing and grasp on to branches.

Crossing Gaps

Branches are not continuous, and any arboreal animal must be able to move between gaps in the branches, or even between trees. This can be accomplished by reaching across gaps, by leaping across them or gliding between them.

Obstructions

Arboreal habitats often contain many obstructions, both in the form of branches emerging from the one being moved on and other branches impinging on the space the animal needs to move through. These obstructions may impede locomotion, or may be used as additional contact points to enhance it. While obstructions tend to impede limbed animals, they benefit snakes by providing anchor points.

Anatomical Specializations

Arboreal organisms display many specializations for dealing with the mechanical challenges of moving through their habitats.

Limb Length

Arboreal animals frequently have elongated limbs that help them cross gaps, reach fruit or other resources, test the firmness of support ahead, and in some cases, to brachiate. However, some species of lizard have reduced limb size that helps them avoid limb movement being obstructed by impinging branches.

Prehensile Tails

Many arboreal species, such as tree porcupines, chameleons, silky anteaters, spider monkeys, and possums, use prehensile tails to grasp branches. In the spider monkey and crested gecko, the tip of the tail has either a bare patch or adhesive pad, which provide increased friction.

Claws

Claws can be used to interact with rough substrates and re-orient the direction of forces the animal applies. This is what allows squirrels to climb tree trunks that are so large as to be essentially flat, from the perspective of such a small animal. However, claws can interfere with an animal's ability to grasp very small branches, as they may wrap too far around and prick the animal's own paw.

The silky anteater uses its prehensile tail as a third arm for stabilization and balance, while its claws help better grasp and climb onto branches.

Adhesion

Adhesion is an alternative to claws, which works best on smooth surfaces. Wet adhesion is common in tree frogs and arboreal salamanders, and functions either by suction or by capillary adhesion. Dry adhesion is best typified by the specialized toes of geckos, which use van der Waals forces to adhere to many substrates, even glass.

Gripping

Frictional gripping is used by primates, relying upon hairless fingertips. Squeezing the branch between the fingertips generates frictional force that holds the animal's hand to the branch. However, this type of grip depends upon the angle of the frictional force, thus upon the diameter of the branch, with larger branches resulting in reduced gripping ability. Animals other than primates that use gripping in climbing include the chameleon, which has mitten-like grasping feet, and many birds that grip branches in perching or moving about.

Reversible Feet

To control descent, especially down large diameter branches, some arboreal animals such as squirrels have evolved highly mobile ankle joints that permit rotating the foot into a 'reversed' posture. This allows the claws to hook into the rough surface of the bark, opposing the force of gravity.

Low Center of Mass

Many arboreal species lower their center of mass to reduce pitching and toppling movement when climbing. This may be accomplished by postural changes, altered body proportions, or smaller size.

Small Size

Small size provides many advantages to arboreal species: such as increasing the relative size of branches to the animal, lower center of mass, increased stability, lower mass (allowing movement on smaller branches), and the ability to move through more cluttered habitat. Size relating to weight affects gliding animals such as the reduced weight per snout-vent length for 'flying' frogs.

Hanging under Perches

Some species of primate, bat, and all species of sloth achieve passive stability by hanging beneath the branch. Both pitching and tipping become irrelevant, as the only method of failure would be losing their grip.

The gecko's toes adhere to surfaces via dry adhesion, to allow them to stay firmly attached to a branch or even a flat wall.

Behavioral Specializations

Arboreal species have behaviors specialized for moving in their habitats, most prominently in terms of posture and gait. Specifically, arboreal mammals take longer steps, extend their limbs further forwards and backwards during a step, adopt a more 'crouched' posture to lower their center of mass, and use a diagonal sequence gait.

Ecological Consequences

Arboreal locomotion allows animals access to different resources, depending upon their abilities. Larger species may be restricted to larger-diameter branches that can support their weight, while smaller species may avoid competition by moving in the narrower branches.

Climbing without Trees

Many animals climb in other habitats, such as in rock piles or mountains, and in those habitats, many of the same principles apply due to inclines, narrow ledges, and balance issues. However, less research has been conducted on the specific demands of locomotion in these habitats.

Perhaps the most exceptional of the animals that move on steep or even near vertical rock faces by careful balancing and leaping are the various types of mountain dwelling caprid such as the Barbary sheep, markhor, yak, ibex, tahr, rocky mountain goat, and chamois. Their adaptations may include a soft rubbery pad between their hooves for grip, hooves with sharp keratin rims for lodging in small footholds, and prominent dew claws. The snow leopard, being a predator of such mountain caprids, also has spectacular balance and leaping abilities; being able to leap up to ≈17m (~50 ft). Other balancers and leapers include the mountain zebra, mountain tapir, and hyraxes.

Brachiation

Arboreal snails use their sticky slime to help in climbing up trees, since they lack limbs to do so.

Brachiation is a specialized form of arboreal locomotion, used by primates to move very rapidly while hanging beneath branches. Arguably the epitome of arboreal locomotion, it involves swinging with the arms from one handhold to another. Only a few species are brachiators, and all of these are primates; it is a major means of locomotion among spider monkeys and gibbons, and is occasionally used by female orangutans. Gibbons are the experts of this mode of locomotion, swinging from branch to branch distances of up to 15 m (50 ft), and traveling at speeds of as much as 56 km/h (35 mph).

Gliding and Parachuting

To bridge gaps between trees, many animals such as the flying squirrel have adapted membranes, such as patagia for gliding flight. Some animals can slow their descent in the air using a method known as parachuting, such as *Rhacophorus* (a "flying frog" species) that has adapted toe membranes allowing it to fall more slowly after leaping from trees.

Limbless Climbing

Many species of snake are highly arboreal, and some have evolved specialized musculature for this habitat. While moving in arboreal habitats, snakes move slowly along bare branches using a specialized form of concertina locomotion, but when secondary branches emerge from the branch being moved on, snakes use lateral undulation, a much faster mode. As a result, snakes perform best on small perches in cluttered environments, while limbed organisms seem to do best on large perches in uncluttered environments.

Arboreal Animals

Many species of animals are arboreal, far too many to list individually. This list is of prominently or predominantly arboreal species and higher taxa.

- Primates
- Cats
- brushtail possums

- opossums
- Sloths
- Anteaters
- Treeshrews
- Goats
- Colugos
- Kinkajous
- Viverrids
- Many rodents
- Parrots
- Geckos
- Chameleons
- Many other lizards
- Mambas
- Brown Tree Snakes
- Many other snakes
- Stick insects
- Many other arthropods
- Tree snails
- Koalas

Terrestrial Locomotion

An example of terrestrial locomotion. A horse - an erect-stanced unguligrade quadruped - with a galloping gait. An animation of photos by Eadweard Muybridge

Terrestrial locomotion has evolved as animals adapted from aquatic to terrestrial environments. Locomotion on land raises different problems than that in water, with reduced friction being replaced by the effects of gravity.

There are three basic forms of locomotion found among terrestrial animals

- Legged - Moving by using appendages
- Limbless locomotion - moving without legs, primarily using the body itself as a propulsive structure.
- Rolling - rotating the body over the substrate

Legged Locomotion

Movement on appendages is the most common form of terrestrial locomotion, it is the basic form of locomotion of two major groups with many terrestrial members, the vertebrates and the arthropods. Important aspects of legged locomotion are posture (the way the body is supported by the legs), the number of legs, and the functional structure of the leg and foot. There are also many gaits, ways of moving the legs to locomote, such as walking, running, or jumping.

Posture

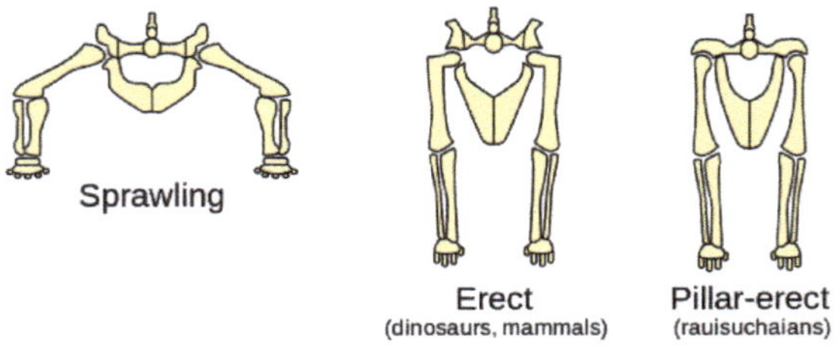

Hip joints and hindlimb postures.

Appendages can be used for movement in a number of ways. The posture, the way the body is supported by the legs, is an important aspect. There are three main ways in which vertebrates support themselves with their legs - sprawling, semi-erect, and fully erect. Some animals may use different postures in different circumstances, depending on the posture's mechanical advantages. Interestingly, there is no detectable difference in energetic cost between stances.

The "sprawling" posture is the most primitive, and is the original limb posture from which the others evolved. The upper limbs are typically held horizontally, while the lower limbs are vertical, though upper limb angle may be substantially increased in large animals. The body may drag along the ground, as in salamanders, or may be substantially elevated, as in monitor lizards. This posture is typically associated with trotting gaits, and the body flexes from side-to-side during movement to increase step length. All limbed reptiles and salamanders use this posture, as does the platypus and several species of frogs that walk. Unusual examples can be found among amphibious fish, such as the mudskipper, which drag themselves across land on their sturdy fins. Among the invertebrates, most arthropods—which includes the most diverse group of animals, the insects—have a stance best described as sprawling. There is also anecdotal evidence that some octopus species (such as the *Pinnoctopus* genus) can also drag themselves across land a short distance by hauling their body along by their tentacles (for example to pursue prey between rockpools) - there may be video evidence of this. The semi-erect posture is more accurately interpreted as an extremely elevated sprawling posture. This mode of locomotion is typically found in large lizards such as monitor lizards and tegus.

Mammals and birds typically have a fully erect posture, though each evolved it independently. In these groups the legs are placed beneath the body. This is often linked with the evolution of endothermy, as it avoids Carrier's constraint and thus allows prolonged periods of activity. The fully erect stance is not necessarily the "most-evolved" stance; evidence suggests that crocodilians evolved a semi-erect stance in their forelimbs from ancestors with fully erect stance as a result of adapting to a mostly aquatic lifestyle, though their hindlimbs are still held fully erect. For example, the mesozoic prehistoric crocodilian *Erpetosuchus* is believed to have had a fully erect stance and been terrestrial.

Number of Legs

The velvet worm (Onychophora)

The number of locomotory appendages varies much between animals, and sometimes the same animal may use different numbers of its legs in different circumstances. The best contender for unipedal movement is the springtail, which while normally hexapedal, hurls itself away from danger using its furcula, a tail-like forked rod that can be rapidly unfurled from the underside of its body.

A number of species move and stand on two legs, that is, they are bipedal. The group that is exclusively bipedal is the birds, which have either an alternating or a hopping gait. There are also a number of bipedal mammals. Most of these move by hopping – including the macropods such as kangaroos and various jumping rodents. Only a few mammals such as humans and the ground pangolin commonly show an alternating bipedal gait. Cockroaches and some lizards may also run on their two hind legs.

With the exception of the birds, terrestrial vertebrate groups with legs are mostly quadrupedal – the mammals, reptiles, and the amphibians usually move on four legs. There are many quadrupedal gaits. The most diverse group of animals on earth, the insects, are included in a larger taxon known as hexapods, most of which are hexapedal, walking and standing on six legs. Exceptions among the insects include praying mantises and water scorpions, which are quadrupeds with their front two legs modified for grasping, some butterflies such as the Lycaenidae (blues and hairstreaks) which use only four legs, and some kinds of insect larvae that may have no legs (e.g., maggots), or additional prolegs (e.g., caterpillars).

Spiders and many of their relatives move on eight legs – they are octopedal. However, some creatures move on many more legs. Terrestrial crustaceans may have a fair number - woodlice having fourteen legs. Also, as previously mentioned, some insect larvae such as caterpillars and sawfly larvae have up to five (caterpillars) or nine (sawflies) additional fleshy prolegs in addition to the six legs normal for insects. Some species of invertebrate have even more legs, the unusual velvet worm having stubby legs under the length of its body, with around several dozen pairs of legs. Centipedes have one pair of legs per body segment, with typically around 50 legs, but some species have over 200. The terrestrial animals with the most legs are the millipedes. They have two pairs of legs per body segment, with common species having between 80 and 400 legs overall – with the rare species *Illacme plenipes* having up to 750 legs. Animals with many legs typically move them in metachronal rhythm, which gives the appearance of waves of motion travelling forwards along their rows of legs.

Leg and Foot Structure

The legs of tetrapods, the main group of terrestrial vertebrates, have internal bones, with externally attached muscles for movement, and the basic form has three key joints: the shoulder joint, the knee joint, and the ankle joint, at which the foot is attached. Within this theme there is much variation in form. An alternative form of vertebrate 'leg' to the tetrapod leg is the fins found on amphibious fish. Also a few tetrapods, such as the macropods, have adapted their tails as additional locomotory appendages.

The basic form of the vertebrate foot has five toes, however some animals will have evolved fewer than this, and some early tetrapods had more; Acanthostega had eight toes. Feet have evolved many forms depending on the animal's needs. One key variation is where on the foot the animal's weight is placed. Most vertebrates—the amphibians, the reptiles, and some mammals such as humans and bears—are plantigrade, walking on the whole of the underside of the foot. Many mammals, such as cats and dogs are digitigrade, walking on their toes, the greater stride length allowing more speed. Digitigrade mammals are also often adept at quiet movement. Birds are also digitigrade. Some animals such as horses are unguligrade, walking on the tips of their toes. This even further increases their stride length and thus their speed. A few mammals are also known to walk on their knuckles, at least for their front legs. Knuckle-walking allows the foot (hand) to specialise for food gathering and/or climbing, as with the great apes and the extinct chalicotheres, or for swimming, as with the platypus. In animals where feet have evolved into functional hands, hand walking is also possible.

Among terrestrial invertebrates there are a number of leg forms. The arthropod legs are jointed and supported by hard external armor, with the muscles attached to the internal surface of this exoskeleton. The other group of legged terrestrial invertebrates, the velvet worms, have soft stumpy legs supported by a hydrostatic skeleton. The prolegs that some caterpillars have in addition to their six more-standard arthropod legs have a similar form to those of velvet worms, and suggest a distant shared ancestry.

Gaits

A jumping kangaroo.

A walking hamster.

Animals show a vast range of gaits, the order that they place and lift their appendages in locomotion. Gaits can be grouped into categories according to their patterns of support sequence. For quadrupeds, there are three main categories: walking gaits, running gaits, and leaping gaits. In one system (relating to horses), there are 60 discrete patterns: 37 walking gaits, 14 running gaits, and 9 leaping gaits.

Walking is the most common gait, where some feet are on the ground at any given time, and found in almost all legged animals. In an informal sense, running is considered to occur when at some points in the stride all feet are off the ground in a moment of suspension. Technically, however, moments of suspension occur in both running gaits (such as trot) and leaping gaits (such as canter and gallop). Gaits involving one or more moments of suspension can be found in many animals, and compared to walking they are faster but more energetically costly forms of locomotion.

Animals will use different gaits for different speeds, terrain, and situations. For example, horses show four natural gaits, the slowest horse gait is the walk, then there are three faster gaits which, from slowest to fastest, are the trot, the canter, and the gallop. Animals may also have unusual gaits that are used occasionally, such as for moving sideways or backwards. For example, the main human gaits are bipedal walking and running, but they employ many other gaits occasionally, including a four-legged crawl in tight spaces.

In walking, and for many animals running, the motion of legs on either side of the body alternates, i.e. is out of phase. Other animals, such as a horse when galloping, or an inchworm, alternate between their front and back legs. An alternative to a gait that alternates between legs is hopping or saltation, where all legs move together. As a main means of locomotion, this is usually found in bipeds or semi-bipeds. Among the mammals saltation is commonly used among macropods (kangaroos and their relatives), jerboas, springhares, kangaroo rats, hopping mice, gerbils, and sportive lemurs. Certain tendons in kangaroo hind legs are very elastic, allowing kangaroos to effectively bounce along conserving energy from hop to hop, making hopping a very energy efficient way to move around in their nutrient poor environment. Saltation is also used by many small birds. Frogs and fleas also hop.

Most animals move in the direction of their head. However, there are some exceptions. Crabs move sideways, and naked mole rats, which live in tight tunnels underground, can move backward or forward with equal facility. Crayfish can move backward much faster than they can move forward.

Gait analysis is the study of gait in humans and other animals. This may involve videoing subjects with markers on particular anatomical landmarks and measuring the forces of their footfall using floor transducers (strain gauges). Skin electrodes may also be used to measure muscle activity.

Limbless Locomotion

A snail moves by slithering.

There are a number of terrestrial and amphibious limbless vertebrates and invertebrates. These animals, due to lack of appendages, use their bodies to generate propulsive force. These movements are sometimes referred to as "slithering" or "crawling", although neither are formally used in the scientific literature and the latter term is also used for some animals moving on all four limbs. All limbless animals come from cold-blooded groups; there are no endothermic limbless animals, i.e. there are no limbless birds or mammals.

Lower Body Surface

Where the foot is important to the legged mammal, for limbless animals the underside of the body is important. Some animals such as snakes or legless lizards move on their smooth dry underside. Other animals have various features that aid movement. Molluscs such as slugs and snails move on a layer of mucus that is secreted from their underside, reducing friction and protecting from injury when moving over sharp objects. Earthworms have small bristles (setae) that hook into the substrate and help them move. Some animals such as leeches have suction cups on either end of the body allowing two anchor movement.

Type of Movement

Some limbless animals, such as leeches, have suction cups on either end of their body, which allow them to move by anchoring the rear end and then moving forward the front end, which is then anchored and then the back end is pulled in, and so on. This is known as two-anchor movement. A legged animal, the inchworm, also moves like this, clasping with appendages at either end of its body.

Limbless animals can also move using pedal locomotary waves, rippling the underside of the body. This is the main method used by molluscs such as slugs and snails, and also large flatworms, some other worms, and even earless seals. The waves may move in the opposite direction to motion, known as retrograde waves, or in the same direction as motion, known as direct waves. Earthworms move by retrograde waves alternatively swelling and contracting down the length of their body, the swollen sections being held in place using setae. Aquatic molluscs such as limpets, which are sometimes out of the water, tend to move using retrograde waves. However terrestrial molluscs such as slugs and snails tend to use direct waves. Lugworms and seals also use direct waves.

Most snakes move using lateral undulation where a lateral wave travels down the snake's body in the opposite direction to the snake's motion and pushes the snake off irregularities in the ground. This mode of locomotion requires these irregularities to function. Another form of locomotion, rectilinear locomotion, is used at times by some snakes, especially large ones such as pythons and boa. Here large scales on the underside of the body, known as scutes are used to push backwards and downwards. This is effective on a flat surface and is used for slow, silent movement, such as when stalking prey. Snakes use concertina locomotion for moving slowly in tunnels, here the snake alternates in bracing parts of its body on it surrounds. Finally the caenophidian snakes use the fast and unusual method of movement known as sidewinding on sand or loose soil. The snake cycles through throwing the front part of its body in the direction of motion and bringing the back part of its body into line crosswise.

Rolling

The pangolin Manis temminckii in defensive position.

Although animals have never evolved wheels for locomotion, a small number of animals will move at times by rolling their whole body. Rolling animals can be divided into those that roll under the force of gravity or wind and those that roll using their own power.

Gravity or Wind Assisted

The web-toed salamander, a 10-centimetre (3.9 in) salamander, lives on steep hills in the Sierra Nevada mountains. When disturbed or startled it coils itself up into a ball, often causing it to roll downhill.

The pebble toad (*Oreophrynella nigra*) lives atop tepui in the Guiana highlands of South America. When threatened, often by tarantulas, it rolls into ball, and typically being on an incline, rolls away under gravity like a loose pebble.

Namib wheeling spiders (*Carparachne spp.*), found in the Namib desert, will actively roll down sand dunes. This action can be used to successfully escape predators such as the *Pompilidae* tarantula wasps, which lay their eggs in a paralyzed spider for their larvae to feed on when they hatch. The spiders flip their body sideways and then cartwheel over their bent legs. The rotation is fast, the golden wheel spider (*Carparachne aureoflava*) moving up to 20 revolutions per second, moving the spider at 1 metre per second.

Coastal tiger beetle larvae when threatened can flick themselves into the air and curl their bodies to form a wheels, which the wind blows, often uphill, as far as 25m and as fast as 11kph. The also may have some ability to steer themselves in this state.

Pangolins, a type of mammal covered in thick scales, roll into a tight ball when threatened. Pangolins have been reported to roll away from danger, by both gravity and self-powered methods. A pangolin in hill country in Sumatra, to flee from the researcher, ran to the edge of a slope and curled into a ball to roll down the slope, crashing through the vegetation, and covering an estimated 30 metres or more in 10 seconds.

Self-powered

Caterpillars of the Mother-Of-Pearl Moth, *Pleuroptya ruralis*, when attacked, will touch their heads to their tails and roll backwards, up to 5 revolutions at about 40 cm per second, which is about 40 times its normal speed.

Nannosquilla decemspinosa, a species of long-bodied, short-legged mantis shrimp, lives in shallow sandy areas along the Pacific coast of Central and South America. When stranded by a low tide the 3 cm stomatopod lies on its back and performs backwards somersaults over and over. The animal moves up to 2 meters at a time by rolling 20–40 times, with speeds of around 72 revolutions per minute. That is 1.5 body lengths per second (3.5 cm/s). Researchers estimate that the stomatopod acts as a true wheel around 40% of the time during this series of rolls. The remaining 60% of the time it has to "jumpstart" a roll by using its body to thrust itself upwards and forwards.

Pangolins have also been reported to roll away from danger by self-powered methods. Witnessed by a lion researcher in the Serengeti in Africa, a group of lions surrounded a pangolin, but could not get purchase on it when it rolled into a ball, and so the lions sat around it waiting and dozing. Surrounded by lions, it would unroll itself slightly and give itself a push to roll some distance, until by doing this multiple times it could get far enough away from the lions to be safe. Moving like this would allow a pangolin to cover distance while still remaining in a protective armoured ball.

Limits and Extremes

The fastest terrestrial animal is the cheetah, which can attain maximal sprint speeds of approximately 104 km/h (64 mph). The fastest running lizard is the Black Iguana, which has been recorded moving at speed of up to 34.9 km/h (21.7 mph)..

Animal Locomotion on the Water Surface

Animal locomotion on the surface layer of water is the study of animal locomotion in the case of small animals that live on the surface layer of water, relying on surface tension to stay afloat.

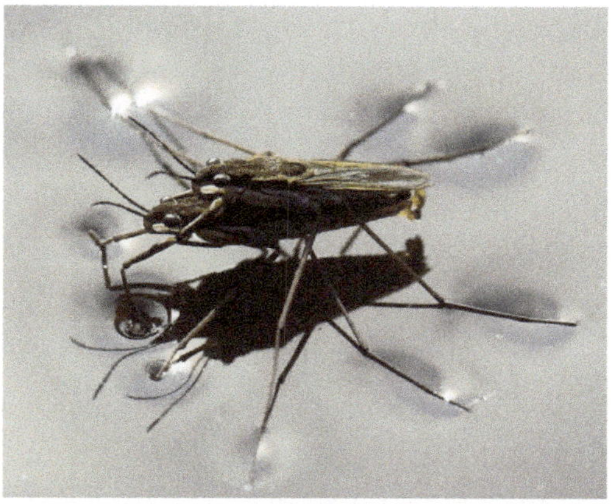

Water striders can move on the surface of water.

There are two types of animal locomotion on water, determined by the ratio of the animal's weight to the water's surface tension: those whose weight is supported by the surface tension at rest, and can therefore easily remain on the water's surface without much exertion, and those whose weight is not supported by the water's surface tension at rest, and must therefore exert additional motion in a direction parallel to the water's surface in order to remain above it. A creature such as the basilisk lizard, often dubbed the 'Jesus lizard', has a weight which is larger than the surface tension can support, and is widely known for running across the surface of water. Another example, the western grebe, performs a mating ritual that includes running across the surface of water.

Surface living animals such as the water strider typically have hydrophobic feet covered in small hairs that prevent the feet from breaking the surface and becoming wet. Another insect known to walk on the water surface is the ant species *Polyrhachis sokolova*. The pygmy gecko (*Coleodactylus amazonicus*), due to its small size and hydrophobic skin is also able to walk on the water surface.

According to biophysicist David L. Hu, there are at least 342 species of water striders. As striders increase in size, their legs become proportionately longer, with *Gigantometra gigas* having a length of over 20 cm requiring a surface tension force of about 40 millinewtons.

Water striders generate thrust by shedding vortices in the water: a series of "U"-shaped vortex filaments is created during the power stroke. The two free ends of the "U" are attached to the water surface. These vortices transfer enough (backward) momentum to the water to propel the animal forwards (note that some momentum is transferred by capillary waves.

Meniscus Climbing

To pass from the water surface to land, a water-walking insect must contend with the slope of the meniscus at the water's edge. Many such insects are unable to climb this meniscus using their usual propulsion mechanism.

David Hu and coworker John W. M. Bush have shown that such insects climb meniscuses by assuming a fixed body posture. This deforms the water surface and generates capillary forces that propels the insect up the slope without moving its appendages.

Hu and Bush conclude that meniscus climbing is an unusual means of propulsion in that the insect propels itself in a quasi-static configuration, without moving its appendages. Biolocomotion is generally characterized by the transfer of muscular strain energy to the kinetic and gravitational potential energy of the creature, and the kinetic energy of the suspending fluid. In contrast, meniscus climbing has a different energy pathway: by deforming the free surface, the insect converts muscular strain to the surface energy that powers its ascent.

Marangoni Propulsion

Many insects, including some terrestrial insects, can release a surfactant and propel themselves using the Marangoni effect. Hu and Bush report that *Microvelia* can attain a peak speed of 17 cm/s, which is twice its peak walking speed, using Marangoni propulsion.

Marangoni propulsion by a wetting arthropod is precisely analogous to a soap boat but the situation for insects such as water striders is more complex. Hu and Bush state that "for nonwetting arthropods, the transfer of chemical to kinetic energy is more subtle, as the Marangoni stress must be communicated across the creature's complex surface layer".

Sailing

Elella, the by-the-wind sailor, is a cnidarian with no means of propulsion other than sailing. A small rigid sail projects into the air and catches the wind. *Velella* sails always align along the direction of the wind where the sail may act as an aerofoil, so that the animals tend to sail downwind at a small angle to the wind.

Undulatory Locomotion

Snakes primarily rely on undulatory locomotion to move through a wide range of environments

Undulatory locomotion is the type of motion characterized by wave-like movement patterns that act to propel an animal forward. Examples of this type of gait include crawling in snakes, or swimming in the lamprey. Although this is typically the type of gait utilized by limbless animals, some creatures with limbs, such as the salamander, choose to forgo use of their legs in certain environments and exhibit undulatory locomotion. This movement strategy is important to study in order to create novel robotic devices capable of traversing a variety of environments.

Environmental Interactions

In limbless locomotion, forward locomotion is generated by propagating flexural waves along the length of the animal's body. Forces generated between the animal and surrounding environment lead to a generation of alternating sideways forces that act to move the animal forward. These forces generate thrust and drag.

Hydrodynamics

Simulation predicts that thrust and drag are dominated by viscous forces at low Reynolds numbers and inertial forces at higher Reynolds numbers. When the animal swims in a fluid, two main forces are thought to play a role:

- Skin Friction: Generated due to the resistance of a fluid to shearing and is proportional to speed of the flow. This dominates undulatory swimming in spermatozoa and the nematode
- Form Force: Generated by the differences in pressure on the surface of the body and it varies with the square of flow speed.

At low Reynolds number ($Re \sim 10^0$), skin friction accounts for nearly all of the thrust and drag. For those animals which undulate at intermediate Reynolds number ($Re \sim 10^1$), such as the Ascidian larvae, both skin friction and form force account for the production of drag and thrust. At high Reynolds number ($Re \sim 10^2$), both skin friction and form force act to generate drag, but only form force produces thrust.

Kinematics

In animals that move without use of limbs, the most common feature of the locomotion is a rostral to caudal wave that travels down their body. However, this pattern can change based on the particular undulating animal, the environment, and the metric in which the animal is optimizing (i.e. speed, energy, etc.). The most common mode of motion is simple undulations in which lateral bending is propagated from head to tail.

Snakes can exhibit 5 different modes of terrestrial locomotion: (1) lateral undulation, (2) sidewinding, (3) concertina, (4) rectilinear, and (5) slide-pushing. Lateral undulation closely resembles the simple undulatory motion observed in many other animals such as in lizards, eels and fish, in which waves of lateral bending propagate down the snakes body.

The American eel typically moves in an aquatic environment, though it can also move on land for short periods of time. It is able to successfully move about in both environments by producing traveling waves of lateral undulations. However, differences between terrestrial and aquatic locomotor strategy suggest that the axial musculature is being activated differently, In terrestrial locomotion, all points along the body move on approximately the same path and, therefore, the lateral displacements along the length of the eel's body is approximately the same. However, in aquatic locomotion, different points along the body follow different paths with increasing lateral amplitude more posteriorly. In general, the amplitude of the lateral undulation and angle of intervertebral flexion is much greater during terrestrial locomotion than that of aquatic.

Musculoskeletal System

Perch filets showing myomere structure

Muscle Architecture

A typical characteristic of many animals that utilize undulatory locomotion is that they have segmented muscles, or blocks of myomeres, running from their head to tails which are separated by connective tissue called myosepta. In addition, some segmented muscle groups, such as the lateral hypaxial musculature in the salamander are oriented at an angle to the longitudinal direction. For these obliquely oriented fibers the strain in the longitudinal direction is greater than the strain in the muscle fiber direction leading to an architectural gear ratio greater than 1. A higher initial angle of orientation and more dorsoventral bulging produces a faster muscle contraction but results in a lower amount of force production. It is hypothesized that animals employ a variable gearing mechanism that allows self-regulation of force and velocity to meet the mechanical demands of the contraction. When a pennate muscle is subjected to a low force, resistance to width changes in the muscle cause it to rotate which consequently produce a higher architectural gear ratio (AGR) (high velocity). However, when subject to a high force, the perpendicular fiber force component overcomes the resistance to width changes and the muscle compresses producing a lower AGR (capable of maintaining a higher force output).

Most fishes bend as a simple, homogenous beam during swimming via contractions of longitudinal red muscle fibers and obliquely oriented white muscle fibers within the segmented axial musculature. The fiber strain (εf) experienced by the longitudinal red muscle fibers is equivalent to the longitudinal strain (εx). The deeper white muscle fibers fishes show diversity in arrangement. These fibers are organized into cone-shaped structures and attach to connective tissue sheets known as myosepta; each fiber shows a characteristic dorsoventral (α) and mediolateral (φ) trajectory. The segmented architecture theory predicts that, $\varepsilon x > \varepsilon f$. This phenomenon results in an architectural gear ratio, determined as longitudinal strain divided by fiber strain (εx / εf), greater than one and longitudinal velocity amplification; furthermore, this emergent velocity amplification may be augmented by variable architectural gearing via mesolateral and dorsoventral shape changes, a pattern seen in pennate muscle contractions. A red-to-white gearing ratio (red εf / white εf) captures the combined effect of the longitudinal red muscle fiber and oblique white muscle fiber strains.

Simple bending behavior in homogenous beams suggests ε increases with distance from the neutral axis (z). This poses a problem to animals, such as fishes and salamanders, which undergo undulatory movement. Muscle fibers are constrained by the length-tension and force-velocity curves. Furthermore, it has been hypothesized that muscle fibers recruited for a particular task must operate within an optimal range of strains (ε) and contractile velocities to generate peak force and power respectively. Non-uniform ε generation during undulatory movement would force differing muscle fibers recruited for the same task to operate on differing portions of the length-tension and force-velocity curves; performance would not be optimal. Alexander predicted that the dorsoventral (α) and mediolateral (φ) orientation of the white fibers of the fish axial musculature may allow more uniform strain across varying mesolateral fiber distances. Unfortunately, the white muscle fiber musculature of fishes is too complex to study uniform strain generation; however, Azizi *et al.* studied this phenomenon using a simplified salamander model.

Siren lacertian, an aquatic salamander, utilizes swimming motions similar to the aforementioned fishes yet contains hypaxial muscle fibers (which generate bending) characterized by a simpler organization. The hypaxial muscle fibers of S. lacertian are obliquely oriented, but have a near zero mediolateral (φ) trajectory and a constant doroslateral (α) trajectory within each segment. Therefore, the effect of doroslateral (α) trajectory and the distance between a given hypaxial muscle layer and the neutral axis of bending (z) on muscle fiber strain (ε) can be studied.

Azizi *et al.* found that longitudinal contractions of the constant volume hypaxial muscles were compensated by an increase in the dorsoventral dimensions. Bulging was accompanied by fiber rotation as well as an increase in both α hypaxial fiber trajectory and architectural gear ratio (AGR), a phenomenon also seen in pennate muscle contractions. Azizi *et al.* constructed a mathematical model to predict the final hypaxial fiber angle, AGR and dorsoventral height, where: λx = longitudinal extension ratio of the segment (portortion of final longitudinal length after contraction to initial longitudinal length), β = final fiber angle, γ = initial fiber angle, f = initial fiber length, and εx and εf = longitudinal and fiber strain respectively.

- $\lambda x = \varepsilon x + 1$
- $\lambda f = \varepsilon f + 1$
- $\varepsilon x = \Delta L/\text{Linital}$
- Architectural gear ratio = $\varepsilon x / \varepsilon f = [\lambda f(\cos\beta / \cos\gamma) -1)]/ (\lambda f - 1)$
- $\beta = \sin{-1} (y2 / \lambda ff)$

This relationship shows that AGR increase with an increase in fiber angle from γ to β. In addition, final fiber angle (β) increases with dorsolateral bulging (y) and fiber contraction, but decreases as a function of initial fiber length.

The application of the latter conclusions can be seen in S. lacertian. This organism undulates as a homogenous beam (just as in fishes) during swimming; thus the distance of a muscle fiber from the neutral axis (z) during bending must be greater for external oblique muscle layers (EO) than internal oblique muscle layers (IO). The relationship between the strains (ε) experienced by the EO and IO and their respective z values is given by the following equation: where εEO and εIO =

strain of the external and internal oblique muscle layers, and zEO and zIO = distance of the external and internal oblique muscle layers respectively from the neutral axis.

$\varepsilon EO = \varepsilon IO\ (zEO / zIO)$

Via this equation, we see that z is directly proportional to ε; the strain experienced by the EO exceeds that of the IO. Azizi *et al.* discovered that the initial hypaxial fiber α trajectory in the EO is greater than that of the IO. Because initial α trajectory is proportional to the AGR, the EO contracts with a greater AGR than the IO. The resulting velocity amplification allows both layers of muscles to operate at similar strains and shortening velocities; this enables the EO and IO to function on comparable portions of the length-tension and force-velocity curves. Muscles recruited for a similar task ought to operate at similar strains and velocities to maximize force and power output. Therefore, variability in AGR within the hypaxial musculature of the Siren lacertian counteracts varying mesolateral fiber distances and optimizes performance. Azizi *et al.* termed this phenomenon as fiber strain homogeneity in segmented musculature.

Muscle Activity

In addition to a rostral to caudal kinematic wave that travels down the animals body during undulatory locomotion, there is also a corresponding wave of muscle activation that travels in the rostro-caudal direction. However, while this pattern is characteristic of undulatory locomotion, it too can vary with environment.

American Eel

Aquatic Locomotion: Electromyogram (EMG) recordings of the American eel reveal a similar pattern of muscle activation during aquatic movement as that of fish. At slow speeds only the most posterior end of the eel's muscles are activated with more anterior muscle recruited at higher speeds. As in many other animals, the muscles activate late in the lengthening phase of the muscle strain cycle, just prior to muscle shortening which is a pattern believed to maximize work output from the muscle.

Terrestrial Locomotion: EMG recordings show a longer absolute duration and duty cycle of muscle activity during locomotion on land. Also, the absolute intensity is much higher while on land which is expected from the increase in gravitational forces acting on the animal. However, the intensity level decreases more posteriorly along the length of the eel's body. Also, the timing of muscle activation shifts to later in the strain cycle of muscle shortening.

Energetics

Animals with elongated bodies and reduced or no legs have evolved differently from their limbed relatives. In the past, some have speculated that this evolution was due to a lower energetic cost associated with limbless locomotion. The biomechanical arguments used to support this rationale include that (1) there is no cost associatied with the vertical displacement of the center of mass typically found with limbed animals, (2) there is no cost associated with accelerating or decelerating limbs, and (3) there is a lower cost for supporting the body. This hypothesis has been studied further by examining the oxygen consumption rates in the snake during different modes of locomotion: lateral undulation, concertina, and sidewinding. The *net cost of transport* (NCT), which indicates the amount of energy required to move a unit of mass a given distance, for a snake moving with a lateral undulatory gait is identical to that of a limbed lizard with the same mass. However, a snake utilizing concertina locomotion produces a much higher net cost of transport, while sidewinding actually produces a lower net cost of transport. Therefore, the different modes of locomotion are of primary importance when determining energetic cost. The reason that lateral undulation has the same energetic efficiency as limbed animals and not less, as hypothesized earlier, might be due to the additional biomechanical cost associated with this type of movement due to the force needed to bend the body laterally, push its sides against a vertical surface, and overcome sliding friction.

Neuromuscular System

Intersegmental Coordination

Wavelike motor pattern typically arise from a series of coupled segmental oscillator. Each segmental oscillator is capable of producing a rhythmic motor output in the absence of sensory feedback. One such example is the half center oscillator which consist of two neurons that are mutually inhibitory and produce activity 180 degrees out of phase. The phase relationships between these oscillators are established by the emergent properties of the oscillators and the coupling between them. Forward swimming can by accomplished by a series of coupled oscillators in which the anterior oscillators have a shorter endogenous frequency than the posterior oscillators. In this case, all oscillators will be driven at the same period but the anterior oscillators will lead in phase. In addition, the phase relations can be established by asymmetries in the couplings between oscillators or by sensory feedback mechanisms.

- Leech

The leech moves by producing dorsoventral undulations. The phase lags between body segments is about 20 degrees and independent of cycle period. Thus, both hemisegments of the oscillator fire synchronously to produce a contraction. Only the ganglia rostral to the midpoint are capable of producing oscillation individually. There is U-shaped gradient in endogenous segment oscillation as well with the highest oscillations frequencies occurring near the middle of the animal. Although the couplings between neurons spans six segments in both the anterior and posterior direction, there are asymmetries between the various interconnections because the oscillators are active at three different phases. Those that are active in the 0 degree phase project only in the descending direction while those projecting in the ascending direction are active at 120 degrees or 240 degrees. In addition, sensory feedback from the environment may contribute to resultant phase lag.

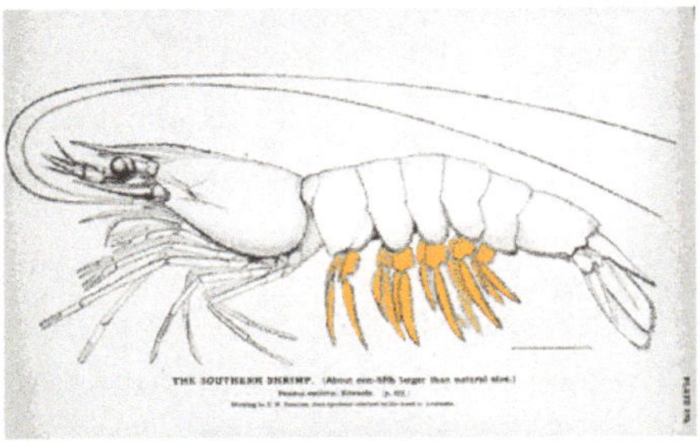

Pleopods

- Lamprey

The lamprey moves using lateral undulation and consequently left and right motor hemisegments are active 180 degrees out of phase. Also, it has been found that the endogenous frequency of the more anterior oscillators is higher than that of the more posterior ganglia. In addition, inhibitory interneurons in the lamprey project 14-20 segments caudally but have short rostral projections. Sensory feedback may be important for appropriately responding to perturbations, but seems to be less important for maintaince of appropriate phase relations.

Robotics

Based on biologically hypothesized connections of the central pattern generator in the salamander, a robotic system has been created which exhibits the same characteristics of the actual animal. Electrophysiology studies have shown that stimulation of the mesencephalic locomotor region (MLR) located in the brain of the salamander produce different gaits, swimming or walking, depending on intensity level. Similarly, the CPG model in the robot can exhibit walking at low levels of tonic drive and swimming at high levels of tonic drive. The model is based on the four assumptions that:

- Tonic stimulation of the body CPG produces spontaneous traveling waves. When the limb CPG is activated it overrides the body CPG.
- The strength of the coupling from the limb to the body CPG is stronger than that from body to limb.
- Limb oscillators saturate and stop oscillating at higher tonic drives.
- Limb oscillators have lower intrinsic frequencies than body CPGs at the same tonic drive.

This model encompasses the basic features of salamander locomotion.

Comparative Foot Morphology

Comparative foot morphology involves comparing the form of distal limb structures of a variety of terrestrial vertebrates. Understanding the role that the foot plays for each type of organism must take account of the differences in body type, foot shape, arrangement of structures, loading conditions and other variables. However, similarities also exist among the feet of many different terrestrial vertebrates. The paw of the dog, the hoof of the horse, the *manus* (foot) and *pes* (foot) of the elephant, and the foot of the human all share some common features of structure, organization and function. Their foot structures function as the load-transmission platform which is essential to balance, standing and types of locomotion (such as walking, trotting, galloping and running).

The discipline of biomimetics applies the information gained by comparing the foot morphology of a variety of terrestrial vertebrates to human-engineering problems. For instance, it may provide insights that make it possible to alter the foot's load transmission in people who wear an external orthosis because of paralysis from spinal-cord injury, or who use a prosthesis following the diabetes-related amputation of a leg. Such knowledge can be incorporated in technology that improves a person's balance when standing; enables them to walk more efficiently, and to exercise; or otherwise enhances their quality of life by improving their mobility.

Structure

Limb and foot structure of representative terrestrial vertebrates:

Variability in Scaling and Limb Coordination

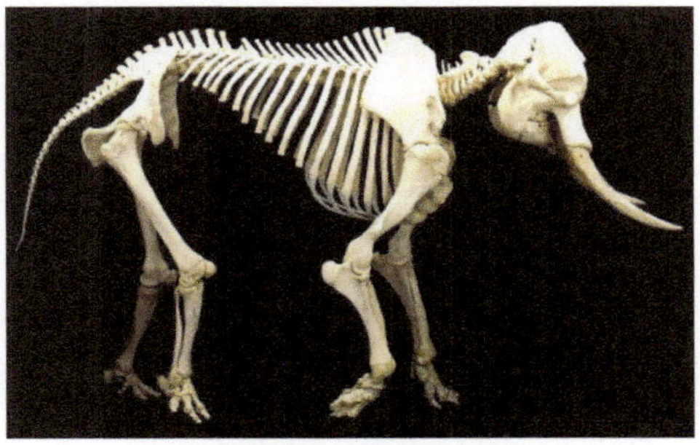

Elephant skeleton

There is considerable variation in the scale and proportions of body and limb, as well as the nature of loading, during standing and locomotion both among and between quadrupeds and bipeds. The anterior-posterior body mass distribution varies considerably among mammalian quadrupeds, which affects limb loading. When standing, many terrestrial quadrupeds support more of their weight on their forelimbs rather than their hind limbs; however, the distribution of body mass and limb loading changes when they move. Humans have a lower-limb mass that is greater than their upper-limb mass. The hind limbs of the dog and horse have a slightly greater mass than the forelimbs, whereas the elephant has proportionally longer limbs. The elephant's forelimbs are longer than its hind limbs.

In the horse and dog, the hind limbs play an important role in primary propulsion. The legged locomotion of humans generally distributes an equal loading on each lower limb. The locomotion of the elephant (which is the largest terrestrial vertebrate) displays a similar loading distribution on its hind limbs and forelimbs. The walking and running gaits of quadrupeds and bipeds show differences in the relative phase of the movements of their forelimbs and hind limbs, as well as of their right-side limbs versus their left-side limbs. Many of the aforementioned variables are connected with differences in the scaling of body and limb dimension as well as in patterns of limb coordination and movement. However, little is understood concerning the functional contribution of the foot and its structures during the weight-bearing phase. The comparative morphology of the distal limb and foot structure of some representative terrestrial vertebrates reveals some interesting similarities.

Columnar Organization of Limb Structures

Even many terrestrial vertebrates exhibit differences in the scaling of limb dimension, limb coordination and magnitude of forelimb-hind limb loading, in the dog, horse and elephant the structure of the distal forelimb is similar to that of the distal hind limb. In the human, the structures of the hand are generally similar in shape and arrangement to those of the foot. Terrestrial vertebrate quadrupeds and bipeds generally possess distal limb and foot endoskeleton structures that are aligned in series, stacked in a relatively vertical orientation and arranged in a quasi-columnar fashion in the extended limb. In the dog and horse, the bones of the proximal limbs are oriented vertically, whereas the distal limb structures of the ankle and foot have an angulated orientation. In humans and elephants, a vertical-column orientation of the bones in the limbs and feet is also evident for associated skeletal muscle-tendon units. The horse's foot contains an external nail (hoof) oriented about the perimeter in the shape of a semicircle. The underlying bones are arranged in a semi-vertical orientation. The dog's paw similarly contains bones arranged in a semi-vertical orientation.

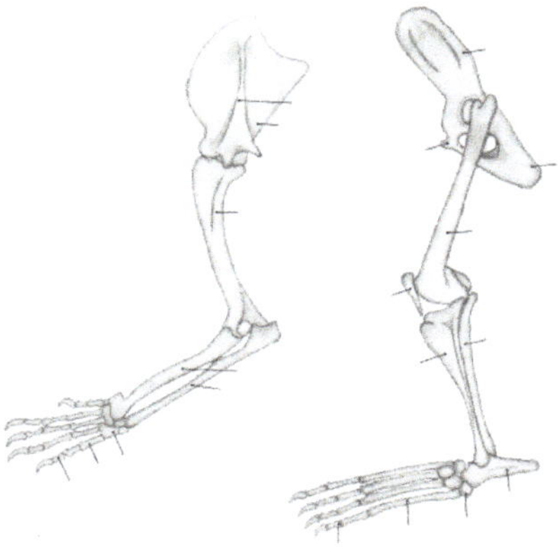

Limb skeleton of a lion, an example of an angulated bony column

In the human and the elephant, the column orientation of the foot complex is replaced in humans by a plantigrade orientation, and in elephants by a semi-plantigrade alignment of the hind limb foot structure. This difference in orientation in the foot bones and joints of humans and elephants helps them to adapt to variations in the terrain.

Distal Cushion

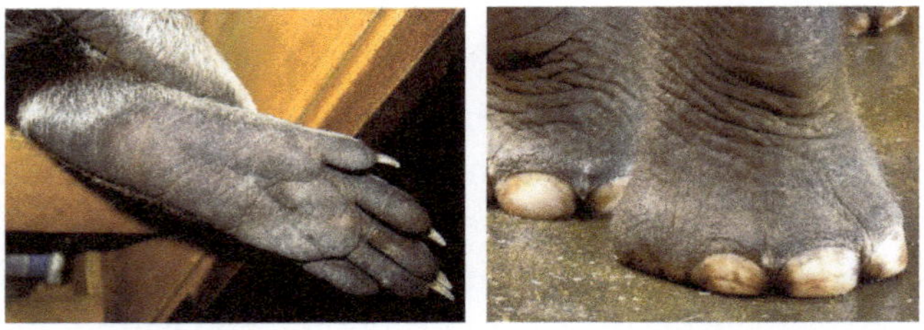

Distal cushions on the foot of a raccoon and an elephant

Many representative terrestrial vertebrates possess a distal cushion on the under-surface of the foot. The dog's paw contains a number of visco-elastic pads oriented along the middle and distal foot. The horse possesses a centralized digital pad known as the frog, which is located at the distal aspect of the foot and surrounded by the hoof. Humans possess a tough fibro and elastic pad of fat that is anchored to the skin and bone of the rear portion of the foot.

The foot of the elephant possesses what is perhaps one of the most unusual distal cushions found in vertebrates. The forefoot (*manus*) and hindfoot (*pes*) contain huge pads of fat that are scaled to cope with the massive loadings imposed by the largest terrestrial vertebrate. In addition, a cartilage-like projection (prepollex in the forelimb and prehallux in the hind limb) appears to anchor the distal cushion to the bones of the elephant's foot.

The distal cushions of all these organisms (dog, horse, human and elephant) are dynamic structures during locomotion, alternating between phases of compression and expansion; it has been suggested that these structures thereby reduce the loads experienced by the skeletal system.

Organization

Arrangement of foot structures:

Because of the wide variety in body types, scaling and morphology of the distal limbs of terrestrial vertebrates, there exists a degree of controversy concerning the nature and organization of foot structures. One organizational approach to understanding foot structures makes distinctions regarding their regional anatomy. The foot structures are divided into segments from proximal to distal and are grouped according to similarity in shape, dimension and function. In this approach, the foot may be described in three segments: as the hindfoot, midfoot and forefoot.

The hindfoot is the most proximal and posterior portion of the foot. Functionally, the structures contained in this region are typically robust, possessing a larger size and girth than the other structures of the foot. The structures of the hindfoot are usually adapted for transmitting large loads between the proximal and distal aspects of the limb when the foot contacts the ground. This is apparent in the human and elephant foot, where the hindfoot undergoes greater loading during initial contact in many forms of locomotion. The hindfoot structures of the dog and horse are located relatively proximally compared to the elephant and human foot.

The midfoot is the intermediate portion of the foot between the hindfoot and forefoot. The structures in this region are intermediate in size, and typically transmit loads from the hindfoot to the forefoot. The human transverse tarsal joint of the midfoot transmits forces from the subtalar joint in the hindfoot to the forefoot joints (metatarsophalangeal and interphalangeal) and associated bones (metatarsals and phalanges). The midfoot of the dog, horse and elephant contains similar intermediate structures having similar functions to those of the human midfoot.

The forefoot represents the most distal portion of the foot. In the human and elephant, the bone structures contained in this region are generally longer and narrower. The structures of the forefoot play a role in providing leverage for terminal stance propulsion and load transfer.

Function

Load transmission of the foot in representative terrestrial vertebrates:

Dog Paw

The paw of the dog has a digitigrade orientation. The vertical columnar orientation of the proximal bones of the limbs, which articulate with distal foot structures that are arranged in quasi-vertical columnar orientation, is well-aligned to transmit loadings during weight-bearing contact of the skeleton with the ground. The angled orientation of the elongated metatarsal and the digits extends the area available for storing and releasing mechanical energy in the muscle tendon units originating proximally to the ankle joint and terminating at the distal aspect of the foot bones. When muscle tendon units lengthen, the load strain facilitates mechanical activity. These muscle tendon unit structures appear well designed to aid in the ground-reaction transmission of forces that is essential for locomotion. In addition, the pads of the distal paw appear to allow load attenuation, by enhancing shock absorption during the paw's contact with the ground.

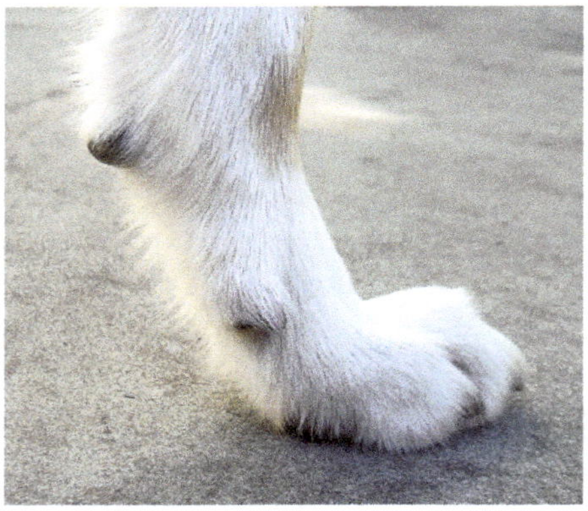

Dog paw

Horse Foot

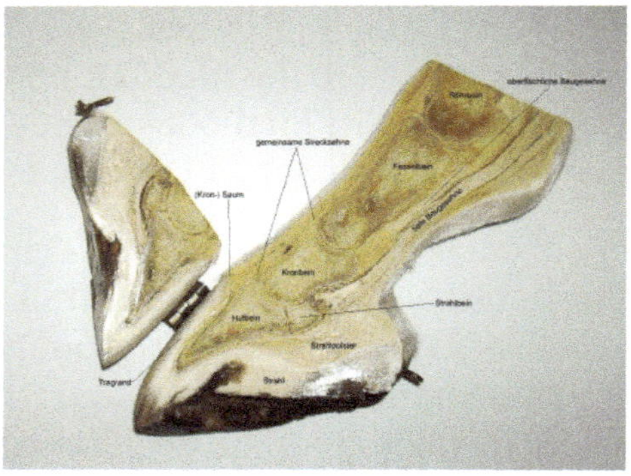

Section of a horse foot

The horse's foot is in an unguligrade orientation. The columnar orientation of bones and connective tissue is similarly well-aligned to transmit loads during the weight-bearing phase of locomotion. The thick keratinized and semicircular hoof changes shape during loading and unloading. Similarly, the cushioned frog situated centrally at the rear ends of the hoof undergoes compression during loading, and expansion when unloaded. Together, the hoof and cushioned frog structures may work in concert with hoof capsule to provide shock absorption. The horse hoof also acts dynamically during loading, which may cushion the endoskeleton from high loads that would otherwise produce critical deformation.

Elephant Foot

The hind limb and foot of the elephant are oriented semi-plantigrade, and closely resemble the structure and function of the human foot. The tarsals and metapodials are arranged so as to form an arch, similarly to the human foot. The six toes of each foot of the elephant are enclosed in a flexible sheath of skin. Similar to the dog's paw, the elephant's phalanges are oriented in a downward direction. The distal phalanges of the elephant do not directly touch the ground, and are attached to the respective nail/hoof. Distal cushions occupy the spaces between the muscle tendon units and ligaments within the hindfoot, midfoot and forefoot bones on the plantar surface. The distal cushion is highly innervated by sensory structures (Meissner's and Pacinian corpuscles), making the distal foot one of the most sensitive structures of the elephant (more so than its trunk). The cushions of the elephant's foot respond to the requirement to store and absorb mechanical loads when they are compressed, and to distribute locomotor loads over a large area in order to keep foot tissue stresses within acceptable levels. In addition, the musculoskeletal foot arch and sole cushion of the elephant act in concert, similarly to the horse's cushioned frog and hoof and the human foot. In the elephant, the nearly half-cupula-shaped arrangement of the bony elements of the metatarsals and toes has interesting similarities to the structure of the arches of human feet.

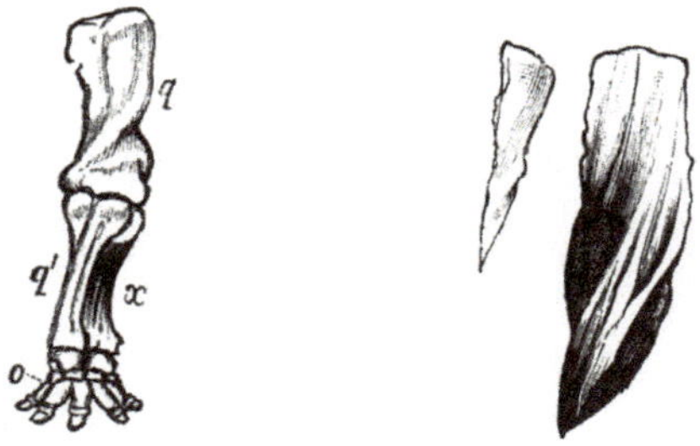

Leg skeleton of the modern elephant

Recently, scientists at the Royal Veterinary College in the United Kingdom have discovered that the elephant possesses a sixth false toe, a sesamoid, located similarly to the giant panda's extra "thumb". They found that this sixth toe acts to support and distribute the weight of the elephant.

Human Foot

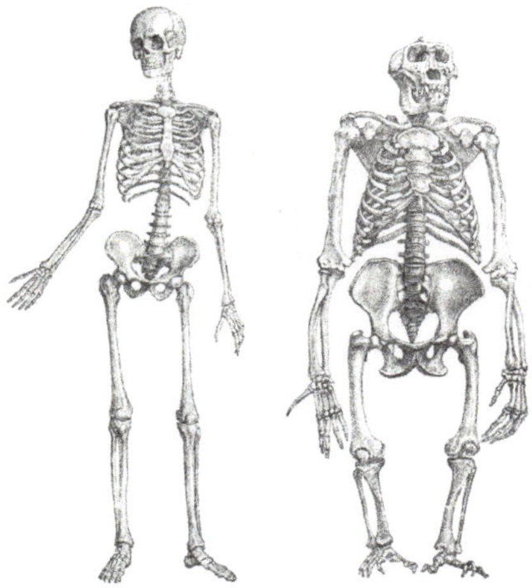

Skeleton of the human and gorilla (gorilla shown in non-natural posture)

The unique plantigrade alignment of the human foot results in a distal-limb structure that can adapt to a variety of conditions. The less mobile and more robust tarsal bones are shaped and aligned to accept and transmit large loads during the early phases of stance (initial contact and loading response phases of walking, and inadvertent heel strikes during running). The tarsals of the midfoot, which are smaller and shorter than the hindfoot tarsals, appear well oriented to transmit loads between the hindfoot and forefoot; this is necessary for load transfer and locking of the foot complex into a rigid lever for late stance phase. Conversely, the midfoot bones and joints also allow for the transmission of loads and inter-joint movement that unlocks the foot to create a loosely packed structure which renders the foot highly compliant over a variety of surfaces. In this configuration, the foot is able to absorb and damp the large loads encountered during heel strike and early weight acceptance. The forefoot, with its long metatarsal and relatively long phalanges, transmits loads during the end-of-stance phase that facilitate the push-off and transfer of forward momentum. The forefoot also serves as a lever to allow balance during standing and jumping. In addition, the arches of the foot that span the hindfoot, midfoot and forefoot play a critical role in the nature of transformation of the foot from a rigid lever to a flexible weight-accepting structure.

With a running gait, the foot-loading order is usually the reverse of walking. The foot strikes the ground with the ball of the foot, and then the heel drops. The heel drop elastically extends the Achilles tendon; this extension is reversed during the push-off.

Clinical Implications

Veterinarian or human healthcare professionals often respond when the foot of a dog, horse, elephant or human develops an abnormality. They typically investigate to understand the nature of the pathology in order to generate and implement a clinical treatment plan. For instance, the paws of the dog and the hindfoot work together to absorb the shock of jumping and running, and to provide flexibility of movement. If the dog's skeletal structures in areas other than the foot are compromised, the foot may be burdened with compensatory loading. Structural faults such as straight or loose shoulders, straight stifles, loose hips, and lack of balance between the forefoot and hindfoot, can all cause gait abnormalities that in turn damage the hindfoot and paws by overloading their foot structures as they compensate for the structural faults.

In the horse, dryness of the hoof may cause stiffening of the external foot structure. The stiffer hoof reduces the foot's load attenuation capacity, rendering the horse unable to bear much weight on the distal limb. Similar characteristic features emerge in the human foot in the form of the *pes cavus* alignment deformity, which is produced by tight connective tissue structures and joint congruency that create a rigid foot complex. Individuals with *pes cavus* display characteristic reduced load-attenuation features, and other structures proximal to the foot may compensate with increased load transfer (i.e., excessive loading to the knees, hips, lumbo-pelvic joints or lumbar vertebrae). Foot disorders are common in captive elephants. However, the cause is poorly understood.

Aquatic Locomotion

Aquatic locomotion is biologically propelled motion through a liquid medium. The simplest propulsive systems are composed of cilia and flagella. Swimming has evolved a number of times in a range of organisms including arthropods, fish, molluscs, reptiles, birds, and mammals.

Evolution of Swimming

Swimming evolved a number of times in unrelated lineages, and the evolutionary pressures leading to its adoption are unknown. Supposed jellyfish fossils occur in the Ediacaran, but the first free-swimming animals appear in the Early to Middle Cambrian. These are mostly related to the arthropods, and include the Anomalocaridids, which swam by means of lateral lobes in a fashion reminiscent of today's cuttlefish. Cephalopods joined the ranks of the nekton in the late Cambrian, and chordates were probably swimming from the Early Cambrian. Many terrestrial animals retain some capacity to swim, however some have returned to the water and developed the capacities for aquatic locomotion.

Micro-organisms

Bacterial

Ciliates

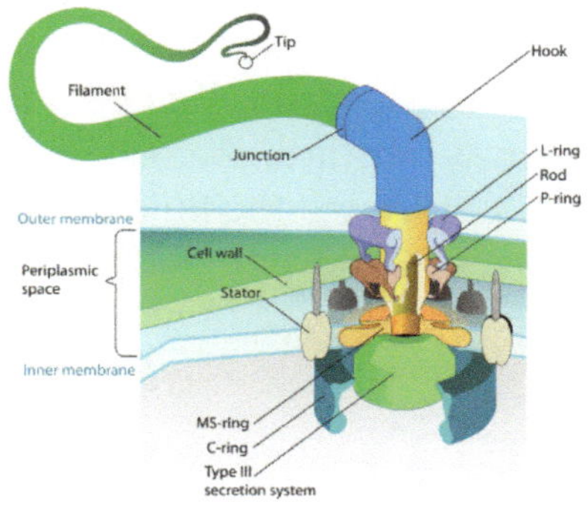

The flagellum of a Gram-negative bacteria is rotated by a molecular motor at its base

Ciliates use small flagella called cilia to move through the water. One ciliate will generally have hundreds to thousands of cilia that are densely packed together in arrays. During movement, an individual cilium deforms using a high-friction power stroke followed by a low-friction recovery stroke. Since there are multiple cilia packed together on an individual organism, they display collective behavior in a metachronal rhythm. This means the deformation of one cilium is in phase with the deformation of its neighbor, causing deformation waves that propagate along the surface of the organism. These propagating waves of cilia are what allow the organism to use the cilia in a coordinated manner to move. A typical example of a ciliated microorganism is the *Paramecium*, a one-celled, ciliated protozoan covered by thousands of cilia. The cilia beating together allow the *Paramecium* to propel through the water at speeds of 500 micrometers per second.

Salmon spermatozoa for artificial propagation

Flagellates

Certain organisms such as bacteria and animal sperm have flagellum which have developed a way to move in liquid environments. A rotary motor model shows that bacteria uses the protons of an electrochemical gradient in order to move their flagella. Torque in the flagella of bacteria is created by particles that conduct protons around the base of the flagellum. The direction of rotation of the flagella in bacteria comes from the occupancy of the proton channels along the perimeter of the flagellar motor.

Movement of sperm is called sperm motility. The middle of the mammalian spermatozoon contains mitochondria that power the movement of the flagellum of the sperm. The motor around the base produces torque, just like in bacteria for movement through the aqueous environment.

Pseudopodia

Movement using a pseudopod is accomplished through increases in pressure at one point on the cell membrane. This pressure increase is the result of actin polymerization between the cortex and the membrane. As the pressure increases the cell membrane is pushed outward creating the pseudopod. When the pseudopod moves outward, the rest of the body is pulled forward by cortical tension. The result is cell movement through the fluid medium. Furthermore, the direction of movement is determined by chemotaxis. When chemoattraction occurs in a particular area of the cell membrane, actin polymerization can begin and move the cell in that direction. An excellent example of an organism that utilizes pseudopods is *Naegleria fowleri*.

A Simple Animation

Invertebrates

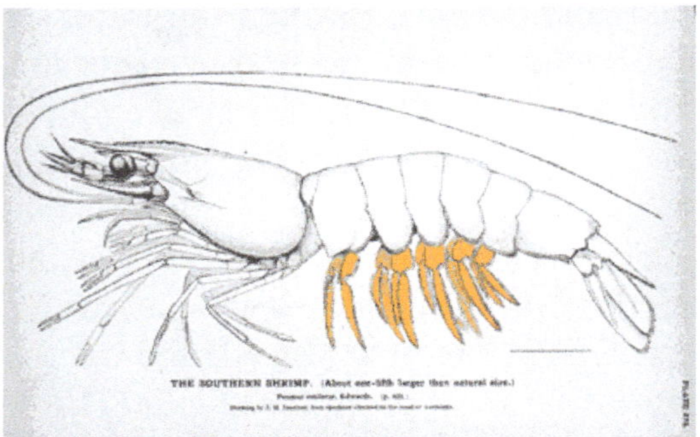

Shrimp paddle with special swimming legs (pleopods)

Among the radiata, jellyfish and their kin, the main form of swimming is to flex their cup shaped bodies. All jellyfish are free-swimming, although many of these spend most of their time swimming passively. Passive swimming is akin to gliding; the organism floats, using currents where it can, and does not exert any energy into controlling its position or motion. Active swimming, in contrast, involves the expenditure of energy to travel to a desired location.

In bilateria, there are many methods of swimming. The arrow worms (chaetognatha) undulate their finned bodies, not unlike fish. Nematodes swim by undulating their fin-less bodies. Some Arthropod groups can swim - including many crustaceans. Most crustaceans, such as shrimp, will usually swim by paddling with special swimming legs (pleopods). Swimming crabs swim with modified walking legs (pereiopods). Daphnia, a crustacean, swims by beating its antennae instead.

There are also a number of forms of swimming molluscs. Many free-swimming sea slugs, such as sea angels, flap fin-like structures. Some shelled molluscs, such as scallops can briefly swim by clapping their two shells open and closed. The molluscs most evolved for swimming are the cephalopods.

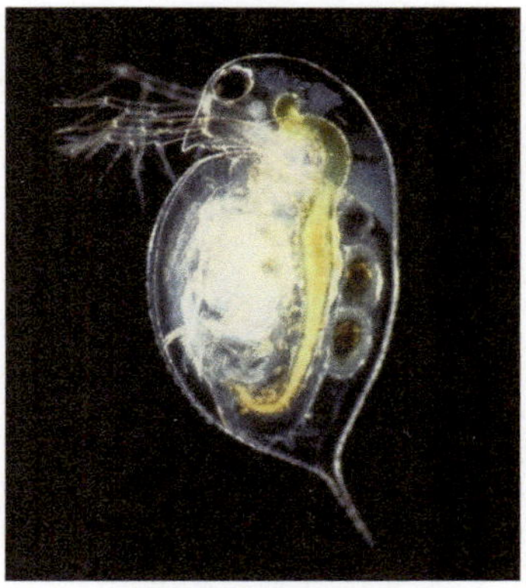

Daphnia swims by beating its antennae

Among the Deuterostomia, there are a number of swimmers as well. Feather stars can swim by undulating their many arms . Salps move by pumping waters through their gelatinous bodies. The deuterosomes most evolved for swimming are found among the vertebrates, notably the fish.

Jet Propulsion

Octopuses swim headfirst, with arms trailing behind

Jet propulsion is a method of aquatic locomotion where animals fill a muscular cavity and squirt out water to propel them in the opposite direction of the squirting water. Most organisms are equipped with one of two designs for jet propulsion; they can draw water from the rear and expel it from the rear, such as jellyfish, or draw water from front and expel it from the rear, such as salps. Filling up the cavity causes an increase in both the mass and drag of the animal. Because of the expanse of the contracting cavity, the animal's velocity fluctuates as it moves through the water, accelerating while expelling water and decelerating while vacuuming water. Even though these fluctuations in drag and mass can be ignored if the frequency of the jet-propulsion cycles is high enough, jet-propulsion is a relatively inefficient method of aquatic locomotion.

Jellyfish pulsate their bell for a type of jet locomotion

Scallops swim by clapping their two shells open and closed

All cephalopods can move by jet propulsion, but this is a very energy-consuming way to travel compared to the tail propulsion used by fish. The relative efficiency of jet propulsion decreases further as animal size increases. Since the Paleozoic, as competition with fish produced an environment where efficient motion was crucial to survival, jet propulsion has taken a back role, with fins and tentacles used to maintain a steady velocity. The stop-start motion provided by the jets, however, continues to be useful for providing bursts of high speed - not least when capturing prey or avoiding predators. Indeed, it makes cephalopods the fastest marine invertebrates, and they can outaccelerate most fish. Oxygenated water is taken into the mantle cavity to the gills and through muscular contraction of this cavity, the spent water is expelled through the hyponome, created by a fold in the mantle. Motion of the cephalopods is usually backward as water is forced out anteriorly through the hyponome, but direction can be controlled somewhat by pointing it in different directions. Most cephalopods float (i.e. are neutrally buoyant), so do not need to swim to remain afloat. Squid swim more slowly than fish, but use more power to generate their speed. The loss in efficiency is due to the amount of water the squid can accelerate out of its mantle cavity.

Jellyfish use a one-way water cavity design which generates a phase of continuous cycles of jet-propulsion followed by a rest phase. The Froude efficiency is about 0.09, which indicates a very costly method of locomotion. The metabolic cost of transport for jellyfish is high when compared to a fish of equal mass.

Other jet-propelled animals have similar problems in efficiency. Scallops, which use a similar design to jellyfish, swim by quickly opening and closing their shells, which draws in water and expels it from all sides. This locomotion is used as a means to escape predators such as starfish. Afterwards, the shell acts as a hydrofoil to counteract the scallop's tendency to sink. The Froude efficiency is low for this type of movement, about 0.3, which is why it's used as an emergency escape mechanism from predators. However, the amount of work the scallop has to do is mitigated by the elastic hinge that connects the two shells of the bivalve. Squids swim by drawing water into their mantle cavity and expelling it through their siphon. The Froude efficiency of their jet-propulsion system is around 0.29, which is much lower than a fish of the same mass.

Much of the work done by scallop muscles to close its shell is stored as elastic energy in abductin tissue, which acts as a spring to open the shell. The elasticity causes the work done against the water to be low because of the large openings the water has to enter and the small openings the water has to leave. The inertial work of scallop jet-propulsion is also low. Because of the low inertial work, the energy savings created by the elastic tissue is so small that it's negligible. Medusae can also use their elastic mesoglea to enlarge their bell. Their mantle contains a layer of muscle sandwiched between elastic fibers. The muscle fibers run around the bell circumferentially while the elastic fibers run through the muscle and along the sides of the bell to prevent lengthening. After making a single contraction, the bell vibrates passively at the resonant frequency to refill the bell. However, in contrast with scallops, the inertial work is similar to the hydrodynamic work due to how medusas expel water - through a large opening at low velocity. Because of this, the negative pressure created by the vibrating cavity is lower than the positive pressure of the jet, meaning that inertial work of the mantle is small. Thus, jet-propulsion is shown as an inefficient swimming technique.

Fish

Open water fish, like this Atlantic bluefin tuna, are usually streamlined for straightline speed, with a deeply forked tail and a smooth body shaped like a spindle tapered at both ends.

Many reef fish, like this queen angelfish, have a body flattened like a pancake, with pectoral and pelvic fins that act with the flattened body to maximize manoeuvrability.

Many fish swim through water by creating undulations with their bodies or oscillating their fins. The undulations create components of forward thrust complemented by a rearward force, side forces which are wasted portions of energy, and a normal force that is between the forward thrust and side force. Different fish swim by undulating different parts of their bodies. Eel-shaped fish undulate their entire body in rhythmic sequences. Streamlined fish, such as salmon, undulate the caudal portions of their bodies. Some fish, such as sharks, use stiff, strong fins to create dynamic lift and propel themselves. It is common for fish to use more than one form of propulsion, although they will display one dominant mode of swimming Gait changes have even been observed in juvenile reef fish of various sizes. Depending on their needs, fish can rapidly alternate between synchronized fin beats and alternating fin beats

According to *Guinness World Records 2009*, *H. zosterae* (the dwarf seahorse) is the slowest moving fish, with a top speed of about 5 feet (150 cm) per hour. They swim very poorly, rapidly fluttering a dorsal fin and using pectoral fins (located behind their eyes) to steer. Seahorses have no caudal fin.

Body-caudal Fin (BCF) Propulsion

- Anguilliform: Anguilliform swimmers are typically slow swimmers. They undulate the majority of their body and use their head as the fulcrum for the load they are moving. At any point during their undulation, their body has an amplitude between 0.5-1.0 wavelengths. The amplitude that they move their body through allows them to swim backwards. Anguilliform locomotion is usually seen in fish with long, slender bodies like eels, lampreys, oarfish, and a number of catfish species.
- Subcarangiform, Carangiform, Thunniform: These swimmers undulate the posterior half of their body and are much faster than anguilliform swimmers. At any point while they are swimming, a wavelength <1 can be seen in the undulation pattern of the body. Some Carangiform swimmers include nurse sharks, bamboo sharks, and reef sharks. Thunniform swimmers are very fast and some common Thunniform swimmers include tuna, white sharks, salmon, jacks, and mako sharks. Thunniform swimmers only undulate their high aspect ratio caudal fin, so they are usually very stiff to push more water out of the way.
- Ostraciiform: Ostraciiform swimmers oscillate their caudal region, making them relatively slow swimmers. Boxfish, torpedo rays, and momyrs employ Ostraciiform locomotion. The cow fish uses Osctraciiform locomotion to hover in the water column.

Median Paired Fin (MPF) Propulsion

- Tetraodoniform, Balistiform, Diodontiform: These swimmers oscillate their median fins. They are typically slow swimmers, and some notable examples include the oceanic sunfish (which has extremely modified anal and dorsal fins), puffer fish, and triggerfish.
- Rajiform, Amiiform, Gymnotiform: This locomotory mode is accomplished by undulation of the pectoral and median fins. During their undulation pattern, a wavelength >1 can be seen in their fins. They are typically slow to moderate swimmers, and some examples include rays, bowfin, and knife fishes. The black ghost knife fish is a Gymnotiform swimmer that has a very long ventral ribbon fin. Thrust is produced by passing waves down the ribbon fin while the body remains rigid. This also allows the ghost knife fish to swim in reverse.
- Labriform: Labriform swimmers are also slow swimmers. They oscillate their pectoral fins to create thrust. Oscillating fins create thrust when a starting vortex is shed from the trailing edge of the fin. As the foil departs from the starting vortex, the effect of that vortex diminishes, while the bound circulation remains, producing lift. Labriform swimming can be viewed as continuously starting and stopping. Wrasses and surf perch are common Labriform swimmers

Hydrofoils

The leopard shark angles its pectoral fins so they behave as hydrofoils to control the animal's pitch

Hydrofoils, or fins, are used to push against the water to create a normal force to provide thrust, propelling the animal through water. Sea turtles and penguins beat their paired hydrofoils to create lift. Some paired fins, such as pectoral fins on leopard sharks, can be angled at varying degrees to allow the animal to rise, fall, or maintain its level in the water column. The reduction of fin surface area helps to minimize drag, and therefore increase efficiency. Regardless of size of the animal, at any particular speed, maximum possible lift is proportional to (wing area) x $(\text{speed})^2$. Dolphins and whales have large, horizontal caudal hydrofoils, while many fish and sharks have vertical caudal hydrofoils. Porpoising (seen in cetaceans, penguins, and pinnipeds) may save energy if they are moving fast. Since drag increases with speed, the work required to swim unit distance is greater at higher speeds, but the work needed to jump unit distance is independent of speed. Seals propel themselves through the water with their caudal tail, while sea lions create thrust solely with their pectoral flippers.

Drag Powered Swimming

The slowest-moving fishes are the sea horses, often found in reefs

As with moving through any fluid, friction is created when molecules of the fluid collide with organism. The collision causes drag against moving fish, which is why many fish are streamlined in shape. Streamlined shapes work to reduce drag by orienting elongated objects parallel to the force of drag, therefore allowing the current to pass over and taper off the end of the fish. This streamlined shape allows for more efficient use of energy locomotion. Some flat-shaped fish can take advantage of pressure drag by having a flat bottom surface and curved top surface. The pressure drag created allows for the upward lift of the fish.

Appendages of aquatic organisms propel them in two main and biomechanically extreme mechanisms. Some use lift powered swimming, which can be compared to flying as appendages flap like wings, and reduce drag on the surface of the appendage. Others use drag powered swimming, which can be compared to oars rowing a boat, with movement in a horizontal plane, or paddling, with movement in the parasagittal plane.

Drag swimmers use a cyclic motion where they push water back in a power stroke, and return their limb forward in the return or recovery stroke. When they push water directly backwards, this moves their body forward, but as they return their limbs to the starting position, they push water forward, which will thus pull them back to some degree, and so opposes the direction that the body is heading. This opposing force is called drag. The return-stroke drag causes drag swimmers to employ different strategies than lift swimmers. Reducing drag on the return stroke is essential for optimizing efficiency. For example, ducks paddle through the water spreading the webs of their feet as they move water back, and then when they return their feet to the front they pull their webs together to reduce the subsequent pull of water forward. The legs of water beetles have little hairs which spread out to catch up and move water back in the power stroke, but lay flat as the appendage moves forward in the return stroke. Also, the water beetle's legs have a side that is wider and is held perpendicular to the motion when pushing backward, but the leg is then rotated when the limb is to return forward, so that the thinner side will catch up less water.

Drag swimmers experience a lessened efficiency in swimming due to resistance which affects their optimum speed. The less drag a fish experiences, the more it will be able to maintain higher speeds. Morphology of the fish can be designed to reduce drag, such as streamlining the body. The cost of transport is much higher for the drag swimmer, and when deviating from its optimum speed, the drag swimmer is energetically strained much more than the lift swimmer. There are natural processes in place to optimize energy use, and it is thought that adjustments of metabolic rates can compensate in part for mechanical disadvantages.

Semi-aquatic animals compared to fully aquatic animals exhibit exacerbation of drag. Design that allows them to function out of the water limits the efficiency possible to be reached when in the water. In water swimming at the surface exposes them to resistive wave drag and is associated with a higher cost than submerged swimming. Swimming below the surface exposes them to resistance due to return strokes and pressure, but primarily friction. Frictional drag is due to fluid viscosity and morphology characteristics. Pressure drag is due to the difference of water flow around the body and is also affected by body morphology. Semi-aquatic organisms encounter increased resistive forces when in or out of the water, as they are not specialized for either habitat. The morphology of otters and beavers, for example, must meet needs for both environments. Their fur decreases streamlining and creates additional drag. The platypus may be a good example of an intermediate between drag and lift swimmers because it has been shown to have a rowing mechanism which is similar to lift-based pectoral oscillation. The limbs of semi-aquatic organisms are reserved for use on land and using them in water not only increases the cost of locomotion, but limits them to drag-based modes. Although they are less efficient, drag swimmers are able to produce more thrust at low speeds than lift swimmers. They are also thought to be better for maneuverability due to the large thrust produced.

Amphibians

Most of the Amphibia have a larval state, which has inherited anguilliform motion, and a laterally compressed tail to go with it, from fish ancestors. The corresponding tetrapod adult forms, even in the tail-retaining sub-class Urodeles, are sometimes aquatic to only a negligible extent (as in the genus Salamandra, whose tail has lost its suitability for aquatic propulsion), but the majority of Urodeles, from the newts to the giant salamander Megalobatrachus, retain a laterally compressed tail for a life that is aquatic to a considerable degree, which can use in a carangiform motion.

Common toad (Bufo bufo) swimming

Of the tailless amphibians (the frogs and toads of the sub-class Anura) the majority are aquatic to an insignificant extent in adult life, but in that considerable minority that are mainly aquatic we encounter for the first time the problem of adapting the tailless-tetrapod structure for aquatic propulsion. The mode that they use is unrelated to any used by fish. With their flexible back legs and webbed feet they execute something close to the leg movements of a human 'breast stroke,' rather more efficiently because the legs are better streamlined.

Reptiles

Nile crocodile (Crocodylus niloticus) swimming

From the point of view of aquatic propulsion, the descent of modern members of the class Reptilia from archaic tailed Amphibia is most obvious in the case of the order Crocodilia (crocodiles and alligators), which use their deep, laterally compressed tails in an essentially carangiform mode of propulsion .

Terrestrial snakes, in spite of their 'bad' hydromechanical shape with roughly circular cross-section and gradual posterior taper, swim fairly readily when required, by an anguilliform propulsion.

Cheloniidae (true turtles) have found a beautiful solution to the problem of tetrapod swimming through the development of their forelimbs into flippers of high-aspect-ratio wing shape, with which they imitate a bird's propulsive mode more accurately than do the eagle-rays themselves.

Immature Hawaiian green sea turtle in shallow waters

Macroplata

Fin and Flipper Locomotion

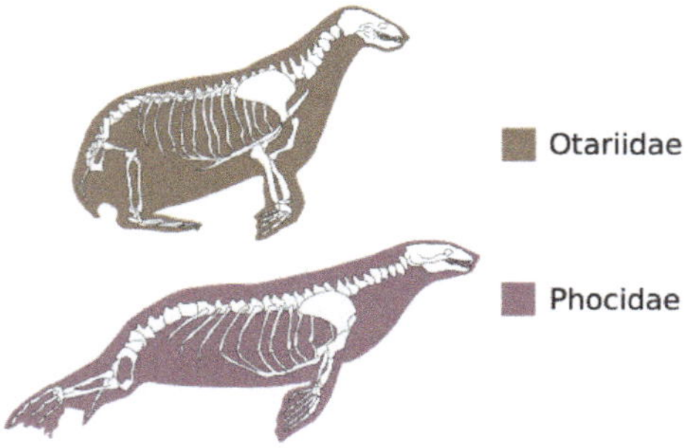

Comparative skeletal anatomy of a typical otariid seal and a typical phocid seal

Aquatic reptiles such as sea turtles and extinct species like Pliosauroids predominantly use their pectoral flippers to propel themselves through the water and their pelvic flippers for maneuvering. During swimming they move their pectoral flippers in a dorso-ventral motion, causing forward motion. During swimming, they rotate their front flippers to decrease drag through the water column and increase efficiency. Newly hatched sea turtles exhibit several behavioral skills that help orientate themselves towards the ocean as well as identifying the transition from sand to water. If rotated in the pitch, yaw or roll direction, the hatchlings are capable of counteracting the forces acting upon them by correcting with either their pectoral or pelvic flippers and redirecting themselves towards the open ocean.

Among mammals otariids (fur seals) swim primarily with their front flippers, using the rear flippers for steering, and phocids (true seals) move the rear flippers laterally, pushing the animal through the water.

Escape Reactions

Animated representation of lobstering

Some arthropods, such as lobsters and shrimps, can propel themselves backwards quickly by flicking their tail, known as lobstering or the caridoid escape reaction.

Varieties of fish, such as teleosts, also use fast-starts to escape from predators. Fast-starts are characterized by the muscle contraction on one side of the fish twisting the fish into a C-shape. Afterwards, muscle contraction occurs on the opposite side to allow the fish to enter into a steady swimming state with waves of undulation traveling alongside the body. The power of the bending motion comes from fast-twitch muscle fibers located in the central region of the fish. The signal to perform this contraction comes from a set of Mauthner cells which simultaneously send a signal to the muscles on one side of the fish. Mauthner cells are activated when something startles the fish and can be activated by visual or sound-based stimuli.

Fast-starts are split up into three stages. Stage one, which is called the preparatory stroke, is characterized by the initial bending to a C-shape with small delay caused by hydrodynamic resistance. Stage two, the propulsive stroke, involves the body bending rapidly to the other side, which may occur multiple times. Stage three, the rest phase, cause the fish to return to normal steady-state swimming and the body undulations begin to cease. Large muscles located closer to the central portion of the fish are stronger and generate more force than the muscles in the tail. This asymmetry in muscle composition causes body undulations that occur in Stage 3. Once the fast-start is completed, the position of the fish has been shown to have a certain level of unpredictability, which helps fish survive against predators.

The rate at which the body can bend is limited by resistance contained in the inertia of each body part. However, this inertia assists the fish in creating propulsion as a result of the momentum created against the water. The forward propulsion created from C-starts, and steady-state swimming in general, is a result of the body of the fish pushing against the water. Waves of undulation create rearward momentum against the water providing the forward thrust required to push the fish forward.

Efficiency

The Froude propulsion efficiency is defined as the ratio of power output to the power input:

$$nf = 2U_1 / (U_1 + U_2)$$

where U1 = free stream velocity and U2 = jet velocity. A good efficiency for carangiform propulsion is between 50 and 80%.

Minimizing Drag

Pressure differences occur outside the boundary layer of swimming organisms due to disrupted flow around the body. The difference on the up- and down-stream surfaces of the body is pressure drag, which creates a downstream force on the object. Frictional drag, on the other hand, is a result of fluid viscosity in the boundary layer. Higher turbulence causes greater frictional drag.

Reynolds number (Re) is the measure of the relationships between inertial and viscous forces in flow ((animal's length x animal's velocity)/kinematic viscosity of the fluid). Turbulent flow can be found at higher Re values, where the boundary layer separates and creates a wake, and laminar flow can be found at lower Re values, when the boundary layer separation is delayed, reducing wake and kinetic energy loss to opposing water momentum.

The body shape of a swimming organism affects the resulting drag. Long, slender bodies reduce pressure drag by streamlining, while short, round bodies reduce frictional drag; therefore, the optimal shape of an organism depends on its niche. Swimming organisms with a fusiform shape are likely to experience the greatest reduction in both pressure and frictional drag.

Wing shape also affects the amount of drag experienced by an organism, as with different methods of stroke, recovery of the pre-stroke position results in the accumulation of drag.

High-speed ram ventilation creates laminar flow of water from the gills along the body of an organism.

The secretion of mucus along the organism's body surface, or the addition of long-chained polymers to the velocity gradient, can reduce frictional drag experienced by the organism.

Buoyancy

Many aquatic/marine organisms have developed organs to compensate for their weight and control their buoyancy in the water. These structures, make the density of their bodies very close to that of the surrounding water. Some hydrozoans, such as siphonophores, has gas-filled floats; the Nautilus, Sepia, and Spirula (Cephalopods) have chambers of gas within their shells; and most teleost fish and many lantern fish (Myctophidae) are equipped with swim bladders. Many aquatic and marine organisms may also be composed of low-density materials. Deep-water teleosts, which do not have a swim bladder, have few lipids and proteins, deeply ossified bones, and watery tissues that maintain their buoyancy. Some sharks' livers are composed of low-density lipids, such as hydrocarbon squalene or wax esters (also found in Myctophidae without swim bladders), which provide buoyancy.

Swimming animals that are denser than water must generate lift or adapt a benthic lifestyle. Movement of the fish to generate hydrodynamic lift is necessary to prevent sinking. Often, their bodies act as hydrofoils, a task that is more effective in flat-bodied fish. At a small tilt angle, the lift is greater for flat fish than it is for fish with narrow bodies. Narrow-bodied fish use their fins as hydrofoils while their bodies remain horizontal. In sharks, the heterocercal tail shape drives water downward, creating a counteracting upward force while thrusting the shark forward. The lift generated is assisted by the pectoral fins and upward-angle body positioning. It is supposed that tunas primarily use their pectoral fins for lift.

Buoyancy maintenance is metabolically expensive. Growing and sustaining a buoyancy organ, adjusting the composition of biological makeup, and exerting physical strain to stay in motion demands large amounts of energy. It is proposed that lift may be physically generated at a lower energy cost by swimming upward and gliding downward, in a "climb and glide" motion, rather than constant swimming on a plane.

Temperature

Temperature can also greatly affect the ability of aquatic organisms to move through water. This is because temperature not only affects the properties of the water, but also the organisms in the water, as most have an ideal range specific to their body and metabolic needs.

Q10 (temperature coefficient), the factor by which a rate increases at a 10 °C increase in temperature, is used to measure how organisms' performance relies on temperature. Most have increased rates as water becomes warmer, but some have limits to this and others find ways to alter such effects, such as by endothermy or earlier recruitment of faster muscle.

For example, *Crocodylus porosus*, or estuarine crocodiles, were found to increase swimming speed from 15 °C to 23 °C and then to have peak swimming speed from 23 °C to 33 °C. However, performance began to decline as temperature rose beyond that point, showing a limit to the range of temperatures at which this species could ideally perform.

Submergence

The more of the animal's body that is submerged while swimming, the less energy it uses. Swimming on the surface requires two to three times more energy than when completely submerged. This is because of the bow wave that is formed at the front when the animal is pushing the surface of the water when swimming, creating extra drag.

Secondary Evolution

Chinstrap penguin leaping over water

While tetrapods lost many of their natural adaptations to swimming when they evolved onto the land, many have re-evolved the ability to swim or have indeed returned to a completely aquatic lifestyle.

Primarily or exclusively aquatic animals have re-evolved from terrestrial tetrapods multiple times: examples include amphibians such as newts, reptiles such as crocodiles, sea turtles, ichthyosaurs, plesiosaurs and mosasaurs, marine mammals such as whales, seals and otters, and birds such as penguins. Many species of snakes are also aquatic and live their entire lives in the water. Among invertebrates, a number of insect species have adaptations for aquatic life and locomotion. Examples of aquatic insects include dragonfly larvae, water boatmen, and diving beetles. There are also aquatic spiders, although they tend to prefer other modes of locomotion under water than swimming proper.

Swimming dog

Even though primarily terrestrial tetrapods have lost many of their adaptations to swimming, the ability to swim has been preserved or re-developed in many of them. It may never have been completely lost.

Examples are: Some breeds of dog swim recreationally. Umbra, a world record-holding dog, can swim 4 miles (6.4 km) in 73 minutes, placing her in the top 25% in human long-distance swimming competitions. Although most cats hate water, adult cats are good swimmers. The fishing cat is one wild species of cat that has evolved special adaptations for an aquatic or semi-aquatic lifestyle – webbed digits. Tigers and some individual jaguars are the only big cats known to go into water readily, though other big cats, including lions, have been observed swimming. A few domestic cat breeds also like swimming, such as the Turkish Van.

Horses, moose, and elk are very powerful swimmers, and can travel long distances in the water. Elephants are also capable of swimming, even in deep waters. Eyewitnesses have confirmed that camels, including dromedary and Bactrian camels, can swim, despite the fact that there is little deep water in their natural habitats.

Both domestic and wild rabbits can swim. Domestic rabbits are sometimes trained to swim as a circus attraction. A wild rabbit famously swam in an apparent attack on U.S. President Jimmy Carter's boat when it was threatened in its natural habitat.

The guinea pig (or cavy) is noted as having an excellent swimming ability. Mice can swim quite well. They do panic when placed in water, but many lab mice are used in the Morris water maze, a test to measure learning. When mice swim, they use their tails like flagella and kick with their legs.

Many snakes are excellent swimmers as well. Large adult anacondas spend the majority of their time in the water, and have difficulty moving on land.

Many monkeys can naturally swim and some, like the proboscis monkey, crab-eating macaque, and rhesus macaque swim regularly.

Human Swimming

Swimming has been known amongst humans since prehistoric times; the earliest record of swimming dates back to Stone Age paintings from around 7,000 years ago. Competitive swimming started in Europe around 1800 and was part of the first modern 1896 Summer Olympics in Athens, though not in a form comparable to the contemporary events. It was not until 1908 that regulations were implemented by the International Swimming Federation to produce competitive swimming.

Fin and Flipper Locomotion

Periophthalmus gracilis (from Malaysia to North Australia)

Fin and flipper locomotion occurs mostly in aquatic locomotion, and rarely in terrestrial locomotion. From the three common states of matter — gas, liquid and solid, these appendages are adapted for liquids, mostly Fresh or Saltwater and used in locomotion, steering and balancing of the body. Locomotion is important in order to escape predators, acquire food, find mates and bury for shelter, nest or food. Aquatic locomotion consists of swimming, whereas terrestrial locomotion encompasses walking, 'crutching', jumping, digging as well as covering. Some animals such as Sea turtles and Mudskippers use these two environments for different purposes, for example using the land for nesting, and the sea to hunt for food.

Aquatic Locomotion with Fins and Flippers

Aquatic Locomotion of Fish

Fish live in Fresh or Saltwater habitats and some exceptions are capable of coming on land (Mudskippers). Most fish have a line of muscle blocks, called myomeres, along each side of the body. To swim, they alternately contract one side and relax the other side in a progression which goes from the head to the tail. In this way, an undulatory locomotion results, first bending the body one way in a wave which travels down the body, and then back the other way, with the contracting and relaxing muscles switching roles. They use their fins to propel themselves through the water in this swimming motion. Actinopterygians, the ray-finned fish show an evolutionary pattern of fine control ability to control the dorsal and ventral lobe of the caudal fin. Through developmental changes, intrinsic caudal muscles were added, which enable fish to exhibit such complex maneuvers such as control during acceleration, braking and backing. Studies have shown that the muscles in the caudal fin, have independent activity patterns from the myotomal musculature. These results show specific kinematic roles for different part of the fish's musculature. A curious example of fish adaption is the Ocean sunfish, also known as the *Mola mola*. These fish have undergone significant developmental changes reducing their spinal cord, giving them a disk like appearance, and investing in two very large fins for propulsion. This adaptation usually gives them the appearance that they are as long as they are tall. They are also amazing fish in that they hold the world record in weight gain from fry to adult (60 million times its weight).

Aquatic Locomotion of Marine Mammals

Swimming mammals, such as whales, dolphins,and sea lions, use their flippers to move forward through the water column. During swimming sea lions have a thrust phase, which lasts about 60% of the full cycle, and the recovery phase lasts the remaining 40%. A full cycle duration lasts about 0.5 to 1.0 seconds. Changing direction is a very rapid maneuver that is initiated by head movement towards the back of the animal that is followed by a spiral turn with the body. Due to their pectoral flippers being so closely located to their center of gravity, sea lions are capable of displaying astounding maneuverability in the pitch, roll, and yaw direction and are therefore not constrained, turning stochastically as they please. It is hypothesized that the increased level of maneuverability is caused by their complex habitat. Hunting occurs in difficult environments containing rocky inshore/kelp forest communities, with many niches for prey to hide, therefore requiring speed and maneuverability for capture. The complex skills of a sea lion are learned early on in ontogeny and most are perfected by the time the pups reach one year. Whales and dolphins are less maneuverable and more constrained in their movements. However, dolphins are capable of accelerating as fast as sea lions, but they are not capable of turning as quickly and as efficiently. For both whales and dolphins, their center of gravity does not line up with their pectoral flippers in a straight line, causing a much more rigid and stable swimming pattern.

Aquatic Locomotion of Marine Reptiles

Aquatic reptiles such as sea turtles predominantly use their pectoral flippers to propulse through the water and their pelvic flippers for maneuvering. During swimming they move their pectoral flippers in a clapping motion underneath their body and pull them back up into an airplane position, causing forward motion. During the swimming motion it is really important that they rotate their front flipper in order to decrease drag through the water column and increase their efficiency. Sea turtles exhibit a natural suite of behavior skills that help them direct themselves towards the ocean as well as identify the transition from sand to water after hatching. If rotated in the pitch, yaw or roll direction the hatchlings are capable of counteracting the forces acting upon them by correcting with either their pectoral or pelvic flippers and redirecting themselves towards the open ocean.

Terrestrial Locomotion

Terrestrial Locomotion of Fish

Terrestrial locomotion poses new obstacles such as gravity and new media, including sand, mudd, twigs, logs, debris, grass and many more. Fins and flippers are aquatically adapted appendages and typically aren't very useful in such an environment. It could be hypothesized that fish would try to "swim" on land, but studies have shown that some fish evolved to cope with the terrestrial environment. Mudskippers, for example demonstrate a 'crutching' gait which enables them to 'walk' over muddy surfaces as well as dig burrows to hide in. Mudskippers are also able to jump up to 3 cm distances. This behavior is described as starting with a J-curvature of the body at about

2/3 of its body length (with its tail wrapped towards the head), followed by a straightening of their body which propulses them like a projectile through the air. This behavior enables them to cope with the new environment and opens their habitat to new food sources as well as new predators.

Mudskippers in The Gambia

Terrestrial Locomotion of Marine Reptiles

Caretta caretta Jekyll Island, GA

Reptiles, such as sea turtles spend most of their lives in the ocean. However, their life cycle requires the females to come on shore and lay their nests on the beach. Consequently, the hatchlings emerge from the sand and have to run toward the water. Depending on their species, sea turtles are described to have either a symmetrical gait (diagonally opposite limbs are moving together) or an asymmetrical gait (Contra-lateral limbs move together). For example, loggerhead sea turtle hatchlings are commonly seen exhibiting symmetrical gait on sand, whereas, leatherback sea turtles employ the asymmetrical gait while on land. Notably, leatherbacks employ their front (pelvic) flippers more during forward terrestrial locomotion. Sea turtles can be seen nesting on subtropical and tropical beaches all around the world and exhibit such behavior such as arribada (Collective animal behavior). This is a phenomenon seen in Kemp's Ridley turtles which emerge all at once in one night only onto the beach to lay their nests.

References

- Weis, J.S. (2012). Walking sideways: the remarkable world of crabs. Ithaca, NY: Cornell University Press. pp. 63–77. ISBN 978-0-8014-5050-1. OCLC 794640315.
- Campbell, Neil A.; Reece, Jane B. (2005). Biology, 7th Edition. San Francisco: Pearson - Benjamin Cummings. pp. 522–523. ISBN 0-8053-7171-0.

- Dorit, R. L.; Walker, W. F.; Barnes, R. D. (1991). Zoology. Saunders College Publishing. p. 782. ISBN 978-0-03-030504-7.
- King, R.S. (2013). "BiLBIQ: A Biologically Inspired Robot with Walking and Rolling Locomotion". Biosystems and Biorobotics. 2. Springer, Verlag, Berlin, Heidelberg. doi:10.1007/978-3-642-34682-8. ISBN 978-3-642-34681-1.
- Jurmain, Robert; Kilgore, Lynn; Trevathan, Wenda (2008). Essentials of Physical Anthropology (7 ed.). Cengage Learning. p. 109. ISBN 9780495509394.
- Davies, S.J.J.F. (2003). "Birds I Tinamous and Ratites to Hoatzins". In Hutchins, Michael. Grzimek's Animal Life Encyclopedia. 8 (2 ed.). Farmington Hills, MI: Gale Group. pp. 99–101. ISBN 0-7876-5784-0.
- Roberts, Tristan David Martin (1995). Understanding Balance: The Mechanics of Posture and Locomotion. Nelson Thornes. p. 211. ISBN 978-1-56593-416-0. Retrieved 18 March 2015.
- Wilbur, Karl M.; Clarke, M.R.; Trueman, E.R., eds. (1985), "11: Evolution of Buoyancy and Locomotion in recent cephalopods", The Mollusca, 12. Paleontology and neontology of Cephalopods, New York: Academic Press, ISBN 0-12-728702-7
- Wilbur, Karl M.; Clarke, M.R.; Trueman, E.R., eds. (1985), The Mollusca, 12. Paleontology and neontology of Cephalopods, New York: Academic Press, ISBN 0-12-728702-7
- Daniel L. Gilbert; William J. Adelman; John M. Arnold (1990). Squid as Experimental Animals. Daniel L. Gilbert, John M. Arnold (illustrated ed.). Springer,. ISBN 978-0-306-43513-3.
- Harkness, John E.; Wagner, Joseph E. (1995). The Biology and Medicine of Rabbits and Rodents. Williams & Wilkins. pp. 30–39. ISBN 0-683-03919-9.
- Kehm, G (2007). Olympic Swimming and Diving:Great Moments in Olympic History. The Rosen Publishing Group. pp. 14. ISBN 1-4042-0970-0.

3

Fish Physiology

Fish physiology is the study of the physiological functions of a fish and how the parts of fishes function together. There is a difference between fish physiology and fish anatomy. Fish jaw, swim bladder and fish fin are some of the features of fish physiology. The major aspects of fish physiology are discussed in this chapter.

Fish Physiology

Fish physiology is the scientific study of how the component parts of fish function together in the living fish. It can be contrasted with fish anatomy, which is the study of the form or morphology of fishes. In practice, fish anatomy and physiology complement each other, the former dealing with the structure of a fish, its organs or component parts and how they are put together, such as might be observed on the dissecting table or under the microscope, and the later dealing with how those components function together in the living fish.

When threatened, the toxic pufferfish fills its extremely elastic stomach with water.

Respiration

Most fish exchange gases using gills on either side of the pharynx (throat). Gills are tissues which consist of threadlike structures called filaments. These filaments have many functions and "are involved in ion and water transfer as well as oxygen, carbon dioxide, acid and ammonia exchange. Each filament contains a capillary network that provides a large surface area for exchanging oxygen and carbon dioxide. Fish exchange gases by pulling oxygen-rich water through their mouths and pumping it over their gills. In some fish, capillary blood flows in the opposite direction to the water, causing countercurrent exchange. The gills push the oxygen-poor water out through openings in the sides of the pharynx.

Fish from multiple groups can live out of the water for extended time periods. Amphibious fish such as the mudskipper can live and move about on land for up to several days, or live in stagnant or otherwise oxygen depleted water. Many such fish can breathe air via a variety of mechanisms. The skin of anguillid eels may absorb oxygen directly. The buccal cavity of the electric eel may breathe air. Catfish of the families Loricariidae, Callichthyidae, and Scoloplacidae absorb air through their digestive tracts. Lungfish, with the exception of the Australian lungfish, and bichirs have paired lungs similar to those of tetrapods and must surface to gulp fresh air through the mouth and pass spent air out through the gills. Gar and bowfin have a vascularized swim bladder that functions in the same way. Loaches, trahiras, and many catfish breathe by passing air through the gut. Mudskippers breathe by absorbing oxygen across the skin (similar to frogs). A number of fish have evolved so-called accessory breathing organs that extract oxygen from the air. Labyrinth fish (such as gouramis and bettas) have a labyrinth organ above the gills that performs this function. A few other fish have structures resembling labyrinth organs in form and function, most notably snakeheads, pikeheads, and the Clariidae catfish family.

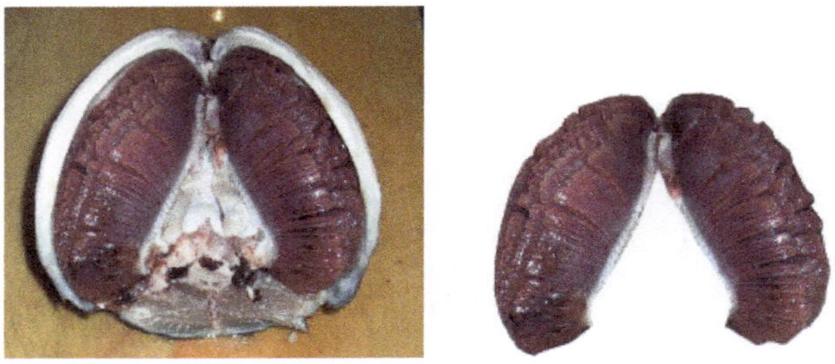

Gills inside the head of a tuna. The head is snout-down, with the view looking towards the mouth. On the right are the detached gills.

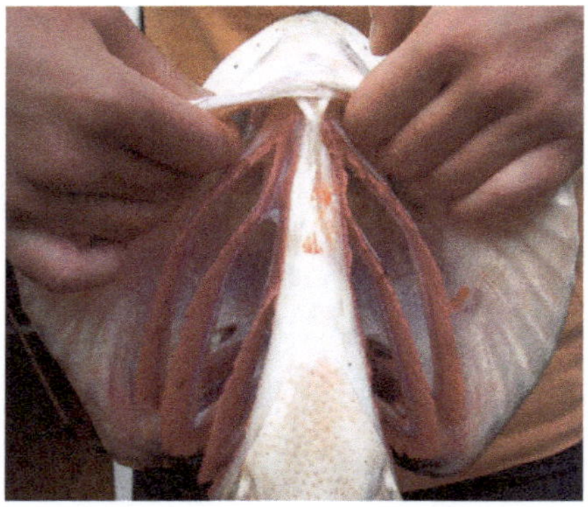

Gill arches bearing gills in a pike

Breathing air is primarily of use to fish that inhabit shallow, seasonally variable waters where the water's oxygen concentration may seasonally decline. Fish dependent solely on dissolved oxygen, such as perch and cichlids, quickly suffocate, while air-breathers survive for much longer, in some cases in water that is little more than wet mud. At the most extreme, some air-breathing fish are able to survive in damp burrows for weeks without water, entering a state of aestivation (summertime hibernation) until water returns.

Air breathing fish can be divided into *obligate* air breathers and *facultative* air breathers. Obligate air breathers, such as the African lungfish, are obligated to breathe air periodically or they suffocate. Facultative air breathers, such as the catfish *Hypostomus plecostomus*, only breathe air if they need to and can otherwise rely on their gills for oxygen. Most air breathing fish are facultative air breathers that avoid the energetic cost of rising to the surface and the fitness cost of exposure to surface predators.

All basal vertebrates breathe with gills. The gills are carried right behind the head, bordering the posterior margins of a series of openings from the esophagus to the exterior. Each gill is supported by a cartilagenous or bony gill arch. The gills of vertebrates typically develop in the walls of the pharynx, along a series of gill slits opening to the exterior. Most species employ a countercurrent exchange system to enhance the diffusion of substances in and out of the gill, with blood and water flowing in opposite directions to each other.

The gills are composed of comb-like filaments, the gill lamellae, which help increase their surface area for oxygen exchange. When a fish breathes, it draws in a mouthful of water at regular intervals. Then it draws the sides of its throat together, forcing the water through the gill openings, so that it passes over the gills to the outside. The bony fish have three pairs of arches, cartilaginous fish have five to seven pairs, while the primitive jawless fish have seven. The vertebrate ancestor no doubt had more arches, as some of their chordate relatives have more than 50 pairs of gills.

Higher vertebrates do not develop gills, the gill arches form during fetal development, and lay the basis of essential structures such as jaws, the thyroid gland, the larynx, the *columella* (corresponding to the stapes in mammals) and in mammals the malleus and incus. Fish gill slits may be the evolutionary ancestors of the tonsils, thymus gland, and Eustachian tubes, as well as many other structures derived from the embryonic branchial pouches.

Scientists have investigated what part of the body is responsible for maintaining the respiratory rhythm. They found that neurons located in the brainstem of fish are responsible for the genesis of the respiratory rhythm. The position of these neurons is slightly different from the centers of respiratory genesis in mammals but they are located in the same brain compartment, which has caused debates about the homology of respiratory centers between aquatic and terrestrial species. In both aquatic and terrestrial respiration, the exact mechanisms by which neurons can generate this involuntary rhythm are still not completely understood.

Another important feature of the respiratory rhythm is that it is modulated to adapt to the oxygen consumption of the body. As observed in mammals, fish "breathe" faster and heavier when they do physical exercise. The mechanisms by which these changes occur have been strongly debated over more than 100 years between scientists. The authors can be classified in 2 schools:

1. Those who think that the major part of the respiratory changes are pre-programmed in the brain, which would imply that neurons from locomotion centers of the brain connect to respiratory centers in anticipation of movements.

2. Those who think that the major part of the respiratory changes result from the detection of muscle contraction, and that respiration is adapted as a consequence of muscular contraction and oxygen consumption. This would imply that the brain possesses some kind of detection mechanisms that would trigger a respiratory response when muscular contraction occurs.

Many now agree that both mechanisms are probably present and complementary, or working alongside a mechanism that can detect changes in oxygen and/or carbon dioxide blood saturation.

Bony Fish

In bony fish, the gills lie in a branchial chamber covered by a bony operculum. The great majority of bony fish species have five pairs of gills, although a few have lost some over the course of evolution. The operculum can be important in adjusting the pressure of water inside of the pharynx to allow proper ventilation of the gills, so that bony fish do not have to rely on ram ventilation (and hence near constant motion) to breathe. Valves inside the mouth keep the water from escaping.

The gill arches of bony fish typically have no septum, so that the gills alone project from the arch, supported by individual gill rays. Some species retain gill rakers. Though all but the most primitive bony fish lack a spiracle, the pseudobranch associated with it often remains, being located at the base of the operculum. This is, however, often greatly reduced, consisting of a small mass of cells without any remaining gill-like structure.

Marine teleosts also use gills to excrete electrolytes. The gills' large surface area tends to create a problem for fish that seek to regulate the osmolarity of their internal fluids. Saltwater is less dilute than these internal fluids, so saltwater fish lose large quantities of water osmotically through their gills. To regain the water, they drink large amounts of seawater and excrete the salt. Freshwater is more dilute than the internal fluids of fish, however, so freshwater fish gain water osmotically through their gills.

In some primitive bony fishes and amphibians, the larvae bear external gills, branching off from the gill arches. These are reduced in adulthood, their function taken over by the gills proper in fishes and by lungs in most amphibians. Some amphibians retain the external larval gills in adulthood, the complex internal gill system as seen in fish apparently being irrevocably lost very early in the evolution of tetrapods.

Cartilaginous Fish

Like other fish, sharks extract oxygen from seawater as it passes over their gills. Unlike other fish, shark gill slits are not covered, but lie in a row behind the head. A modified slit called a spiracle lies just behind the eye, which assists the shark with taking in water during respiration and plays a major role in bottom–dwelling sharks. Spiracles are reduced or missing in active pelagic sharks. While the shark is moving, water passes through the mouth and over the gills in a process known as "ram ventilation". While at rest, most sharks pump water over their gills to ensure a constant supply of oxygenated water. A small number of species have lost the ability to pump water through their gills and must swim without rest. These species are *obligate ram ventilators* and would presumably asphyxiate if unable to move. Obligate ram ventilation is also true of some pelagic bony fish species.

The respiration and circulation process begins when deoxygenated blood travels to the shark's two-chambered heart. Here the shark pumps blood to its gills via the ventral aorta artery where it branches into afferent brachial arteries. Reoxygenation takes place in the gills and the reoxygenated blood flows into the efferent brachial arteries, which come together to form the dorsal aorta. The blood flows from the dorsal aorta throughout the body. The deoxygenated blood from the body then flows through the posterior cardinal veins and enters the posterior cardinal sinuses. From there blood enters the heart ventricle and the cycle repeats.

Sharks and rays typically have five pairs of gill slits that open directly to the outside of the body, though some more primitive sharks have six or seven pairs. Adjacent slits are separated by a cartilaginous gill arch from which projects a long sheet-like septum, partly supported by a further piece of cartilage called the *gill ray*. The individual lamellae of the gills lie on either side of the septum. The base of the arch may also support gill rakers, small projecting elements that help to filter food from the water.

A smaller opening, the spiracle, lies in the back of the first gill slit. This bears a small *pseudobranch* that resembles a gill in structure, but only receives blood already oxygenated by the true gills. The spiracle is thought to be homologous to the ear opening in higher vertebrates.

Most sharks rely on ram ventilation, forcing water into the mouth and over the gills by rapidly swimming forward. In slow-moving or bottom dwelling species, especially among skates and rays, the spiracle may be enlarged, and the fish breathes by sucking water through this opening, instead of through the mouth.

Chimaeras differ from other cartilagenous fish, having lost both the spiracle and the fifth gill slit. The remaining slits are covered by an operculum, developed from the septum of the gill arch in front of the first gill.

Lampreys and Hagfish

Lampreys and hagfish do not have gill slits as such. Instead, the gills are contained in spherical pouches, with a circular opening to the outside. Like the gill slits of higher fish, each pouch contains two gills. In some cases, the openings may be fused together, effectively forming an operculum. Lampreys have seven pairs of pouches, while hagfishes may have six to fourteen, depending on the species. In the hagfish, the pouches connect with the pharynx internally. In adult lampreys, a separate respiratory tube develops beneath the pharynx proper, separating food and water from respiration by closing a valve at its anterior end.

Circulation

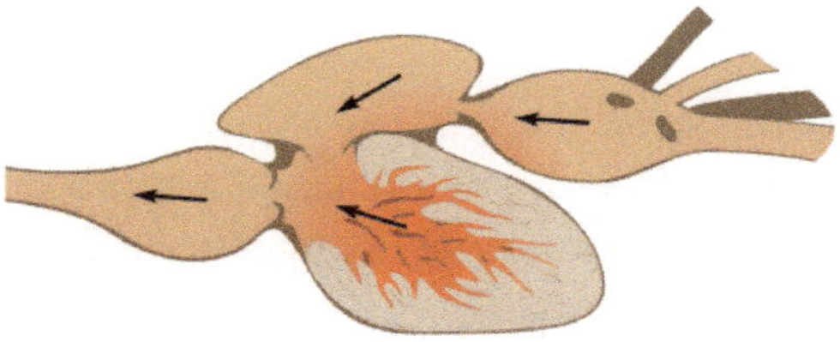

Two-chambered heart of a fish

The circulatory systems of all vertebrates are *closed*, just as in humans. Still, the systems of fish, amphibians, reptiles, and birds show various stages of the evolution of the circulatory system. In fish, the system has only one circuit, with the blood being pumped through the capillaries of the gills and on to the capillaries of the body tissues. This is known as *single cycle* circulation. The heart of fish is therefore only a single pump (consisting of two chambers). Fish have a closed-loop circulatory system. The heart pumps the blood in a single loop throughout the body. In most fish, the heart consists of four parts, including two chambers and an entrance and exit. The first part is the sinus venosus, a thin-walled sac that collects blood from the fish's veins before allowing it to flow to the second part, the atrium, which is a large muscular chamber. The atrium serves as a one-way antechamber, sends blood to the third part, ventricle. The ventricle is another thick-walled, muscular chamber and it pumps the blood, first to the fourth part, bulbus arteriosus, a large tube, and then out of the heart. The bulbus arteriosus connects to the aorta, through which blood flows to the gills for oxygenation.

In amphibians and most reptiles, a double circulatory system is used, but the heart is not always completely separated into two pumps. Amphibians have a three-chambered heart.

Digestion

Jaws allow fish to eat a wide variety of food, including plants and other organisms. Fish ingest food through the mouth and break it down in the esophagus. In the stomach, food is further digested and, in many fish, processed in finger-shaped pouches called pyloric caeca, which secrete digestive enzymes and absorb nutrients. Organs such as the liver and pancreas add enzymes and various chemicals as the food moves through the digestive tract. The intestine completes the process of digestion and nutrient absorption.

In most vertebrates, digestion is a four-stage process involving the main structures of the digestive tract, starting with ingestion, placing food into the mouth, and concluding with the excretion of undigested material through the anus. From the mouth, the food moves to the stomach, where as bolus it is broken down chemically. It then moves to the intestine, where the process of breaking the food down into simple molecules continues and the results are absorbed as nutrients into the circulatory and lymphatic system.

Although the precise shape and size of the stomach varies widely among different vertebrates, the relative positions of the oesophageal and duodenal openings remain relatively constant. As a result, the organ always curves somewhat to the left before curving back to meet the pyloric sphincter. However, lampreys, hagfishes, chimaeras, lungfishes, and some teleost fish have no stomach at all, with the oesophagus opening directly into the intestine. These animals all consume diets that either require little storage of food, or no pre-digestion with gastric juices, or both.

The small intestine is the part of the digestive tract following the stomach and followed by the large intestine, and is where much of the digestion and absorption of food takes place. In fish, the divisions of the small intestine are not clear, and the terms *anterior* or *proximal* intestine may be used instead of duodenum. The small intestine is found in all teleosts, although its form and length vary enormously between species. In teleosts, it is relatively short, typically around one and a half times the length of the fish's body. It commonly has a number of *pyloric caeca*, small pouch-like structures along its length that help to increase the overall surface area of the organ for digesting food. There is no ileocaecal valve in teleosts, with the boundary between the small intestine and the rectum being marked only by the end of the digestive epithelium.

There is no small intestine as such in non-teleost fish, such as sharks, sturgeons, and lungfish. Instead, the digestive part of the gut forms a spiral intestine, connecting the stomach to the rectum. In this type of gut, the intestine itself is relatively straight, but has a long fold running along the inner surface in a spiral fashion, sometimes for dozens of turns. This valve greatly increases both the surface area and the effective length of the intestine. The lining of the spiral intestine is similar to that of the small intestine in teleosts and non-mammalian tetrapods. In lampreys, the spiral valve is extremely small, possibly because their diet requires little digestion. Hagfish have no spiral valve at all, with digestion occurring for almost the entire length of the intestine, which is not subdivided into different regions.

The large intestine is the last part of the digestive system normally found in vertebrate animals. Its function is to absorb water from the remaining indigestible food matter, and then to pass useless waste material from the body. In fish, there is no true large intestine, but simply a short rectum connecting the end of the digestive part of the gut to the cloaca. In sharks, this includes a *rectal gland* that secretes salt to help the animal maintain osmotic balance with the seawater. The gland somewhat resembles a caecum in structure, but is not a homologous structure.

As with many aquatic animals, most fish release their nitrogenous wastes as ammonia. Some of the wastes diffuse through the gills. Blood wastes are filtered by the kidneys.

Saltwater fish tend to lose water because of osmosis. Their kidneys return water to the body. The reverse happens in freshwater fish: they tend to gain water osmotically. Their kidneys produce dilute urine for excretion. Some fish have specially adapted kidneys that vary in function, allowing them to move from freshwater to saltwater.

In sharks, digestion can take a long time. The food moves from the mouth to a J-shaped stomach, where it is stored and initial digestion occurs. Unwanted items may never get past the stomach, and instead the shark either vomits or turns its stomachs inside out and ejects unwanted items from its mouth. One of the biggest differences between the digestive systems of sharks and mammals is that sharks have much shorter intestines. This short length is achieved by the spiral valve with multiple turns within a single short section instead of a long tube-like intestine. The valve provides a long surface area, requiring food to circulate inside the short gut until fully digested, when remaining waste products pass into the cloaca.

Endocrine System

Regulation of Social Behaviour

Neuropeptides similar to oxytocin are found in virtually all vertebrates, including fish. In humans, oxytocin is linked with prosocial behaviours.

Research into cichlids from the social species *Neolamprologus pulcher* indicates that the oxycotin analog isotocin increases responsiveness to social information, which suggests that it is a key regulator of social behaviour in all fish, that has evolved from a prehistoric ancestor.

Osmoregulation

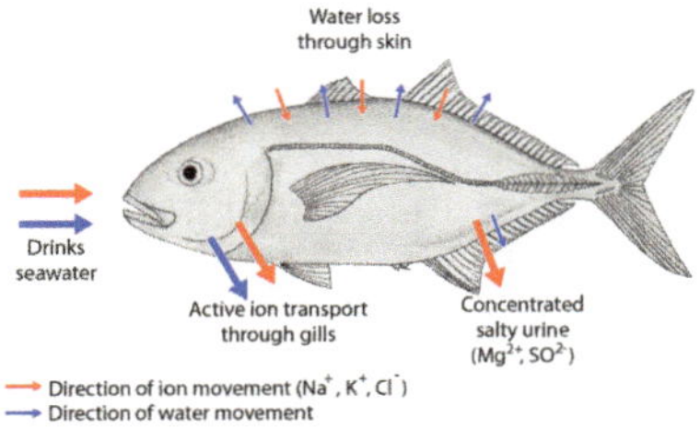

Movement of water and ions in saltwater fish

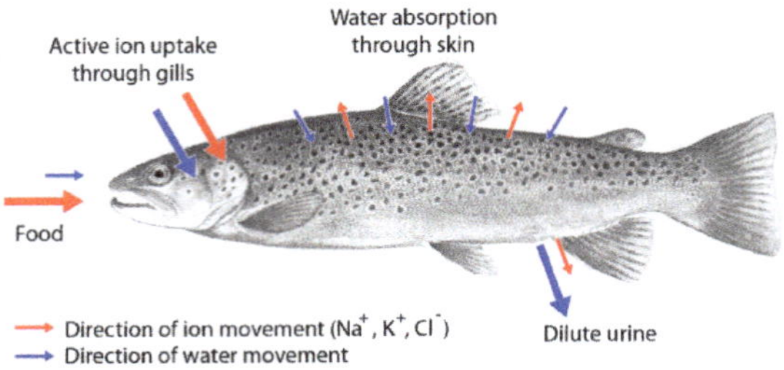

Movement of water and ions in freshwater fish

Two major types of osmoregulation are osmoconformers and osmoregulators. Osmoconformers match their body osmolarity to their environment actively or passively. Most marine invertebrates are osmoconformers, although their ionic composition may be different from that of seawater.

Osmoregulators tightly regulate their body osmolarity, which always stays constant, and are more common in the animal kingdom. Osmoregulators actively control salt concentrations despite the salt concentrations in the environment. An example is freshwater fish. The gills actively uptake salt from the environment by the use of mitochondria-rich cells. Water will diffuse into the fish, so it excretes a very hypotonic (dilute) urine to expel all the excess water. A marine fish has an internal osmotic concentration lower than that of the surrounding seawater, so it tends to lose water and gain salt. It actively excretes salt out from the gills. Most fish are stenohaline, which means they are restricted to either salt or fresh water and cannot survive in water with a different salt concentration than they are adapted to. However, some fish show a tremendous ability to effectively osmoregulate across a broad range of salinities; fish with this ability are known as euryhaline species, e.g., salmon. Salmon has been observed to inhabit two utterly disparate environments — marine and fresh water — and it is inherent to adapt to both by bringing in behavioral and physiological modifications.

In contrast to bony fish, with the exception of the coelacanth, the blood and other tissue of sharks and Chondrichthyes is generally isotonic to their marine environments because of the high concentration of urea and trimethylamine N-oxide (TMAO), allowing them to be in osmotic balance with the seawater. This adaptation prevents most sharks from surviving in freshwater, and they are therefore confined to marine environments. A few exceptions exist, such as the bull shark, which has developed a way to change its kidney function to excrete large amounts of urea. When a shark dies, the urea is broken down to ammonia by bacteria, causing the dead body to gradually smell strongly of ammonia.

Sharks have adopted a different, efficient mechanism to conserve water, i.e., osmoregulation. They retain urea in their blood in relatively higher concentration. Urea is damaging to living tissue so, to cope with this problem, some fish retain *trimethylamine oxide*. This provides a better solution to urea's toxicity. Sharks, having slightly higher solute concentration (i.e., above 1000 mOsm which is sea solute concentration), do not drink water like fresh water fish.

Thermoregulation

Homeothermy and poikilothermy refer to how stable an organism's temperature is. Most endothermic organisms are homeothermic, like mammals. However, animals with facultative endothermy are often poikilothermic, meaning their temperature can vary considerably. Similarly, most fish are ectotherms, as all of their heat comes from the surrounding water. However, most are homeotherms because their temperature is very stable.

Most organisms have a preferred temperature range, however some can be acclimated to temperatures colder or warmer than what they are typically used to. An organism's preferred temperature is typically the temperature at which the organism's physiological processes can act at optimal rates. When fish become acclimated to other temperatures, the efficiency of their physiological processes may decrease but will continue to function. This is called the thermal neutral zone at which an organism can survive indefinitely.

H.M. Vernon has done work on the death temperature and paralysis temperature (temperature of heat rigor) of various animals. He found that species of the same class showed very similar temperature values, those from the Amphibia examined being 38.5 °C, fish 39 °C, Reptilia 45 °C, and various Molluscs 46 °C.

To cope with low temperatures, some fish have developed the ability to remain functional even when the water temperature is below freezing; some use natural antifreeze or antifreeze proteins to resist ice crystal formation in their tissues.

Most sharks are "cold-blooded" or, more precisely, poikilothermic, meaning that their internal body temperature matches that of their ambient environment. Members of the family Lamnidae (such as the shortfin mako shark and the great white shark) are homeothermic and maintain a higher body temperature than the surrounding water. In these sharks, a strip of aerobic red muscle located near the center of the body generates the heat, which the body retains via a countercurrent exchange mechanism by a system of blood vessels called the rete mirabile ("miraculous net"). The common thresher shark has a similar mechanism for maintaining an elevated body temperature, which is thought to have evolved independently.

Tuna can maintain the temperature of certain parts of their body above the temperature of ambient seawater. For example, bluefin tuna maintain a core body temperature of 25–33 °C (77–91 °F), in water as cold as 6 °C (43 °F). However, unlike typical endothermic creatures such as mammals and birds, tuna do not maintain temperature within a relatively narrow range. Tuna achieve endothermy by conserving the heat generated through normal metabolism. The rete mirabile ("wonderful net"), the intertwining of veins and arteries in the body's periphery, transfers heat from venous blood to arterial blood via a counter-current exchange system, thus mitigating the effects of surface cooling. This allows the tuna to elevate the temperatures of the highly aerobic tissues of the skeletal muscles, eyes and brain, which supports faster swimming speeds and reduced energy expenditure, and which enables them to survive in cooler waters over a wider range of ocean environments than those of other fish. In all tunas, however, the heart operates at ambient temperature, as it receives cooled blood, and coronary circulation is directly from the gills.

- Homeothermy: Although most fish are exclusively ectothermic, there are exceptions. Certain species of fish maintain elevated body temperatures. Endothermic teleosts (bony fish) are all in the suborder Scombroidei and include the billfishes, tunas, including a "primitive" mackerel species, *Gasterochisma melampus*. All sharks in the family Lamnidae – shortfin mako, long fin mako, white, porbeagle, and salmon shark – are endothermic, and evidence suggests the trait exists in family Alopiidae (thresher sharks). The degree of endothermy varies from the billfish, which warm only their eyes and brain, to bluefin tuna and porbeagle sharks who maintain body temperatures elevated in excess of 20 °C above ambient water temperatures. Endothermy, though metabolically costly, is thought to provide advantages such as increased muscle strength, higher rates of central nervous system processing, and higher rates of digestion.

In some fish, a rete mirabile allows for an increase in muscle temperature in regions where this network of vein and arteries is found. The fish is able to thermoregulate certain areas of their body. Additionally, this increase in temperature leads to an increase in basal metabolic temperature. The fish is now able to split ATP at a higher rate and ultimately can swim faster.

The eye of a swordfish can generate heat to better cope with detecting their prey at depths of 2000 feet.

Muscular System

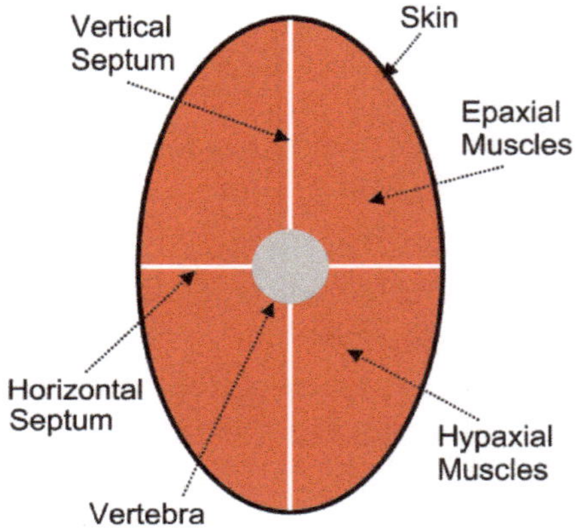

Cross section of a teleost

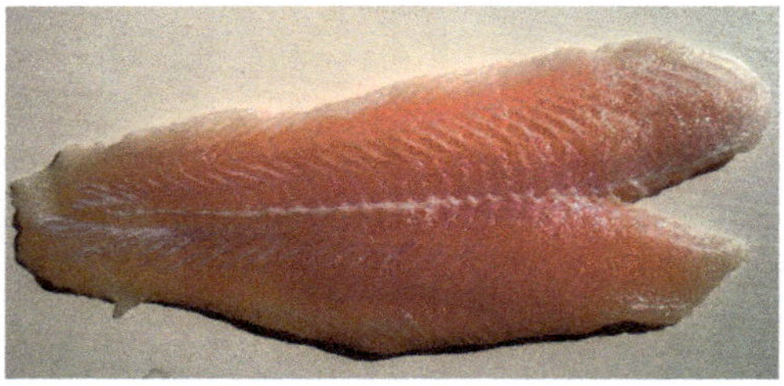
Iridescent shark filets showing myomere structure

Fish swim by contracting longitudinal red muscle and obliquely oriented white muscles. The red muscle is aerobic and needs oxygen which is supplied by myoglobin. The white muscle is anaerobic and it does not need oxygen. Red muscles are used for sustained activity such as cruising at slow speeds on ocean migrations. White muscles are used for bursts of activity, such as jumping or sudden bursts of speed for catching prey.

Mostly fish have white muscles, but the muscles of some fishes, such as scombroids and salmonids, range from pink to dark red. The red myotomal muscles derive their colour from myoglobin, an oxygen-binding molecule, which tuna express in quantities far higher than most other fish. The oxygen-rich blood further enables energy delivery to their muscles.

Most fish move by alternately contracting paired sets of muscles on either side of the backbone. These contractions form S-shaped curves that move down the body. As each curve reaches the back fin, backward force is applied to the water, and in conjunction with the fins, moves the fish forward. The fish's fins function like an airplane's flaps. Fins also increase the tail's surface area, increasing speed. The streamlined body of the fish decreases the amount of friction from the water.

A typical characteristic of many animals that utilize undulatory locomotion is that they have segmented muscles, or blocks of myomeres, running from their head to tails which are separated by connective tissue called myosepta. In addition, some segmented muscle groups, such the lateral hypaxial musculature in the salamander are oriented at an angle to the longitudinal direction. For these obliquely oriented fibers the strain in the longitudinal direction is greater than the strain in the muscle fiber direction leading to an architectural gear ratio greater than 1. A higher initial angle of orientation and more dorsoventral bulging produces a faster muscle contraction but results in a lower amount of force production. It is hypothesized that animals employ a variable gearing mechanism that allows self-regulation of force and velocity to meet the mechanical demands of the contraction. When a pennate muscle is subjected to a low force, resistance to width changes in the muscle cause it to rotate which consequently produce a higher architectural gear ratio (AGR) (high velocity). However, when subject to a high force, the perpendicular fiber force component overcomes the resistance to width changes and the muscle compresses producing a lower AGR (capable of maintaining a higher force output).

Most fishes bend as a simple, homogenous beam during swimming via contractions of longitudinal red muscle fibers and obliquely oriented white muscle fibers within the segmented axial musculature. The fiber strain (εf) experienced by the longitudinal red muscle fibers is equivalent to the longitudinal strain (εx). The deeper white muscle fibers fishes show diversity in arrangement. These fibers are organized into cone-shaped structures and attach to connective tissue sheets known as myosepta; each fiber shows a characteristic dorsoventral (α) and mediolateral (φ) trajectory. The segmented architecture theory predicts that, $\varepsilon x > \varepsilon f$. This phenomenon results in an architectural gear ratio, determined as longitudinal strain divided by fiber strain ($\varepsilon x / \varepsilon f$), greater than one and longitudinal velocity amplification; furthermore, this emergent velocity amplification may be augmented by variable architectural gearing via mesolateral and dorsoventral shape changes, a pattern seen in pennate muscle contractions. A red-to-white gearing ratio (red εf / white εf) captures the combined effect of the longitudinal red muscle fiber and oblique white muscle fiber strains.

Buoyancy

Swim bladder of a common rudd

The body of a fish is denser than water, so fish must compensate for the difference or they will sink. Many bony fishes have an internal organ called a swim bladder, or gas bladder, that adjusts their buoyancy through manipulation of gases. In this way, fish can stay at the current water depth, or ascend or descend without having to waste energy in swimming. The bladder is only found in bony fishes. In the more primitive groups like some minnows, bichirs and lungfish, the bladder is open to the esophagus and double as a lung. It is often absent in fast swimming fishes such as the tuna and mackerel families. The condition of a bladder open to the esophagus is called physostome, the closed condition physoclist. In the latter, the gas content of the bladder is controlled through the rete mirabilis, a network of blood vessels effecting gas exchange between the bladder and the blood.

Sharks, like this three tonne great white shark, don't have swim bladders. Most sharks need to keep swimming to avoid sinking.

In some fish, a rete mirabile fills the swim bladder with oxygen. A countercurrent exchange system is utilized between the venous and arterial capillaries. By lowering the pH levels in the venous capillaries, oxygen unbinds from blood hemoglobin. This causes an increase in venous blood oxygen concentration, allowing the oxygen to diffuse through the capillary membrane and into the arterial capillaries, where oxygen is still sequestered to hemoglobin. The cycle of diffusion continues until the concentration of oxygen in the arterial capillaries is supersaturated (larger than the concentration of oxygen in the swim bladder). At this point, the free oxygen in the arterial capillaries diffuses into the swim bladder via the gas gland.

Unlike bony fish, sharks do not have gas-filled swim bladders for buoyancy. Instead, sharks rely on a large liver filled with oil that contains squalene, and their cartilage, which is about half the normal density of bone. Their liver constitutes up to 30% of their total body mass. The liver's effectiveness is limited, so sharks employ dynamic lift to maintain depth when not swimming. Sand tiger sharks store air in their stomachs, using it as a form of swim bladder. Most sharks need to constantly swim in order to breathe and cannot sleep very long without sinking (if at all). However, certain species, like the nurse shark, are capable of pumping water across their gills, allowing them to rest on the ocean bottom.

Sensory Systems

Most fish possess highly developed sense organs. Nearly all daylight fish have color vision that is at least as good as a human's Many fish also have chemoreceptors that are responsible for extraordinary senses of taste and smell. Although they have ears, many fish may not hear very well. Most fish have sensitive receptors that form the lateral line system, which detects gentle currents and vibrations, and senses the motion of nearby fish and prey. Sharks can sense frequencies in the range of 25 to 50 Hz through their lateral line.

Fish orient themselves using landmarks and may use mental maps based on multiple landmarks or symbols. Fish behavior in mazes reveals that they possess spatial memory and visual discrimination.

Vision

Vision is an important sensory system for most species of fish. Fish eyes are similar to those of terrestrial vertebrates like birds and mammals, but have a more spherical lens. Their retinas generally have both rod cells and cone cells (for scotopic and photopic vision), and most species have colour vision. Some fish can see ultraviolet and some can see polarized light. Amongst jawless fish, the lamprey has well-developed eyes, while the hagfish has only primitive eyespots. Fish vision shows adaptation to their visual environment, for example deep sea fishes have eyes suited to the dark environment.

Hearing

Hearing is an important sensory system for most species of fish. Hearing threshold and the ability to localize sound sources are reduced underwater, in which the speed of sound is faster than in air. Underwater hearing is by bone conduction, and localization of sound appears to depend on differences in amplitude detected by bone conduction. Aquatic animals such as fish, however, have a more specialized hearing apparatus that is effective underwater.

Fish can sense sound through their lateral lines and their otoliths (ears). Some fishes, such as some species of carp and herring, hear through their swim bladders, which function rather like a hearing aid.

Hearing is well-developed in carp, which have the Weberian organ, three specialized vertebral processes that transfer vibrations in the swim bladder to the inner ear.

Although it is hard to test sharks' hearing, they may have a sharp sense of hearing and can possibly hear prey many miles away. A small opening on each side of their heads (not the spiracle) leads directly into the inner ear through a thin channel. The lateral line shows a similar arrangement, and is open to the environment via a series of openings called lateral line pores. This is a reminder of the common origin of these two vibration- and sound-detecting organs that are grouped together as the acoustico-lateralis system. In bony fish and tetrapods the external opening into the inner ear has been lost.

Chemoreception

The shape of the hammerhead shark's head may enhance olfaction by spacing the nostrils further apart.

Sharks have keen olfactory senses, located in the short duct (which is not fused, unlike bony fish) between the anterior and posterior nasal openings, with some species able to detect as little as one part per million of blood in seawater.

Sharks have the ability to determine the direction of a given scent based on the timing of scent detection in each nostril. This is similar to the method mammals use to determine direction of sound.

They are more attracted to the chemicals found in the intestines of many species, and as a result often linger near or in sewage outfalls. Some species, such as nurse sharks, have external barbels that greatly increase their ability to sense prey.

Magnetoception

Electroreception

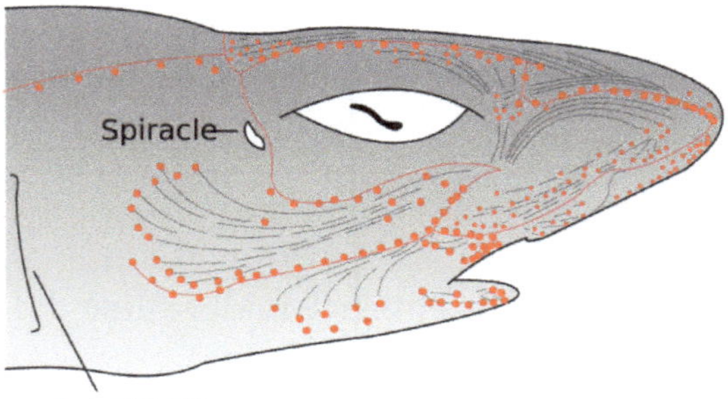

Electromagnetic field receptors (ampullae of Lorenzini) and motion detecting canals in the head of a shark

Some fish, such as catfish and sharks, have organs that detect weak electric currents on the order of millivolt. Other fish, like the South American electric fishes Gymnotiformes, can produce weak electric currents, which they use in navigation and social communication. In sharks, the ampullae of Lorenzini are electroreceptor organs. They number in the hundreds to thousands. Sharks use the ampullae of Lorenzini to detect the electromagnetic fields that all living things produce. This helps sharks (particularly the hammerhead shark) find prey. The shark has the greatest electrical sensitivity of any animal. Sharks find prey hidden in sand by detecting the electric fields they produce. Ocean currents moving in the magnetic field of the Earth also generate electric fields that sharks can use for orientation and possibly navigation.

- The ampullae of Lorenzini allow sharks to sense electrical discharges.
- Electric fish are able to produce electric fields by modified muscles in their body.

Pain

Experiments done by William Tavolga provide evidence that fish have pain and fear responses. For instance, in Tavolga's experiments, toadfish grunted when electrically shocked and over time they came to grunt at the mere sight of an electrode.

In 2003, Scottish scientists at the University of Edinburgh and the Roslin Institute concluded that rainbow trout exhibit behaviors often associated with pain in other animals. Bee venom and acetic acid injected into the lips resulted in fish rocking their bodies and rubbing their lips along the sides and floors of their tanks, which the researchers concluded were attempts to relieve pain, similar to what mammals would do. Neurons fired in a pattern resembling human neuronal patterns.

Professor James D. Rose of the University of Wyoming claimed the study was flawed since it did not provide proof that fish possess "conscious awareness, particularly a kind of awareness that is meaningfully like ours". Rose argues that since fish brains are so different from human brains, fish are probably not conscious in the manner humans are, so that reactions similar to human reactions to pain instead have other causes. Rose had published a study a year earlier arguing that fish cannot feel pain because their brains lack a neocortex. However, animal behaviorist Temple Grandin argues that fish could still have consciousness without a neocortex because "different species can use different brain structures and systems to handle the same functions."

Animal welfare advocates raise concerns about the possible suffering of fish caused by angling. Some countries, such as Germany have banned specific types of fishing, and the British RSPCA now formally prosecutes individuals who are cruel to fish.

Reproductive Processes

Oogonia development in teleosts fish varies according to the group, and the determination of oogenesis dynamics allows the understanding of maturation and fertilisation processes. Changes in the nucleus, ooplasm, and the surrounding layers characterize the oocyte maturation process.

Postovulatory follicles are structures formed after oocyte release; they do not have endocrine function, present a wide irregular lumen, and are rapidly reabsorbed in a process involving the apoptosis of follicular cells. A degenerative process called follicular atresia reabsorbs vitellogenic oocytes not spawned. This process can also occur, but less frequently, in oocytes in other development stages.

Some fish are hermaphrodites, having both testes and ovaries either at different phases in their life cycle or, as in hamlets, have them simultaneously.

Over 97% of all known fish are oviparous, that is, the eggs develop outside the mother's body. Examples of oviparous fish include salmon, goldfish, cichlids, tuna, and eels. In the majority of these species, fertilisation takes place outside the mother's body, with the male and female fish shedding their gametes into the surrounding water. However, a few oviparous fish practice internal fertilisation, with the male using some sort of intromittent organ to deliver sperm into the genital opening of the female, most notably the oviparous sharks, such as the horn shark, and oviparous rays, such as skates. In these cases, the male is equipped with a pair of modified pelvic fins known as claspers.

Marine fish can produce high numbers of eggs which are often released into the open water column. The eggs have an average diameter of 1 millimetre (0.039 in). The eggs are generally surrounded by the extraembryonic membranes but do not develop a shell, hard or soft, around these membranes. Some fish have thick, leathery coats, especially if they must withstand physical force or desiccation. These type of eggs can also be very small and fragile.

Egg of lamprey	Egg of catshark (mermaids' purse)	Egg of bullhead shark	Egg of chimaera

The newly hatched young of oviparous fish are called larvae. They are usually poorly formed, carry a large yolk sac (for nourishment) and are very different in appearance from juvenile and adult specimens. The larval period in oviparous fish is relatively short (usually only several weeks), and larvae rapidly grow and change appearance and structure (a process termed metamorphosis) to become juveniles. During this transition larvae must switch from their yolk sac to feeding on zooplankton prey, a process which depends on typically inadequate zooplankton density, starving many larvae.

In ovoviviparous fish the eggs develop inside the mother's body after internal fertilisation but receive little or no nourishment directly from the mother, depending instead on the yolk. Each embryo develops in its own egg. Familiar examples of ovoviviparous fish include guppies, angel sharks, and coelacanths.

Some species of fish are viviparous. In such species the mother retains the eggs and nourishes the embryos. Typically, viviparous fish have a structure analogous to the placenta seen in mammals connecting the mother's blood supply with that of the embryo. Examples of viviparous fish include the surf-perches, splitfins, and lemon shark. Some viviparous fish exhibit oophagy, in which the developing embryos eat other eggs produced by the mother. This has been observed primarily among sharks, such as the shortfin mako and porbeagle, but is known for a few bony fish as well, such as the halfbeak *Nomorhamphus ebrardtii.* Intrauterine cannibalism is an even more unusual mode of vivipary, in which the largest embryos eat weaker and smaller siblings. This behavior is also most commonly found among sharks, such as the grey nurse shark, but has also been reported for *Nomorhamphus ebrardtii.*

In many species of fish, fins have been modified to allow Internal fertilisation.

Aquarists commonly refer to ovoviviparous and viviparous fish as livebearers.

- Many fish species are hermaphrodites. *Synchronous hermaphrodites* possess both ovaries and testes at the same time. *Sequential hermaphrodites* have both types of tissue in their gonads, with one type being predominant while the fish belongs to the corresponding gender.

Fish Anatomy

Fish anatomy is the study of the form or morphology of fishes. It can be contrasted with fish physiology, which is the study of how the component parts of fish function together in the living fish.

In practice, fish anatomy and fish physiology complement each other, the former dealing with the structure of a fish, its organs or component parts and how they are put together, such as might be observed on the dissecting table or under the microscope, and the latter dealing with how those components function together in the living fish.

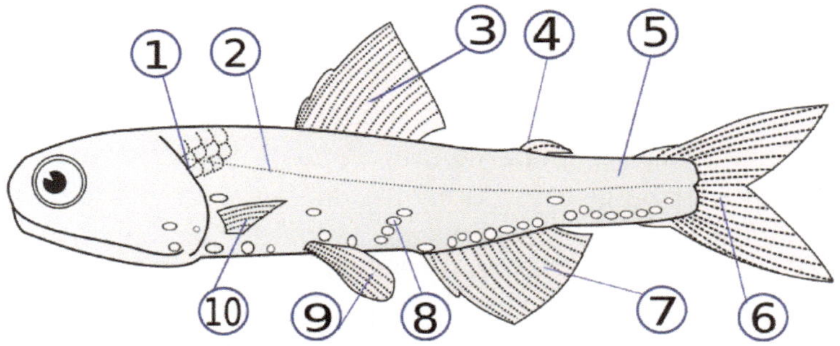

External anatomy of a bony fish (Hector's lanternfish)1. operculum (gill cover), 2. lateral line, 3. dorsal fin, 4. adipose fin,5. caudal peduncle, 6. caudal fin, 7. anal fin, 8. photophores
9. pelvic fins (paired), 10. pectoral fins (paired)

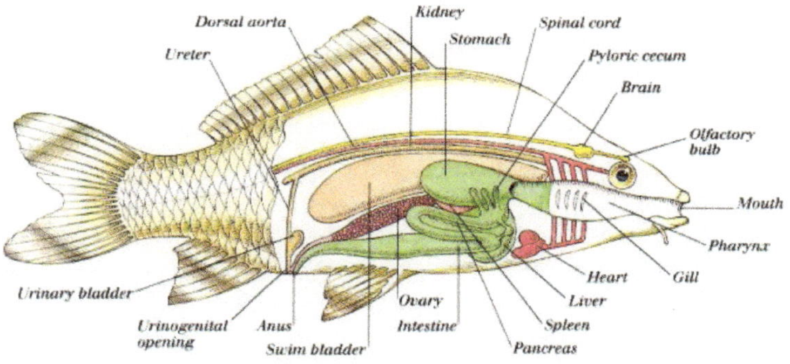

Internal anatomy of a bony fish

The anatomy of fish is often shaped by the physical characteristics of water, the medium in which fish live. Water is much denser than air, holds a relatively small amount of dissolved oxygen, and absorbs more light than air does. The body of a fish is divided into a head, trunk and tail, although the divisions between the three are not always externally visible. The skeleton, which forms the support structure inside the fish, is either made of cartilage, in cartilaginous fish, or bone in bony fish. The main skeletal element is the vertebral column, composed of articulating vertebrae which are lightweight yet strong. The ribs attach to the spine and there are no limbs or limb girdles. The main external features of the fish, the fins, are composed of either bony or soft spines called rays which, with the exception of the caudal fins, have no direct connection with the spine. They are supported by the muscles which compose the main part of the trunk. The heart has two chambers and pumps the blood through the respiratory surfaces of the gills and on round the body in a single circulatory loop. The eyes are adapted for seeing underwater and have only local vision. There is an inner ear but no external or middle ear. Low frequency vibrations are detected by the lateral line system of sense organs that run along the length of the sides of fish, and these respond to nearby movements and to changes in water pressure.

Sharks and rays are basal fish with numerous primitive anatomical features similar to those of ancient fish, including skeletons composed of cartilage. Their bodies tend to be dorso-ventrally flattened, they usually have five pairs of gill slits and a large mouth set on the underside of the head. The dermis is covered with separate dermal placoid scales. They have a cloaca into which the urinary and genital passages open, but not a swim bladder. Cartilaginous fish produce a small number of large, yolky eggs. Some species are ovoviviparous and the young develop internally but others are oviparous and the larvae develop externally in egg cases.

The bony fish lineage shows more derived anatomical traits, often with major evolutionary changes from the features of ancient fish. They have a bony skeleton, are generally laterally flattened, have five pairs of gills protected by an operculum, and a mouth at or near the tip of the snout. The dermis is covered with overlapping scales. Bony fish have a swim bladder which helps them maintain a constant depth in the water column, but not a cloaca. They mostly spawn a large number of small eggs with little yolk which they broadcast into the water column.

Body

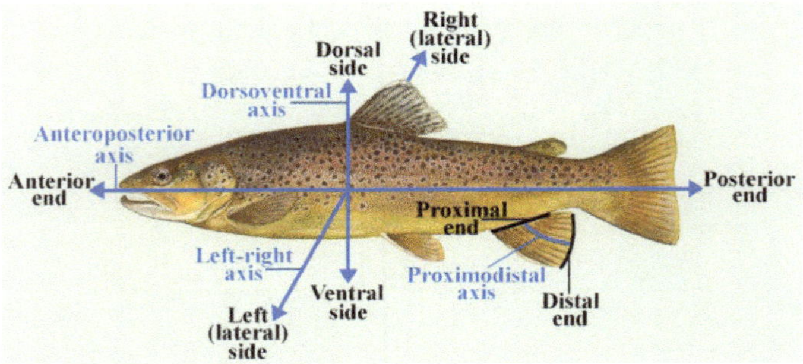

Anatomical directions and axes

In many respects fish anatomy is different from humans and mammals, yet it shares the same basic vertebrate body plan from which all vertebrates have evolved: a notochord, rudimentary vertebrae, and a well-defined head and tail.

Fish have a variety of different body plans. At the broadest level their body is divided into head, trunk, and tail, although the divisions are not always externally visible. The body is often fusiform, a streamlined body plan often found in fast-moving fish. They may also be filiform (eel-shaped) or vermiform (worm-shaped). Also, fish are often either compressed (laterally thin) or depressed (dorso-ventrally flat).

Skeleton

Skeleton of a bony fish

There are two different skeletal types: the exoskeleton, which is the stable outer shell of an organism, and the endoskeleton, which forms the support structure inside the body. The skeleton of the fish is either made of cartilage (cartilaginous fishes) or bones (bony fishes). The main features of the fish, the fins, are bony fin rays and, except for the caudal fin, have no direct connection with the spine. They are supported only by the muscles. The ribs attach to the spine.

Bones are rigid organs that form part of the endoskeleton of vertebrates. They function to move, support, and protect the various organs of the body, produce red and white blood cells and store minerals. Bone tissue is a type of dense connective tissue. Because bones come in a variety of shapes and have a complex internal and external structure they are lightweight, yet strong and hard, in addition to fulfilling their many other functions.

Fish bones have been used to bioremediate lead from contaminated soil.

Vertebrae

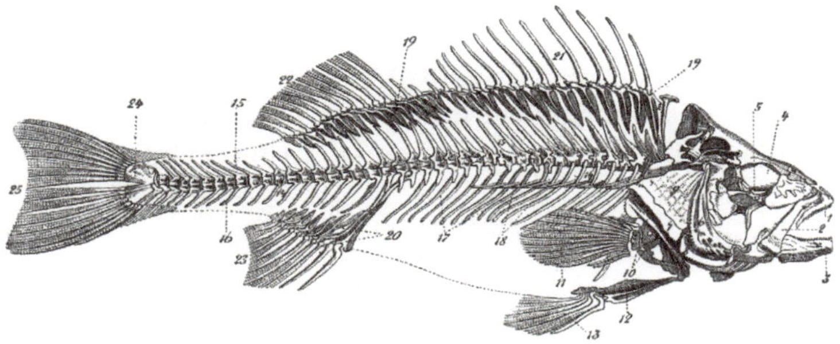

Skeletal structure of a bass showing the vertebral column running from the head to the tail

Fish are vertebrates. All vertebrates are built along the basic chordate body plan: a stiff rod running through the length of the animal (vertebral column or notochord), with a hollow tube of nervous tissue (the spinal cord) above it and the gastrointestinal tract below. In all vertebrates, the mouth is found at, or right below, the anterior end of the animal, while the anus opens to the exterior before the end of the body. The remaining part of the body continuing aft of the anus forms a tail with vertebrae and spinal cord, but no gut.

The defining characteristic of a vertebrate is the vertebral column, in which the notochord (a stiff rod of uniform composition) found in all chordates has been replaced by a segmented series of stiffer elements (vertebrae) separated by mobile joints (intervertebral discs, derived embryonically and evolutionarily from the notochord). However, a few fish have secondarily lost this anatomy, retaining the notochord into adulthood, such as the sturgeon.

Skeletal structure of an Atlantic cod

The X-ray tetra *(Pristella maxillaris)* has a visible backbone. The spinal cord is housed within its backbone

The vertebral column consists of a centrum (the central body or spine of the vertebra), vertebral arches which protrude from the top and bottom of the centrum, and various processes which project from the centrum or arches. An arch extending from the top of the centrum is called a neural arch, while the hemal arch or chevron is found underneath the centrum in the caudal (tail) vertebrae of fish. The centrum of a fish is usually concave at each end (amphicoelous), which limits the motion of the fish. This can be contrasted with the centrum of a mammal, which is flat at each end (acoelous), shaped in a manner that can support and distribute compressive forces.

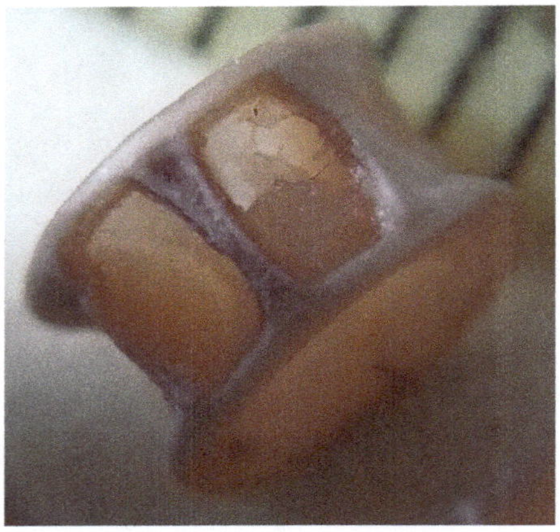

A vertebra (diameter 5 mm) of a small ray-finned fish

The vertebrae of lobe-finned fishes consist of three discrete bony elements. The vertebral arch surrounds the spinal cord, and is of broadly similar form to that found in most other vertebrates. Just beneath the arch lies a small plate-like *pleurocentrum*, which protects the upper surface of the notochord, and below that, a larger arch-shaped *intercentrum* to protect the lower border. Both of these structures are embedded within a single cylindrical mass of cartilage. A similar arrangement was found in primitive tetrapods, but, in the evolutionary line that led to reptiles (and hence, also to mammals and birds), the intercentrum became partially or wholly replaced by an enlarged pleurocentrum, which in turn became the bony vertebral body.

In most ray-finned fishes, including all teleosts, these two structures are fused with, and embedded within, a solid piece of bone superficially resembling the vertebral body of mammals. In living amphibians, there is simply a cylindrical piece of bone below the vertebral arch, with no trace of the separate elements present in the early tetrapods.

In cartilagenous fish, such as sharks, the vertebrae consist of two cartilagenous tubes. The upper tube is formed from the vertebral arches, but also includes additional cartilagenous structures filling in the gaps between the vertebrae, and so enclosing the spinal cord in an essentially continuous sheath. The lower tube surrounds the notochord, and has a complex structure, often including multiple layers of calcification.

Lampreys have vertebral arches, but nothing resembling the vertebral bodies found in all higher vertebrates. Even the arches are discontinuous, consisting of separate pieces of arch-shaped cartilage around the spinal cord in most parts of the body, changing to long strips of cartilage above and below in the tail region. Hagfishes lack a true vertebral column, and are therefore not properly considered vertebrates, but a few tiny neural arches are present in the tail. Hagfishes do, however, possess a cranium. For this reason, the vertebrate subphylum is sometimes referred to as "Craniata" when discussing morphology. Molecular analysis since 1992 has suggested that the hagfishes are most closely related to lampreys, and so also are vertebrates in a monophyletic sense. Others consider them a sister group of vertebrates in the common taxon of Craniata.

Head

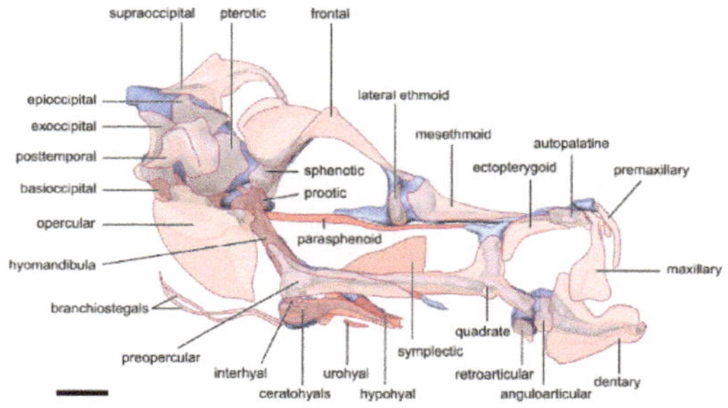

Skull bones as they appear in a seahorse

The head or skull includes the skull roof (a set of bones covering the brain, eyes and nostrils), the snout (from the eye to the forward most point of the upper jaw), the operculum or gill cover (absent in sharks and jawless fish), and the cheek, which extends from the eye to preopercle. The operculum and preopercle may or may not have spines. In sharks and some primitive bony fish a spiracle, small extra gill opening, is found behind each eye.

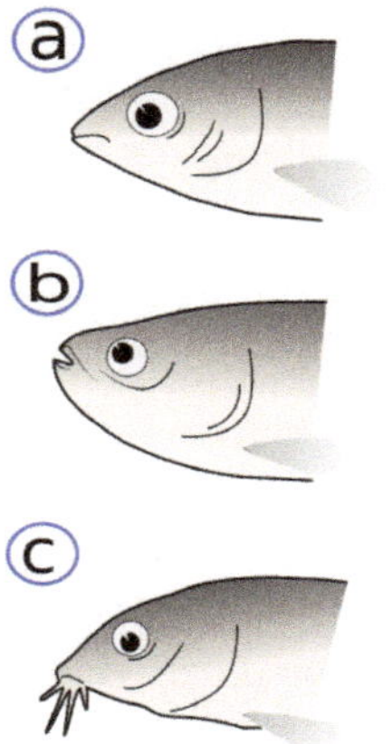

Positions of fish mouths:(a) - terminal,(b) - superior,(c) - subterminal, inferior

Barbels

The skull in fishes is formed from a series of only loosely connected bones. Jawless fish and sharks only possess a cartilaginous endocranium, with the upper and lower jaws of cartilaginous fish being separate elements not attached to the skull.Bony fishes have additional dermal bone, forming a more or less coherent skull roof in lungfish and holost fish. The lower jaw defines a chin.

In lampreys, the mouth is formed into an oral disk. In most jawed fish, however, there are three general configurations. The mouth may be on the forward end of the head (*terminal*), may be upturned (*superior*), or may be turned downwards or on the bottom of the fish (*subterminal* or *inferior*). The mouth may be modified into a suckermouth adapted for clinging onto objects in fast-moving water.

The simpler structure is found in jawless fish, in which the cranium is represented by a trough-like basket of cartilaginous elements only partially enclosing the brain, and associated with the capsules for the inner ears and the single nostril. Distinctively, these fish have no jaws.

Cartilaginous fish, such as sharks, also have simple, and presumably primitive, skull structures. The cranium is a single structure forming a case around the brain, enclosing the lower surface and the sides, but always at least partially open at the top as a large fontanelle. The most anterior part of the cranium includes a forward plate of cartilage, the rostrum, and capsules to enclose the olfactory organs. Behind these are the orbits, and then an additional pair of capsules enclosing the structure of the inner ear. Finally, the skull tapers towards the rear, where the foramen magnum lies immediately above a single condyle, articulating with the first vertebra. There are, in addition, at various points throughout the cranium, smaller foramina for the cranial nerves. The jaws consist of separate hoops of cartilage, almost always distinct from the cranium proper.

In the ray-finned fishes, there has also been considerable modification from the primitive pattern. The roof of the skull is generally well formed, and although the exact relationship of its bones to those of tetrapods is unclear, they are usually given similar names for convenience. Other elements of the skull, however, may be reduced; there is little cheek region behind the enlarged orbits, and little, if any bone in between them. The upper jaw is often formed largely from the premaxilla, with the maxilla itself located further back, and an additional bone, the symplectic, linking the jaw to the rest of the cranium.

Although the skulls of fossil lobe-finned fish resemble those of the early tetrapods, the same cannot be said of those of the living lungfishes. The skull roof is not fully formed, and consists of multiple, somewhat irregularly shaped bones with no direct relationship to those of tetrapods. The upper jaw is formed from the pterygoids and vomers alone, all of which bear teeth. Much of the skull is formed from cartilage, and its overall structure is reduced.

The head may have several fleshy structures known as barbels, which may be very long and resemble whiskers. Many fish species also have a variety of protrusions or spines on the head. The nostrils or nares of almost all fishes do not connect to the oral cavity, but are pits of varying shape and depth.

Skull of a northern pike

Skull of *Tiktaalik*, a genus of extinct sarcopterygian (lobe-finned "fish") from the late Devonian period

External Organs

Jaw

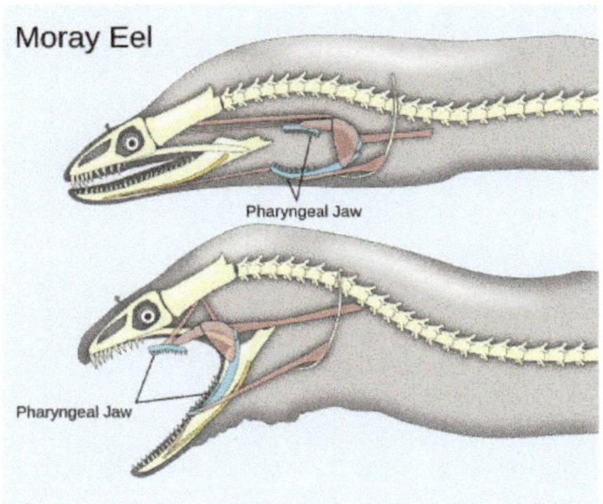

Moray eels have two sets of jaws: the oral jaws that capture prey and the pharyngeal jaws that advance into the mouth and move prey from the oral jaws to the esophagus for swallowing

The vertebrate jaw probably originally evolved in the Silurian period and appeared in the Placoderm fish which further diversified in the Devonian. Jaws are thought to derive from the pharyngeal arches that support the gills in fish. The two most anterior of these arches are thought to have become the jaw itself and the hyoid arch, which braces the jaw against the braincase and increases mechanical efficiency. While there is no fossil evidence directly to support this theory, it makes sense in light of the numbers of pharyngeal arches that are visible in extant jawed (the Gnathostomes), which have seven arches, and primitive jawless vertebrates (the Agnatha), which have nine.

Jaws of great white shark

It is thought that the original selective advantage garnered by the jaw was not related to feeding, but to increased respiration efficiency. The jaws were used in the buccal pump (observable in modern fish and amphibians) that pumps water across the gills of fish or air into the lungs in the case of amphibians. Over evolutionary time the more familiar use of jaws (to humans), in feeding, was selected for and became a very important function in vertebrates.

Linkage systems are widely distributed in animals. The most thorough overview of the different types of linkages in animals has been provided by M. Muller, who also designed a new classification system, which is especially well suited for biological systems. Linkage mechanisms are especially frequent and manifold in the head of bony fishes, such as wrasses, which have evolved many specialized feeding mechanisms. Especially advanced are the linkage mechanisms of jaw protrusion. For suction feeding a system of linked four-bar linkages is responsible for the coordinated opening of the mouth and 3-D expansion of the buccal cavity. Other linkages are responsible for protrusion of the premaxilla.

Eyes

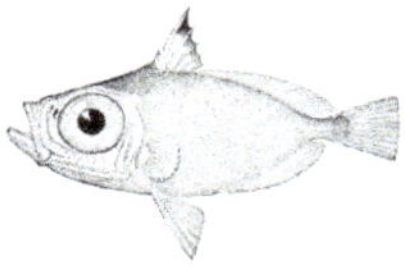

Zenion hololepis is a small deep water fish with large eyes

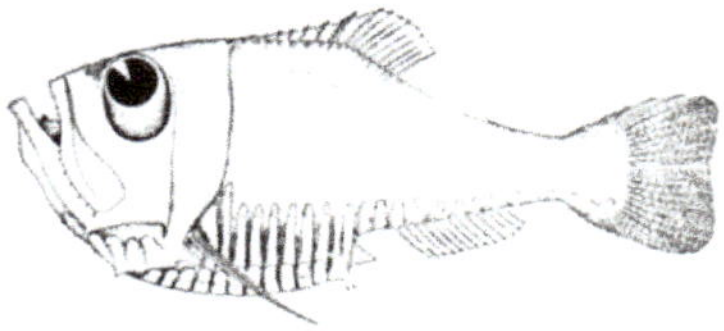

The deep sea half-naked hatchetfish has eyes which look overhead where it can see the silhouettes of prey

Fish eyes are similar to terrestrial vertebrates like birds and mammals, but have a more spherical lens. Their retinas generally have both rod cells and cone cells (for scotopic and photopic vision), and most species have colour vision. Some fish can see ultraviolet and some can see polarized light. Amongst jawless fish, the lamprey has well-developed eyes, while the hagfish has only primitive eyespots. The ancestors of modern hagfish, thought to be the protovertebrate were evidently pushed to very deep, dark waters, where they were less vulnerable to sighted predators, and where it is advantageous to have a convex eye-spot, which gathers more light than a flat or concave one. Unlike humans, fish normally adjust focus by moving the lens closer to or further from the retina.

Gills

Gill of a rainbow trout

The gills, located under the operculum, are a respiratory organ for the extraction of oxygen from water and for the excretion of carbon dioxide. They are not usually visible, but can be seen in some species, such as the frilled shark. The labyrinth organ of Anabantoidei and Clariidae is used to allow the fish to extract oxygen from the air. Gill rakers are bony or cartilaginous, finger-like projections off the gill arch which function in filter-feeders to retain filtered prey.

Skin

The epidermis of fish consists entirely of live cells, with only minimal quantities of keratin in the cells of the superficial layer. It is generally permeable. The dermis of bony fish typically contains relatively little of the connective tissue found in tetrapods. Instead, in most species, it is largely replaced by solid, protective bony scales. Apart from some particularly large dermal bones that form parts of the skull, these scales are lost in tetrapods, although many reptiles do have scales of a different kind, as do pangolins. Cartilaginous fish have numerous tooth-like denticles embedded in their skin, in place of true scales.

Sweat glands and sebaceous glands are both unique to mammals, but other types of skin glands are found in fish. Fish typically have numerous individual mucus-secreting skin cells that aid in insulation and protection, but may also have poison glands, photophores, or cells that produce a more watery, serous fluid. Melanin colours the skin of many species, but in fish the epidermis is often relatively colourless. Instead, the colour of the skin is largely due to chromatophores in the dermis, which, in addition to melanin, may contain guanine or carotenoid pigments. Many species, such as flounders, change the colour of their skin by adjusting the relative size of their chromatophores.

Scales

The outer body of many fish is covered with scales, which are part of the fish's integumentary system. The scales originate from the mesoderm (skin), and may be similar in structure to teeth. Some species are covered instead by scutes. Others have no outer covering on the skin. Most fish are covered in a protective layer of slime (mucus).

Cycloid scales covering rohu

There are four principal types of fish scales.

1. Placoid scales, also called dermal denticles, are similar to teeth in that they are made of dentin covered by enamel. They are typical of sharks and rays.
2. Ganoid scales are flat, basal-looking scales that cover a fish body with little overlapping. They are typical of gar and bichirs.
3. Cycloid scales are small oval-shaped scales with growth rings. Bowfin and remora have cycloid scales.
4. Ctenoid scales are similar to the cycloid scales, with growth rings. They are distinguished by spines that cover one edge. Halibut have this type of scale.

Another, less common, type of scale is the scute, which is:

- an external shield-like bony plate, or
- a modified, thickened scale that often is keeled or spiny, or
- a projecting, modified (rough and strongly ridged) scale, usually associated with the lateral line, or on the caudal peduncle forming caudal keels, or along the ventral profile. Some fish, such as pineconefish, are completely or partially covered in scutes.

Lateral Line

The lateral line is clearly visible as a line of receptors running along the side of this Atlantic cod

The lateral line is a sense organ used to detect movement and vibration in the surrounding water. For example, fish can use their lateral line system to follow the vortices produced by fleeing prey. In most species, it consists of a line of receptors running along each side of the fish.

Photophores

Photophores are light-emitting organs which appears as luminous spots on some fishes. The light can be produced from compounds during the digestion of prey, from specialized mitochondrial cells in the organism called photocytes, or associated with symbiotic bacteria, and are used for attracting food or confusing predators.

Fins

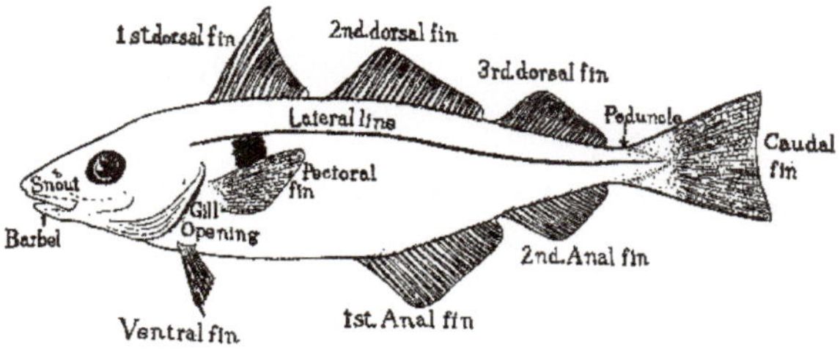

The haddock, a type of cod, is ray-finned. It has three dorsal and two anal fins

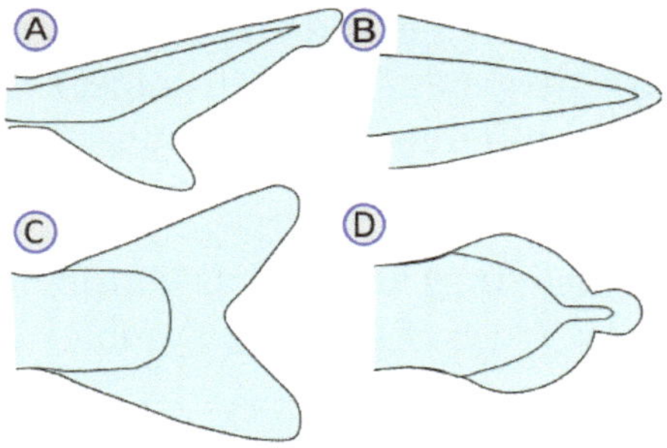

Types of caudal (tail) fin:
(A) - Heterocercal, (B) - Protocercal,(C) - Homocercal, (D) - Diphycercal

Fins are the most distinctive features of fish. They are either composed of bony spines or rays protruding from the body with skin covering them and joining them together, either in a webbed fashion as seen in most bony fish or similar to a flipper as seen in sharks. Apart from the tail or caudal fin, fins have no direct connection with the spine and are supported by muscles only. Their principal function is to help the fish swim. Fins can also be used for gliding or crawling, as seen in the flying fish and frogfish. Fins located in different places on the fish serve different purposes, such as moving forward, turning, and keeping an upright position. For every fin, there are a number of fish species in which this particular fin has been lost during evolution.

Sharks possess a heterocercal caudal fin. The dorsal portion is usually larger than the ventral portion

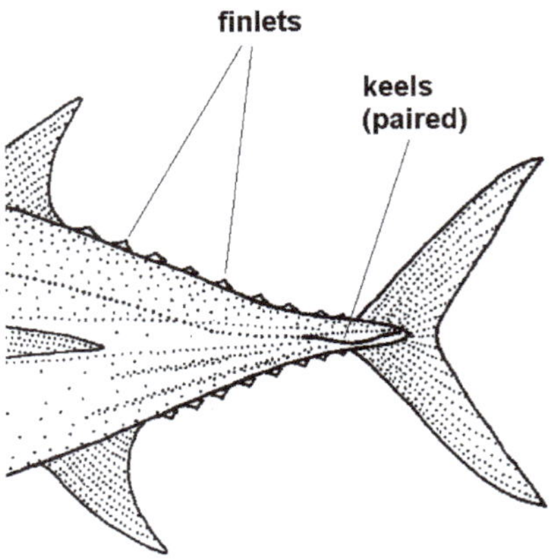

The high performance bigeye tuna is equipped with a homocercal caudal fin and finlets and keels. Drawn by Dr Tony Ayling

Spines and rays

In bony fish, most fins may have spines or rays. A fin may contain only spiny rays, only soft rays, or a combination of both. If both are present, the spiny rays are always anterior. Spines are generally stiff, sharp and unsegmented. Rays are generally soft, flexible, segmented, and may be branched. This segmentation of rays is the main difference that distinguishes them from spines; spines may be flexible in certain species, but never segmented.

Spines have a variety of uses. In catfish, they are used as a form of defense; many catfish have the ability to lock their spines outwards. Triggerfish also use spines to lock themselves in crevices to prevent them being pulled out.

Lepidotrichia are bony, bilaterally-paired, segmented fin rays found in bony fishes. They develop around actinotrichia as part of the dermal exoskeleton. Lepidotrichia may have some cartilage or bone in them as well. They are actually segmented and appear as a series of disks stacked one on top of another. The genetic basis for the formation of the fin rays is thought to be genes coding for the proteins actinodin 1 and actinodin 2.

Types of fin

- Dorsal fins are located on the back. Most fishes have one dorsal fin, but some fishes have two or three . The dorsal fins serve to protect the fish against rolling, and assists in sudden turns and stops. In anglerfish, the anterior of the dorsal fin is modified into an *illicium* and *esca*, a biological equivalent to a fishing rod and lure. The bones that support the dorsal fin are called *Pterygiophore*. There are two to three of them: “proximal”, “middle”, and “distal”. In spinous fins the distal is often fused to the middle, or not present at all.
- The caudal fin is the tail fin, located at the end of the caudal peduncle and is used for propulsion. The caudal peduncle is the narrow part of the fish’s body to which the caudal or tail fin is attached. The hypural joint is the joint between the caudal fin and the last of the vertebrae. The hypural is often fan-shaped. The tail is called:
    - *Heterocercal* if the vertebrae extend into the upper lobe of the tail, making it longer (as in sharks)

- *Reversed heterocercal* if the vertebrae extend into the lower lobe of the tail, making it longer (as in the Anaspida)
- *Protocercal* if the vertebrae extend to the tip of the tail and the tail is symmetrical but not expanded (as in amphioxus)
- *Diphycercal* if the vertebrae extend to the tip of the tail and the tail is symmetrical and expanded (as in the bichir, lungfish, lamprey and coelacanth. Most Palaeozoic fishes had a diphycercal heterocercal tail.)
- Most fish have a *homocercal* tail, where the fin appears superficially symmetric but the vertebrae extend for a very short distance into the upper lobe of the fin. This can be expressed in a variety of shapes. The tail fin can be:
    - rounded at the end
    - truncated: or end in a more-or-less vertical edge, such as in salmon
    - forked: or end in two prongs
    - emarginate: or with a slight inward curve.
    - continuous: with dorsal, caudal and anal fins attached, such as in eels

- The anal fin is located on the ventral surface behind the anus. This fin is used to stabilize the fish while swimming.
- The paired pectoral fins are located on each side, usually just behind the operculum, and are homologous to the forelimbs of tetrapods. A peculiar function of pectoral fins, highly developed in some fish, is the creation of the dynamic lifting force that assists some fish, such as sharks, in maintaining depth and also enables the "flight" for flying fish. In many fish, the pectoral fins aid in walking, especially in the lobe-like fins of some anglerfish and in the mudskipper. Certain rays of the pectoral fins may be adapted into finger-like projections, such as in sea robins and flying gurnards. The "horns" of manta rays and their relatives are called *cephalic fins*; this is actually a modification of the anterior portion of the pectoral fin.
- The paired pelvic or ventral fins are located ventrally below the pectoral fins. They are homologous to the hindlimbs of tetrapods. The pelvic fin assists the fish in going up or down through the water, turning sharply, and stopping quickly. In gobies, the pelvic fins are often fused into a single sucker disk. This can be used to attach to objects.
- The adipose fin is a soft, fleshy fin found on the back behind the dorsal fin and just forward of the caudal fin. It is absent in many fish families, but is found in Salmonidae, characins and catfishes. Its function has remained a mystery, and is frequently clipped off to mark hatchery-raised fish, though data from 2005 showed that trout with their adipose fin removed have an 8% higher tailbeat frequency. Additional research published in 2011 has suggested that the fin may be vital for the detection of and response to stimuli such as touch, sound and changes in pressure. Canadian researchers identified a neural network in the fin, indicating that it likely has a sensory function, but are still not sure exactly what the consequences of removing it are.

- Some types of fast-swimming fish have a horizontal caudal keel just forward of the tail fin. Much like the keel of a ship, this is a lateral ridge on the caudal peduncle, usually composed of scutes that provides stability and support to the caudal fin. There may be a single paired keel, one on each side, or two pairs above and below.
- Finlets are small fins, generally between the dorsal and the caudal fins also between the anal fin and the caudal fin (in bichirs, there are only finlets on the dorsal surface and no dorsal fin). In some fish such as tuna or sauries, they are rayless, non-retractable, and found between the last dorsal and/or anal fin and the caudal fin.

Internal Organs

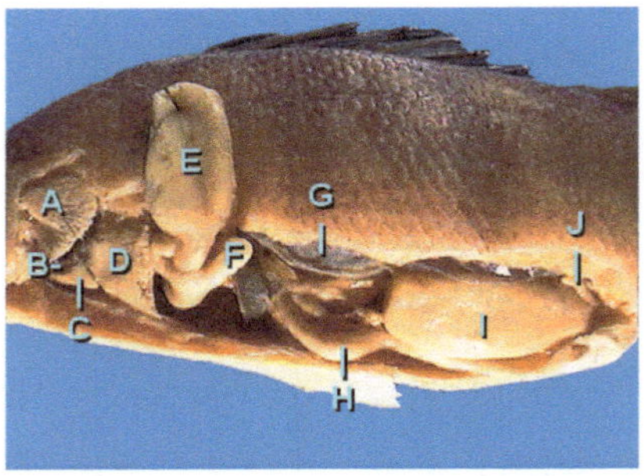

Internal organs of a male yellow perchA = gill, B = heart atrium,C: heart ventricle, D: liver (cut),E: stomach, F = pyloric caeca,G = swim bladder, H = intestine,I = testis, J = urinary bladder

Intestines

As with other vertebrates, the intestines of fish consist of two segments, the small intestine and the large intestine. In most higher vertebrates, the small intestine is further divided into the duodenum and other parts. In fish, the divisions of the small intestine are not as clear, and the terms *anterior intestine* or *proximal intestine* may be used instead of duodenum. In bony fish, the intestine is relatively short, typically around one and a half times the length of the fish's body. It commonly has a number of *pyloric caeca*, small pouch-like structures along its length that help to increase the overall surface area of the organ for digesting food. There is no ileocaecal valve in teleosts, with the boundary between the small intestine and the rectum being marked only by the end of the digestive epithelium. There is no small intestine as such in non-teleost fish, such as sharks, sturgeons, and lungfish. Instead, the digestive part of the gut forms a *spiral intestine*, connecting the stomach to the rectum. In this type of gut, the intestine itself is relatively straight, but has a long fold running along the inner surface in a spiral fashion, sometimes for dozens of turns. This fold creates a valve-like structure that greatly increases both the surface area and the effective length of the intestine. The lining of the spiral intestine is similar to that of the small intestine in teleosts and non-mammalian tetrapods. In lampreys, the spiral valve is extremely small, possibly because their diet requires little digestion. Hagfish have no spiral valve at all, with digestion occurring for almost the entire length of the intestine, which is not subdivided into different regions.

Pyloric Caeca

The pyloric caecum is a pouch, usually peritoneal, at the beginning of the large intestine. It receives faecal material from the ileum, and connects to the ascending colon of the large intestine.

It is present in most amniotes, and also in lungfish. Many fish in addition have a number of small outpocketings, also called pyloric caeca, along their intestine; despite the name they are not homologous with the caecum of amniotes, and their purpose is to increase the overall area of the digestive epithelium.

The black swallower is a species of deep sea fish with an extensible stomach which allows it to swallow fish larger than itself

Internal organs of a female Atlantic cod1. Liver, 2. Gas bladder, 3. ovary,4. Pyloric caeca, 5. Stomach, 6. Intestine

Stomach

As with other vertebrates, the relative positions of the esophageal and duodenal openings to the stomach remain relatively constant. As a result, the stomach always curves somewhat to the left before curving back to meet the pyloric sphincter. However, lampreys, hagfishes, chimaeras, lungfishes, and some teleost fish have no stomach at all, with the esophagus opening directly into the intestine. These fish consume diets that either require little storage of food, or no pre-digestion with gastric juices, or both.

Kidneys

The kidneys of fish are typically narrow, elongated organs, occupying a significant portion of the trunk. They are similar to the mesonephros of higher vertebrates (reptiles, birds and mammals). The kidneys contain clusters of nephrons, serviced by collecting ducts which usually drain into a mesonephric duct. However, the situation is not always so simple. In cartilaginous fish there is also a shorter duct which drains the posterior (metanephric) parts of the kidney, and joins with the mesonephric duct at the bladder or cloaca. Indeed, in many cartilaginous fish, the anterior portion of the kidney may degenerate or cease to function altogether in the adult. Hagfish and lamprey kidneys are unusually simple. They consist of a row of nephrons, each emptying directly into the mesonephric duct.

Spleen

The spleen is found in nearly all vertebrates. It is a non-vital organ, similar in structure to a large lymph node. It acts primarily as a blood filter, and plays important roles in regard to red blood cells and the immune system. In cartilaginous and bony fish it consists primarily of red pulp and is normally a somewhat elongated organ as it actually lies inside the serosal lining of the intestine. The only vertebrates lacking a spleen are the lampreys and hagfishes. Even in these animals, there is a diffuse layer of haematopoeitic tissue within the gut wall, which has a similar structure to red pulp, and is presumed to be homologous with the spleen of higher vertebrates.

Liver

The liver is a large vital organ present in all fish. It has a wide range of functions, including detoxification, protein synthesis, and production of biochemicals necessary for digestion.

Heart

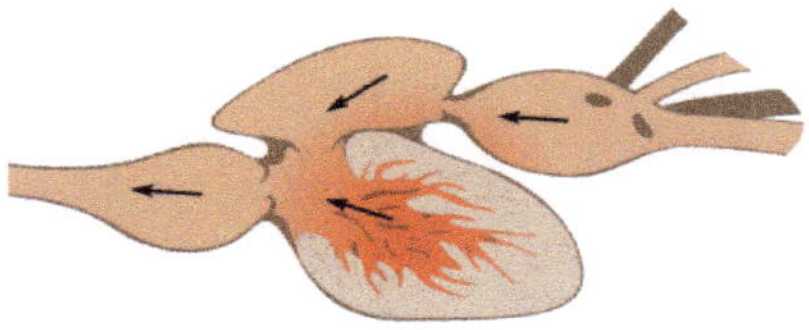

Blood flow through the heart: sinus venosus, atrium, ventricle, and outflow tract

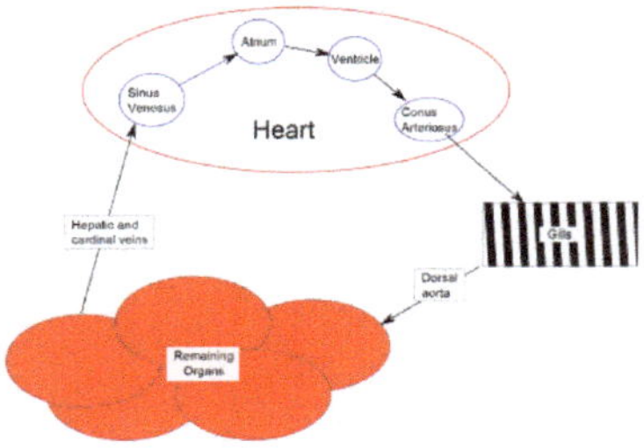

Cardiovascular cycle in a fish

Fish have what is often described as a two-chambered heart, consisting of one atrium to receive blood and one ventricle to pump it, in contrast to three chambers (two atria, one ventricle) of amphibian and most reptile hearts and four chambers (two atria, two ventricles) of mammal and bird hearts. However, the fish heart has entry and exit compartments that may be called chambers, so it is also sometimes described as three-chambered or four-chambered, depending on what is counted as a chamber. The atrium and ventricle are sometimes considered "true chambers", while the others are considered "accessory chambers".

The four compartments are arranged sequentially:

- Sinus venosus, a thin-walled sac or reservoir with some cardiac muscle that collects deoxygenated blood through the incoming hepatic and cardinal veins.
- Atrium, a thicker-walled, muscular chamber that sends blood to the ventricle.
- Ventricle, a thick-walled, muscular chamber that pumps the blood to the fourth part, the outflow tract. The shape of the ventricle varies considerably, usually tubular in fish with elongated bodies, pyramidal with a triangular base in others, or sometimes sac-like in some marine fish.

- The outflow tract (OFT) to the ventral aorta, consisting of the tubular conus arteriosus, bulbus arteriosus, or both. The conus arteriosus, typically found in more primitive species of fish, contracts to assist blood flow to the aorta, while the bulbus anteriosus does not.

Ostial valves, consisting of flap-like connective tissues, prevent blood from flowing backward through the compartments. The ostial valve between the sinus venosus and atrium is called the sino-atrial valve, which closes during ventricular contraction. Between the atrium and ventricle is an ostial valve called the atrio-ventricular valve, and between the bulbus arteriosus and ventricle is an ostial valve called the bulbo-ventricular valve. The conus arteriosus has a variable number of semilunar valves.

The ventral aorta delivers blood to the gills where it is oxygenated and flows, through the dorsal aorta, into the rest of the body. (In tetrapods, the ventral aorta has divided in two; one half forms the ascending aorta, while the other forms the pulmonary artery).

The circulatory systems of all vertebrates, are *closed*. Fish have the simplest circulatory system, consisting of only one circuit, with the blood being pumped through the capillaries of the gills and on to the capillaries of the body tissues. This is known as *single cycle* circulation.

In the adult fish, the four compartments are not arranged in a straight row but, instead form an S-shape with the latter two compartments lying above the former two. This relatively simpler pattern is found in cartilaginous fish and in the ray-finned fish. In teleosts, the conus arteriosus is very small and can more accurately be described as part of the aorta rather than of the heart proper. The conus arteriosus is not present in any amniotes, presumably having been absorbed into the ventricles over the course of evolution. Similarly, while the sinus venosus is present as a vestigial structure in some reptiles and birds, it is otherwise absorbed into the right atrium and is no longer distinguishable.

Swim Bladder

The swim bladder (or gas bladder) is an internal organ that contributes to the ability of a fish to control its buoyancy, and thus to stay at the current water depth, ascend, or descend without having to waste energy in swimming. The bladder is found only in the bony fishes. In the more primitive groups like some minnows, bichirs and lungfish, the bladder is open to the esophagus and doubles as a lung. It is often absent in fast swimming fishes such as the tuna and mackerel families. The condition of a bladder open to the esophagus is called physostome, the closed condition physoclist. In the latter, the gas content of the bladder is controlled through a rete mirabilis, a network of blood vessels effecting gas exchange between the bladder and the blood.

Weberian Apparatus

Fishes of the superorder Ostariophysi possess a structure called the Weberian apparatus, a modification which allow them to hear better. This ability which may well explain the marked success of otophysian fishes. The apparatus is made up of a set of bones known as *Weberian ossicles*, a chain of small bones that connect the auditory system to the swim bladder of fishes. The ossicles connect the gas bladder wall with Y-shaped lymph sinus that abuts the lymph-filled transverse canal joining the sacculi of the right and left ears. This allows the transmission of vibrations to the inner ear. A fully functioning Weberian apparatus consists of the swim bladder, the Weberian ossicles, a portion of the anterior vertebral column, and some muscles and ligaments.

Reproductive Organs

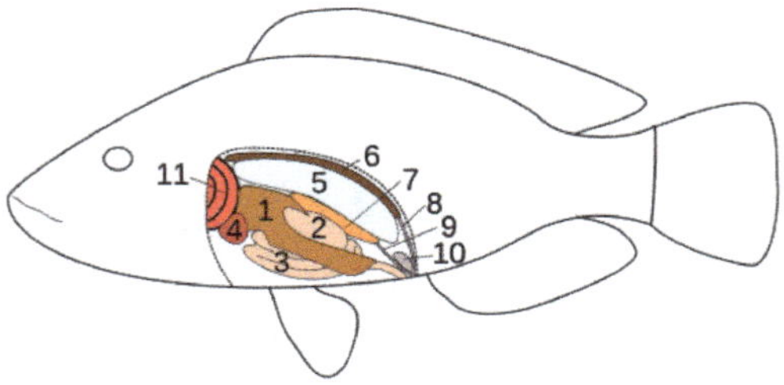

7 = testicles or ovaries

Fish reproductive organs include testes and ovaries. In most species, gonads are paired organs of similar size, which can be partially or totally fused. There may also be a range of secondary organs that increase reproductive fitness. The genital papilla is a small, fleshy tube behind the anus in some fishes, from which the sperm or eggs are released; the sex of a fish often can be determined by the shape of its papilla.

Testes

Most male fish have two testes of similar size. In the case of sharks, the testis on the right side is usually larger. The primitive jawless fish have only a single testis, located in the midline of the body, although even this forms from the fusion of paired structures in the embryo.

Under a tough membranous shell, the tunica albuginea, the testis of some teleost fish, contains very fine coiled tubes called seminiferous tubules. The tubules are lined with a layer of cells (germ cells) that from puberty into old age, develop into sperm cells (also known as spermatozoa or male gametes). The developing sperm travel through the seminiferous tubules to the rete testis located in the mediastinum testis, to the efferent ducts, and then to the epididymis where newly created sperm cells mature. The sperm move into the vas deferens, and are eventually expelled through the urethra and out of the urethral orifice through muscular contractions.

However, most fish do not possess seminiferous tubules. Instead, the sperm are produced in spherical structures called *sperm ampullae*. These are seasonal structures, releasing their contents during the breeding season, and then being reabsorbed by the body. Before the next breeding season, new sperm ampullae begin to form and ripen. The ampullae are otherwise essentially identical to the seminiferous tubules in higher vertebrates, including the same range of cell types.

In terms of spermatogonia distribution, the structure of teleosts testes has two types: in the most common, spermatogonia occur all along the seminiferous tubules, while in Atherinomorph fish they are confined to the distal portion of these structures. Fish can present cystic or semi-cystic spermatogenesis in relation to the release phase of germ cells in cysts to the seminiferous tubules lumen.

Ovaries

Many of the features found in ovaries are common to all vertebrates, including the presence of follicular cells and tunica albuginea There may be hundreds or even millions of fertile eggs present in the ovary of a fish at any given time. Fresh eggs may be developing from the germinal epithelium throughout life. Corpora lutea are found only in mammals, and in some elasmobranch fish; in other species, the remnants of the follicle are quickly resorbed by the ovary. The ovary of teleosts is often contains a hollow, lymph-filled space which opens into the oviduct, and into which the eggs are shed. Most normal female fish have two ovaries. In some elasmobranchs, only the right ovary develops fully. In the primitive jawless fish, and some teleosts, there is only one ovary, formed by the fusion of the paired organs in the embryo.

Fish ovaries may be of three types: gymnovarian, secondary gymnovarian or cystovarian. In the first type, the oocytes are released directly into the coelomic cavity and then enter the ostium, then through the oviduct and are eliminated. Secondary gymnovarian ovaries shed ova into the coelom from which they go directly into the oviduct. In the third type, the oocytes are conveyed to the exterior through the oviduct. Gymnovaries are the primitive condition found in lungfish, sturgeon, and bowfin. Cystovaries characterize most teleosts, where the ovary lumen has continuity with the oviduct. Secondary gymnovaries are found in salmonids and a few other teleosts.

Nervous System

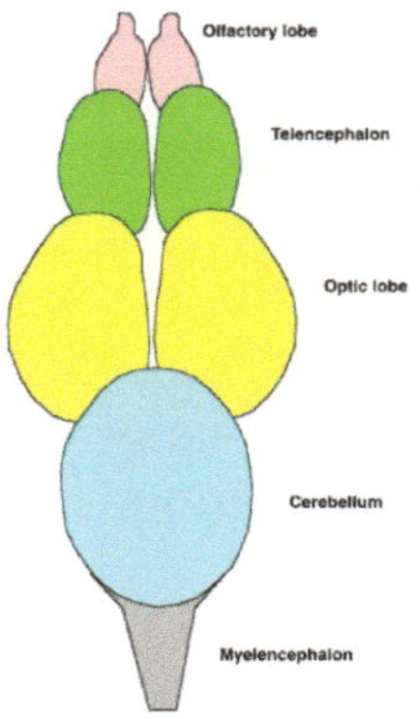

Dorsal view of the brain of the rainbow trout

Central Nervous System

Fish typically have quite small brains relative to body size compared with other vertebrates, typically one-fifteenth the brain mass of a similarly sized bird or mammal. However, some fish have relatively large brains, most notably mormyrids and sharks, which have brains about as massive relative to body weight as birds and marsupials.

Fish brains are divided into several regions. At the front are the olfactory lobes, a pair of structures that receive and process signals from the nostrils via the two olfactory nerves. The olfactory lobes are very large in fish that hunt primarily by smell, such as hagfish, sharks, and catfish. Behind the olfactory lobes is the two-lobed telencephalon, the structural equivalent to the cerebrum in higher vertebrates. In fish the telencephalon is concerned mostly with olfaction. Together these structures form the forebrain.

The forebrain is connected to the midbrain via the diencephalon (in the diagram, this structure is below the optic lobes and consequently not visible). The diencephalon performs functions associated with hormones and homeostasis. The pineal body lies just above the diencephalon. This structure detects light, maintains circadian rhythms, and controls color changes. The midbrain or mesencephalon contains the two optic lobes. These are very large in species that hunt by sight, such as rainbow trout and cichlids.

The hindbrain or metencephalon is particularly involved in swimming and balance. The cerebellum is a single-lobed structure that is typically the biggest part of the brain. Hagfish and lampreys have relatively small cerebellae, while the mormyrid cerebellum is massive and apparently involved in their electrical sense.

The brain stem or myelencephalon is the brain's posterior. As well as controlling some muscles and body organs, in bony fish at least, the brain stem governs respiration and osmoregulation.

Vertebrates are the only chordate group to exhibit a proper brain. A slight swelling of the anterior end of the dorsal nerve cord is found in the lancelet, though it lacks the eyes and other complex sense organs comparable to those of vertebrates. Other chordates do not show any trends towards cephalisation. The central nervous system is based on a hollow nerve tube running along the length of the animal, from which the peripheral nervous system branches out to innervate the various systems. The front end of the nerve tube is expanded by a thickening of the walls and expansion of the central canal of spinal cord into three primary brain vesicles: The prosencephalon (forebrain), mesencephalon (midbrain) and rhombencephalon (hindbrain), further differentiated in the various vertebrate groups. Two laterally placed eyes form around outgrows from the midbrain, except in hagfish, though this may be a secondary loss. The forebrain is well developed and subdivided in most tetrapods, while the midbrain dominate in many fish and some salamanders. Vesicles of the forebrain are usually paired, giving rise to hemispheres like the cerebral hemispheres in mammals. The resulting anatomy of the central nervous system, with a single, hollow ventral nerve cord topped by a series of (often paired) vesicles is unique to vertebrates.

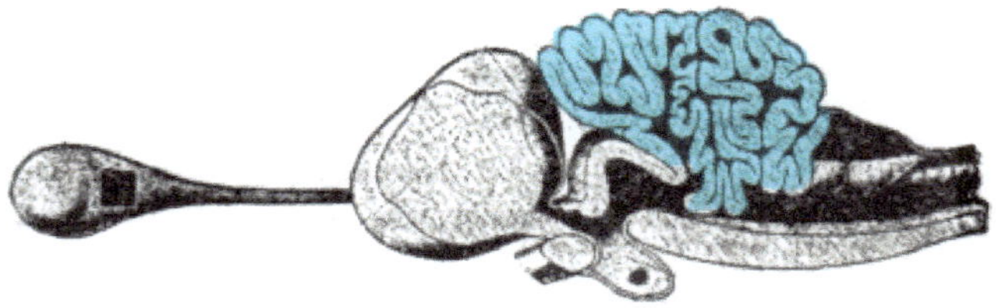

Cross-section of the brain of a porbeagle shark, with the cerebellum highlighted

Cerebellum

The circuits in the cerebellum are similar across all classes of vertebrates, including fish, reptiles, birds, and mammals. There is also an analogous brain structure in cephalopods with well-developed brains, such as octopuses. This has been taken as evidence that the cerebellum performs functions important to all animal species with a brain.

There is considerable variation in the size and shape of the cerebellum in different vertebrate species. In amphibians, lampreys, and hagfish, the cerebellum is little developed; in the latter two groups, it is barely distinguishable from the brain-stem. Although the spinocerebellum is present in these groups, the primary structures are small paired nuclei corresponding to the vestibulocerebellum.

The cerebellum of cartilaginous and bony fishes is extraordinarily large and complex. In at least one important respect, it differs in internal structure from the mammalian cerebellum: The fish cerebellum does not contain discrete deep cerebellar nuclei. Instead, the primary targets of Purkinje cells are a distinct type of cell distributed across the cerebellar cortex, a type not seen in mammals. In mormyrid fish (a family of weakly electrosensitive freshwater fish), the cerebellum is considerably larger than the rest of the brain put together. The largest part of it is a special structure called the *valvula*, which has an unusually regular architecture and receives much of its input from the electrosensory system.

Most species of fish and amphibians possess a lateral line system that senses pressure waves in water. One of the brain areas that receives primary input from the lateral line organ, the medial octavolateral nucleus, has a cerebellum-like structure, with granule cells and parallel fibers. In electrosensitive fish, the input from the electrosensory system goes to the dorsal octavolateral nucleus, which also has a cerebellum-like structure. In ray-finned fishes (by far the largest group), the optic tectum has a layer — the marginal layer — that is cerebellum-like.

Identified Neurons

A neuron is called *identified* if it has properties that distinguish it from every other neuron in the same animal—properties such as location, neurotransmitter, gene expression pattern, and connectivity—and if every individual organism belonging to the same species has one and only one neuron with the same set of properties. In vertebrate nervous systems very few neurons are "identified" in this sense—in humans, there are believed to be none—but in simpler nervous systems, some or all neurons may be thus unique.

In vertebrates, the best known identified neurons are the gigantic Mauthner cells of fish. Every fish has two Mauthner cells, located in the bottom part of the brainstem, one on the left side and one on the right. Each Mauthner cell has an axon that crosses over, innervating neurons at the same brain level and then travelling down through the spinal cord, making numerous connections as it goes. The synapses generated by a Mauthner cell are so powerful that a single action potential gives rise to a major behavioral response: within milliseconds the fish curves its body into a C-shape, then straightens, thereby propelling itself rapidly forward. Functionally this is a fast escape response, triggered most easily by a strong sound wave or pressure wave impinging on the lateral line organ of the fish. Mauthner cells are not the only identified neurons in fish—there are about 20 more types, including pairs of "Mauthner cell analogs" in each spinal segmental nucleus. Although a Mauthner cell is capable of bringing about an escape response all by itself, in the context of ordinary behavior other types of cells usually contribute to shaping the amplitude and direction of the response.

Mauthner cells have been described as command neurons. A command neuron is a special type of identified neuron, defined as a neuron that is capable of driving a specific behavior all by itself. Such neurons appear most commonly in the fast escape systems of various species—the squid giant axon and squid giant synapse, used for pioneering experiments in neurophysiology because of their enormous size, both participate in the fast escape circuit of the squid. The concept of a command neuron has, however, become controversial, because of studies showing that some neurons that initially appeared to fit the description were really only capable of evoking a response in a limited set of circumstances.

- Brainstem

Immune System

Immune organs vary by type of fish. In the jawless fish (lampreys and hagfish), true lymphoid organs are absent. These fish rely on regions of lymphoid tissue within other organs to produce immune cells. For example, erythrocytes, macrophages and plasma cells are produced in the anterior kidney (or pronephros) and some areas of the gut (where granulocytes mature.) They resemble primitive bone marrow in hagfish. Cartilaginous fish (sharks and rays) have a more advanced immune system. They have three specialized organs that are unique to chondrichthyes; the epigonal organs (lymphoid tissue similar to mammalian bone) that surround the gonads, the Leydig's organ within the walls of their esophagus, and a spiral valve in their intestine. These organs house typical immune cells (granulocytes, lymphocytes and plasma cells). They also possess an identifiable thymus and a well-developed spleen (their most important immune organ) where various lymphocytes, plasma cells and macrophages develop and are stored. Chondrostean fish (sturgeons, paddlefish and bichirs) possess a major site for the production of granulocytes within a mass that is associated with the meninges (membranes surrounding the central nervous system.) Their heart is frequently covered with tissue that contains lymphocytes, reticular cells and a small number of macrophages. The chondrostean kidney is an important hemopoietic organ; where erythrocytes, granulocytes, lymphocytes and macrophages develop.

Like chondrostean fish, the major immune tissues of bony fish (or teleostei) include the kidney (especially the anterior kidney), which houses many different immune cells. In addition, teleost fish possess a thymus, spleen and scattered immune areas within mucosal tissues (e.g. in the skin, gills, gut and gonads). Much like the mammalian immune system, teleost erythrocytes, neutrophils and granulocytes are believed to reside in the spleen whereas lymphocytes are the major cell type found in the thymus. In 2006, a lymphatic system similar to that in mammals was described in one species of teleost fish, the zebrafish. Although not confirmed as yet, this system presumably will be where naive (unstimulated) T cells accumulate while waiting to encounter an antigen.

Aquatic Respiration

Sea slugs breathe through a gill (or ctenidium)

Aquatic respiration is the process whereby an aquatic animal obtains oxygen from water.

Respiratory Systems

Fish

In most fish respiration takes place through gills. Lungfish, however, possess one or two lungs. The labyrinth fish have developed a special organ that allows them to take advantage of the oxygen of the air, but is not a true lung. Fish use the process known as countercurrent flow, in which water and blood flow in opposite directions across the gills, maximizing the diffusion of oxygen.

Molluscs

Molluscs generally possess gills that allow exchange of oxygen from an aqueous environment into the circulatory system. These animals also possess a heart that pumps blood which contains hemocyaninine as its oxygen-capturing molecule. Therefore, this respiratory system is similar to that of vertebrate fish. The respiratory system of gastropods can include either gills or a lung.

Arthropods

Aquatic arthropods generally possess some form of gills in which gas exchange takes place by diffusing through the exoskeleton. Others may breathe atmospheric air while remaining submerged, via breathing tubes or trapped air bubbles, though some aquatic insects may remain submerged indefinitely and respire using a plastron.

Aquatic Reptiles

The anatomical structure of the lungs is less complex in reptiles than in mammals, with reptiles lacking the very extensive airway tree structure found in mammalian lungs. Gas exchange in reptiles still occurs in alveoli however, reptiles do not possess a diaphragm. Thus, breathing occurs via a change in the volume of the body cavity which is controlled by contraction of intercostal muscles in all reptiles except turtles. In turtles, contraction of specific pairs of flank muscles governs inspiration or expiration.

Amphibians

Both the lungs and the skin serve as respiratory organs in amphibians. The skin of these animals is highly vascularized and moist, with moisture maintained via secretion of mucus from specialized cells. While the lungs are of primary importance to breathing control, the skin's unique properties aid rapid gas exchange when amphibians are submerged in oxygen-rich water.

Aquatic Birds

The respiratory system of birds differs significantly from that found in mammals, containing unique anatomical features such as air sacs. The lungs of birds also do not have the capacity to inflate as birds lack a diaphragm and a pleural cavity. Gas exchange in birds occurs between air capillaries and blood capillaries, rather than in alveoli.

Gills

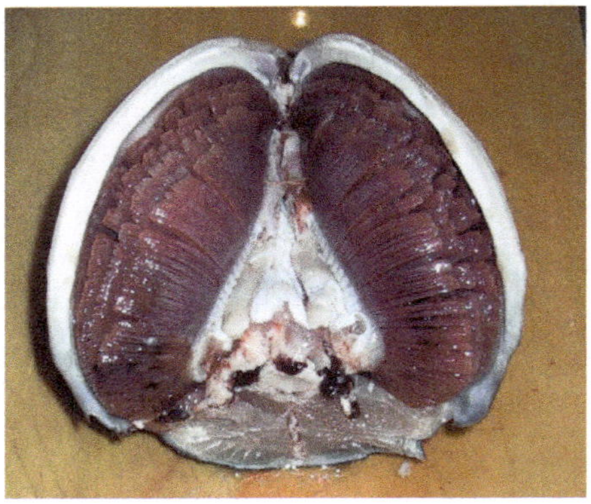

Posterior view of the gills of a tuna

Large aquatic animals have developed gills for respiration which are specifically adapted to their function, for example, they have:

- A large surface area to allow as much oxygen to enter the gills as possible because more of the gas comes into contact with the membrane
- Good blood supply to maintain the concentration gradient needed
- Thin membrane to allow for a short diffusion pathway
- each gill arch has two rows (hemibranchs) of gill filaments
- each gill filament has many lamellae

In osteichthyes, the gills contain 4 gill arches on each side of the head, two on each side for chondrichthyes or 7 gill baskets on each side of the fish's head in Lampreys. In fish, the long bony cover for the gill (the operculum) can be used for pushing water. Some fish pump water using the operculum. Without an operculum, other methods, such as ram ventilation, are required. Some species of sharks use this system. When they swim, water flows into the mouth and across the gills. Because these sharks rely on this technique, they must keep swimming in order to respire.

Bony fish use countercurrent flow to maximize the intake of oxygen that can diffuse through the gill. Countercurrent flow occurs when deoxygenated blood moves through the gill in one direction while oxygenated water moves through the gill in the opposite direction. This mechanism maintains the concentration gradient thus increasing the efficiency of the respiration process as well and prevents the oxygen levels from reaching an equilibrium. Cartilaginous fish do not have a countercurrent flow system as they lack bones which are needed to have the opened out gill that bony fish have.

Control of Respiration

Scientists have investigated what part of the body is responsible for maintaining the respiratory rhythm. They found that neurons located in the brainstem of fish are responsible for the genesis of the respiratory rhythm. The position of these neurons is slightly different from the centers of respiratory genesis in mammals but they are located in the same brain compartment, which has caused debates about the homology of respiratory centers between aquatic and terrestrial species. In both aquatic and terrestrial respiration, the exact mechanisms by which neurons can generate this involuntary rhythm are still not completely understood.

Another important feature of the respiratory rhythm is that it is modulated to adapt to the oxygen consumption of the body. As observed in mammals, fish "breathe" faster and heavier when they do physical exercise. The mechanisms by which these changes occur have been strongly debated over more than 100 years between scientists. The authors can be classified in 2 schools:

1. Those who think that the major part of the respiratory changes are pre-programmed in the brain, which would imply that neurons from locomotion centers of the brain connect to respiratory centers in anticipation of movements.

2. Those who think that the major part of the respiratory changes result from the detection of muscle contraction, and that respiration is adapted as a consequence of muscular contraction and oxygen consumption. This would imply that the brain possesses some kind of detection mechanisms that would trigger a respiratory response when muscular contraction occurs.

Many now agree that both mechanisms are probably present and complementary, or working alongside a mechanism that can detect changes in oxygen and/or carbon dioxide blood saturation.

Gill

The red gills of this common carp are visible as a result of a gill flap birth defect.

A gill is a respiratory organ found in many aquatic organisms that extracts dissolved oxygen from water and excretes carbon dioxide. The gills of some species, such as hermit crabs, have adapted to allow respiration on land provided they are kept moist. The microscopic structure of a gill presents a large surface area to the external environment.

Many microscopic aquatic animals, and some larger but inactive ones, can absorb adequate oxygen through the entire surface of their bodies, and so can respire adequately without a gill. However, more complex or more active aquatic organisms usually require a gill or gills.

Gills usually consist of thin filaments of tissue, branches, or slender, tufted processes that have a highly folded surface to increase surface area. A high surface area is crucial to the gas exchange of aquatic organisms, as water contains only a small fraction of the dissolved oxygen that air does. A cubic meter of air contains about 250 grams of oxygen at STP. The concentration of oxygen in water is lower than in air and it diffuses more slowly. In fresh water, the dissolved oxygen content is approximately 8 cm^3/L compared to that of air which is 210 cm^3/L. Water is 777 times more dense than air and is 100 times more viscous. Oxygen has a diffusion rate in air 10,000 times greater than in water. The use of sac-like lungs to remove oxygen from water would not be efficient enough to sustain life. Rather than using lungs, "[g]aseous exchange takes place across the surface of highly vascularised gills over which a one-way current of water is kept flowing by a specialised pumping mechanism. The density of the water prevents the gills from collapsing and lying on top of each other, which is what happens when a fish is taken out of water."

With the exception of some aquatic insects, the filaments and lamellae (folds) contain blood or coelomic fluid, from which gases are exchanged through the thin walls. The blood carries oxygen to other parts of the body. Carbon dioxide passes from the blood through the thin gill tissue into the water. Gills or gill-like organs, located in different parts of the body, are found in various groups of aquatic animals, including mollusks, crustaceans, insects, fish, and amphibians.

Vertebrate Gills

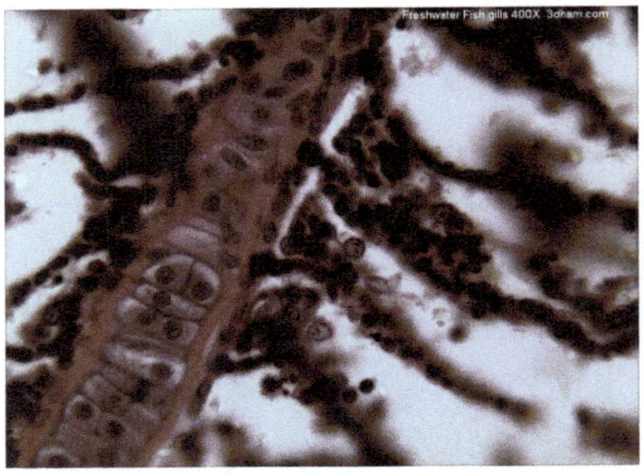

Freshwater Fish Gills magnified 400 times

The gills of vertebrates typically develop in the walls of the pharynx, along a series of gill slits opening to the exterior. Most species employ a countercurrent exchange system to enhance the diffusion of substances in and out of the gill, with blood and water flowing in opposite directions to each other. The gills are composed of comb-like filaments, the gill lamellae, which help increase their surface area for oxygen exchange.

When a fish breathes, it draws in a mouthful of water at regular intervals. Then it draws the sides of its throat together, forcing the water through the gill openings, so it passes over the gills to the outside. Fish gill slits may be the evolutionary ancestors of the tonsils, thymus glands, and Eustachian tubes, as well as many other structures derived from the embryonic branchial pouches.

Fish

Cartilaginous Fish

Sharks and rays typically have five pairs of gill slits that open directly to the outside of the body, though some more primitive sharks have six pairs and the Broadnose sevengill shark being the only cartilaginous fish exceeding this number. Adjacent slits are separated by a cartilaginous gill arch from which projects a cartilaginous gill ray. This gill ray is the support for the sheet-like interbranchial septum, which the individual lamellae of the gills lie on either side of. The base of the arch may also support gill rakers, projections into the pharyngeal cavity that help to prevent large pieces of debris from damaging the delicate gills.

A smaller opening, the spiracle, lies in the back of the first gill slit. This bears a small pseudobranch that resembles a gill in structure, but only receives blood already oxygenated by the true gills. The spiracle is thought to be homologous to the ear opening in higher vertebrates.

Most sharks rely on ram ventilation, forcing water into the mouth and over the gills by rapidly swimming forward. In slow-moving or bottom-dwelling species, especially among skates and rays, the spiracle may be enlarged, and the fish breathes by sucking water through this opening, instead of through the mouth.

Chimaeras differ from other cartilagenous fish, having lost both the spiracle and the fifth gill slit. The remaining slits are covered by an operculum, developed from the septum of the gill arch in front of the first gill.

Bony Fish

In bony fish, the gills lie in a branchial chamber covered by a bony operculum. The great majority of bony fish species have five pairs of gills, although a few have lost some over the course of evolution. The operculum can be important in adjusting the pressure of water inside of the pharynx to allow proper ventilation of the gills, so bony fish do not have to rely on ram ventilation (and hence near constant motion) to breathe. Valves inside the mouth keep the water from escaping.

The gill arches of bony fish typically have no septum, so the gills alone project from the arch, supported by individual gill rays. Some species retain gill rakers. Though all but the most primitive bony fish lack spiracles, the pseudobranch associated with them often remains, being located at the base of the operculum. This is, however, often greatly reduced, consisting of a small mass of cells without any remaining gill-like structure.

Marine teleosts also use gills to excrete electrolytes. The gills' large surface area tends to create a problem for fish that seek to regulate the osmolarity of their internal fluids. Salt water is less dilute than these internal fluids, so saltwater fish lose large quantities of water osmotically through their gills. To regain the water, they drink large amounts of sea water and excrete the salt. Fresh water is more dilute than the internal fluids of fish, however, so freshwater fish gain water osmotically through their gills.

Other Vertebrates

An Alpine newt larva showing the external gills, which flare just behind the head

Lampreys and hagfish do not have gill slits as such. Instead, the gills are contained in spherical pouches, with a circular opening to the outside. Like the gill slits of higher fish, each pouch contains two gills. In some cases, the openings may be fused together, effectively forming an operculum. Lampreys have seven pairs of pouches, while hagfishes may have six to fourteen, depending on the species. In the hagfish, the pouches connect with the pharynx internally and a separate tube which has no respiratory tissue (the pharyngocutaneous duct) develops beneath the pharynx proper, expelling ingested debris by closing a valve at its anterior end.

Tadpoles of amphibians have from three to five gill slits that do not contain actual gills. Usually no spiracle or true operculum is present, though many species have operculum-like structures. Instead of internal gills, they develop three feathery external gills that grow from the outer surface of the gill arches. Sometimes, adults retain these, but they usually disappear at metamorphosis. Lungfish larvae also have external gills, as does the primitive ray-finned fish *Polypterus*, though the latter has a structure different from amphibians. Some salamanders, such as the olm, and the mudpuppy, retain their external gills upon reaching adulthood.

Branchia

Branchia (pl. branchiae) is the naturalists' name for gills. Galen observed that fish had multitudes of openings (*foramina*), big enough to admit gases, but too fine to give passage to water. Pliny the Elder held that fish respired by their gills, but observed that Aristotle was of another opinion.

Invertebrate Gills

A live sea slug, *Pleurobranchaea meckelii*: The gill (or ctenidium) is visible in this view of the right-hand side of the animal.

Respiration in the echinoderms (such as starfish and sea urchins) is carried out using a very primitive version of gills called papulae. These thin protuberances on the surface of the body contain diverticula of the water vascular system. Crustaceans, molluscs, and some aquatic insects have tufted gills or plate-like structures on the surfaces of their bodies.

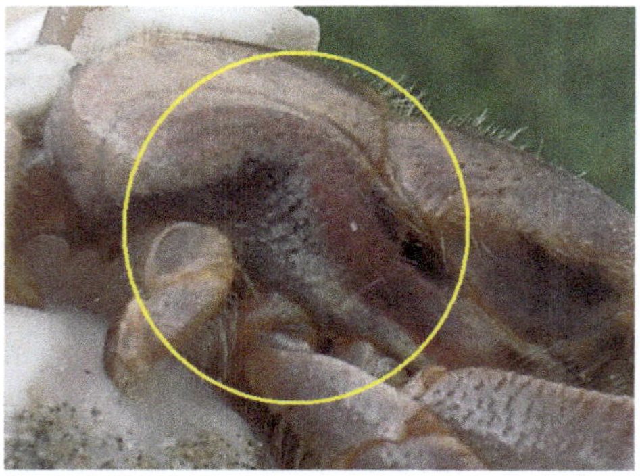

Caribbean hermit crabs have modified gills that allow them to live in humid conditions.

The gills of aquatic insects are tracheal, but the air tubes are sealed, commonly connected to thin external plates or tufted structures that allow diffusion. The oxygen in these tubes is renewed through the gills. In the larval dragon fly, the wall of the caudal end of the alimentary tract (rectum) is richly supplied with tracheae as a rectal gill, and water pumped into and out of the rectum provides oxygen to the closed tracheae.

Plastron

A plastron is a type of structural adaptation occurring among some aquatic arthropods (primarily insects), a form of inorganic gill which holds a thin film of atmospheric oxygen in an area with small openings called spiracles that connect to the tracheal system. The plastron typically consists of dense patches of hydrophobic setae on the body, which prevent water entry into the spiracles, but may also involve scales or microscopic ridges projecting from the cuticle. The physical properties of the interface between the trapped air film and surrounding water allow gas exchange through the spiracles, almost as if the insect were in atmospheric air. Carbon dioxide diffuses into the surrounding water due to its high solubility, while oxygen diffuses into the film as the concentration within the film has been reduced by respiration, and nitrogen also diffuses out as its tension has been increased. Oxygen diffuses into the air film at a higher rate than nitrogen diffuses out. However, water surrounding the insect can become oxygen-depleted if there is no water movement, so many such insects in still water actively direct a flow of water over their bodies.

The inorganic gill mechanism allows aquatic insects with plastrons to remain constantly submerged. Examples include many beetles in the family Elmidae, aquatic weevils, and true bugs in the family Aphelocheiridae, as well as at least one species of ricinuleid arachnid. A somewhat similar mechanism is used by the diving bell spider, which maintains an underwater bubble that exchanges gas like a plastron. Other diving insects (such as backswimmers, and hydrophilid beetles) may carry trapped air bubbles, but deplete the oxygen more quickly, and thus need constant replenishment.

Osteichthyes

Osteichthyes popularly referred to as the bony fish, is a diverse taxonomic group of fish that have skeletons primarily composed of bone tissue, as opposed to cartilage. The vast majority of fish are members of Osteichthyes, which is an extremely diverse and abundant group consisting of 45 orders, and over 435 families and 28,000 species. It is the largest class of vertebrates in existence today. The group Osteichthyes is divided into the ray-finned fish (Actinopterygii) and lobe-finned fish (Sarcopterygii). The oldest known fossils of bony fish are about 420 million years ago, which are also transitional fossils, showing a tooth pattern that is in between the tooth rows of sharks and bony fishes.

Osteichthyes can be compared to Euteleostomi. In paleontology, the terms are synonymous. In ichthyology, the difference is that Euteleostomi presents a cladistic view which includes the terrestrial tetrapods that evolved from lobe-finned fish, whereas on a traditional view, Osteichthyes includes only fishes and is therefore paraphyletic. However, recently published phylogenetic trees treat the Osteichthyes as a clade.

Characteristics

Guiyu oneiros, the earliest known bony fish, lived during the Late Silurian, 419 million years ago). It has the combination of both ray-finned and lobe-finned features, although analysis of the totality of its features place it closer to lobe-finned fish.

Bony fish are characterized by a relatively stable pattern of cranial bones, rooted, medial insertion of mandibular muscle in the lower jaw. The head and pectoral girdles are covered with large dermal bones. The eyeball is supported by a sclerotic ring of four small bones, but this characteristic has been lost or modified in many modern species. The labyrinth in the inner ear contains large otoliths. The braincase, or neurocranium, is frequently divided into anterior and posterior sections divided by a fissure.

Early bony fish had simple lungs (a pouch on either side of the esophagus) which helped them breathe in low-oxygen water. In many bony fish these have evolved into swim bladders, which help the body create a neutral balance between sinking and floating. (The lungs of amphibians, reptiles, birds, and mammals were inherited from their bony fish ancestors.) They do not have fin spines, but instead support the fin with lepidotrichia (bone fin rays). They also have an operculum, which helps them breathe without having to swim.

Bony fish have no placoid scales. Mucus glands coat the body. Most have smooth and overlapping ganoid, cycloid or ctenoid scales.

Classification

Traditionally, Osteichthyes is considered a class, recognised on having a swim bladder, only three pairs of gill arches, hidden behind a bony operculum and a predominately bony skeleton. Under this classification systems, the Osteichthyes are paraphyletic with regard to land vertebrates as the common ancestor of all Osteichthyes includes tetrapods amongst its descendants. The largest subclass, the Actinopterygii (ray-finned fish) are monophyletic, but with the inclusion of the smaller sub-class Sarcopterygii, Osteichthyes is paraphyletic.

This has led to an alternative classification, splitting the Osteichthyes into two full classes. Paradoxically, Sarcopterygii is under this scheme considered monophyletic, as it includes the four classes of tetrapods. Most bony fish belong to the ray-finned fish (Actinopterygii); there are only eight living species of lobe-finned fish (Sarcopterygii), including the lungfish and coelacanths.

Actinopte-rygii	 ray-finned fish	Actinopterygii, or ray-finned fishes, constitute a class or subclass of the bony fishes. The ray-finned fishes are so called because they possess lepidotrichia or "fin rays", their fins being webs of skin supported by bony or horny spines ("rays"), as opposed to the fleshy, lobed fins that characterize the class Sarcopterygii which also, however, possess lepidotrichia. These actinopterygian fin rays attach directly to the proximal or basal skeletal elements, the radials, which represent the link or connection between these fins and the internal skeleton (e.g., pelvic and pectoral girdles). In terms of numbers, actinopterygians are the dominant class of vertebrates, comprising nearly 99% of the over 30,000 species of fish (Davis, Brian 2010). They are ubiquitous throughout freshwater and marine environments from the deep sea to the highest mountain streams. Extant species can range in size from *Paedocypris*, at 8 mm (0.3 in), to the massive ocean sunfish, at 2,300 kg (5,070 lb), and the long-bodied oarfish, to at least 11 m (36 ft).
Sarcoptery-gii	 lobe-finned fish	Sarcopterygii *(fleshy fin)* or lobe-finned fish constitute a clade (traditionally a class or subclass) of the bony fish, though a strict cladistic view includes the terrestrial vertebrates. The living sarcopterygians are the coelacanths, lungfish, and the tetrapods. Early lobe-finned fishes have fleshy, lobed, paired fins, which are joined to the body by a single bone. Their fins differ from those of all other fish in that each is borne on a fleshy, lobelike, scaly stalk extending from the body. Pectoral and pelvic fins have articulations resembling those of tetrapod limbs. These fins evolved into legs of the first tetrapod land vertebrates, amphibians. They also possess two dorsal fins with separate bases, as opposed to the single dorsal fin of actinopterygians (ray-finned fish). The braincase of sarcoptergygians primitively has a hinge line, but this is lost in tetrapods and lungfish. Many early lobe-finned fishes have a symmetrical tail. All lobe-finned fishes possess teeth covered with true enamel.

Phylogeny

The phylogeny of living bony fishes is shown in the cladogram.

Biology

All bony fish possess gills. For the majority this is their sole or main means of respiration. Lungfish and other osteichthyan species are capable of respiration through lungs or vascularized swim bladders. Other species can respire through their skin, intestines, and/or stomach.

Osteichthyes are primitively ectothermic (cold blooded), meaning that their body temperature is dependent on that of the water. But some of the larger marine osteichthyids, such as the opah, swordfish and tuna have independently evolved various levels of endothermy. Bony fish can be any type of heterotroph: numerous species of omnivore, carnivore, herbivore, filter-feeder or detritivore are documented.

Some bony fish are hermaphrodites, and a number of species exhibit parthenogenesis. Fertilization is usually external, but can be internal. Development is usually oviparous (egg-laying) but can be ovoviviparous, or viviparous. Although there is usually no parental care after birth, before birth parents may scatter, hide, guard or brood eggs, with sea horses being notable in that the males undergo a form of "pregnancy", brooding eggs deposited in a ventral pouch by a female.

Examples

The ocean sunfish is the heaviest bony fish in the world, while the longest is the king of herrings, a type of oarfish. Specimens of ocean sunfish have been observed up to 3.3 metres (11 ft) in length and weighing up to 2,303 kilograms (5,077 lb). Other very large bony fish include the Atlantic blue marlin, some specimens of which have been recorded as in excess of 820 kilograms (1,810 lb), the black marlin, some sturgeon species, and the giant and goliath grouper, which both can exceed 300 kilograms (660 lb) in weight. In contrast, the dwarf pygmy goby measures a minute 15 millimetres (0.59 in).

Arapaima gigas is the largest species of freshwater bony fish. The largest bony fish ever was *Leedsichthys*, which dwarfed the beluga sturgeon, ocean sunfish, giant grouper, and all the other giant bony fishes alive today.

Comparison with Cartilaginous Fishes

Cartilaginous fishes can be further divided into sharks, rays and chimaeras. In the table below, the comparison is made between sharks and bony fishes. For the further differences with rays.

Comparison of cartilaginous and bony fishes		
Characteristic	**Sharks (cartilaginous)**	**Bony fishes**
Habitat	Mainly marine	Marine and freshwater
Shape	Usually dorso-ventrally flattened	Usually bilaterally flattened
Exoskeleton	Separate dermal placoid scales	Overlapping dermal cosmoid, ganoid, cycloid or ctenoid scales
Endoskeleton	Cartilaginous	Mostly bony
Caudal fin	Heterocercal	Heterocercal or diphycercal
Pelvic fins	Usually posterior.	Mostly anterior, occasionally posterior.
Intromittent organ	Males use pelvic fins as claspers for transferring sperm to a female	Do not use claspers, though some species use their anal fins as gonopodium for the same purpose
Mouth	Large, crescent shaped on the ventral side of the head	Variable shape and size at the tip or terminal part of the head
Jaw suspension	Hyostylic	Hyostylic and autostylic
Gill openings	Usually five pairs of gill slits which are not protected by an operculum.	Five pairs of gill slits protected by an operculum (a lateral flap of skin).
Type of gills	Larnellibranch with long interbranchial septum	Filiform with reduced interbranchial septum
Spiracles	The first gill slit usually becomes spiracles opening behind the eyes.	No spiracles
Afferent branchial vessels	Five pairs from ventral aorta to gills	Only four pairs
Efferent branchial vessels	Nine pairs	Four pairs
Conus arteriosus	Present in heart	Absent

Cloaca	A true cloaca is present only in cartilaginous fishes and lobe-finned fishes.	In most bony fishes, the cloaca is absent, and the anus, urinary and genital apertures open separately
Stomach	Typically J-shaped	Shape variable. Absent in some.
Intestine	Short with spiral valve in lumen	Long with no spiral valve
Rectal gland	Present	Absent
Liver	Usually has two lobes	Usually has three lobes
Swim bladder	Absent	Usually present
Brain	Has large olfactory lobes and cerebrum with small optic lobes and cerebellum	Has small olfactory lobes and cerebrum and large optic lobes and cerebellum
Restiform bodies	Present in brain	Absent
Ductus endolymphaticus	Opens on top of head	Does not open to exterior
Retina	Lacks cones	Most fish have double cones, a pair of cone cells joined to each other.
Accommodation of eye	Accommodate for near vision by moving the lens closer to the retina	Accommodate for distance vision by moving the lens further from the retina
Ampullae of Lorenzini	Present	Absent
Male genital duct	Connects to the anterior part of the genital kidney	No connection to kidney
Oviducts	Not connected to ovaries	Connected to ovaries
Urinary and genital apertures	United and urinogenital apertures lead into common cloaca	Separate and open independently to exterior
Eggs	A small number of large eggs with plenty of yolk	A large number of small eggs with little yolk
Fertilisation	Internal	Usually external
Development	Ovoviviparous types develop internally. Oviparous types develop externally using egg cases	Normally develop externally without an egg case

Fish Jaw

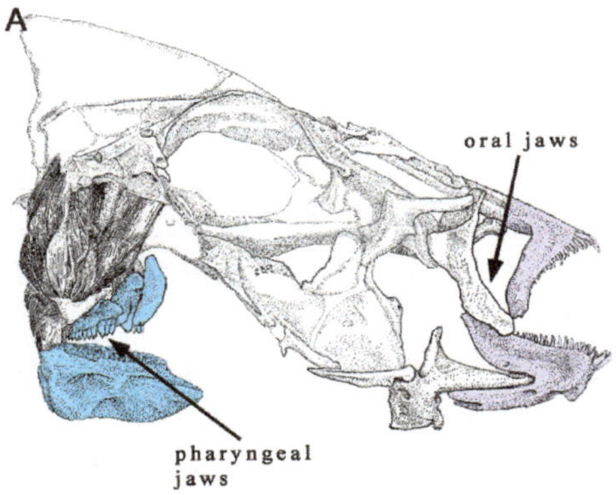

Skull of a generalized cichlid, showing a lateral view of the oral jaws (purple) and the pharyngeal jaws (blue)

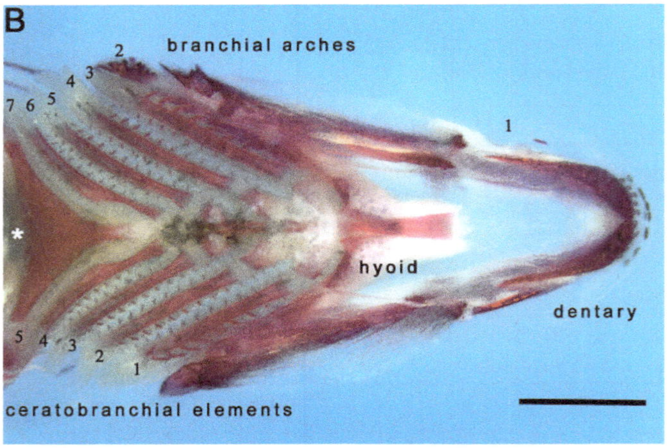

Dorsal view of the lower pharyngeal and oral jaws of a juvenile Malawi eyebiter showing the branchial (pharyngeal) arches and ceratobrachial elements (arch bones). The white asterisk indicates the toothed pharyngeal jaw. Scale bar represents 500 μm.

Most bony fishes have two sets of jaws made mainly of bone. The primary oral jaws open and close the mouth, and a second set of pharyngeal jaws are positioned at the back of the throat. The oral jaws are used to capture and manipulate prey by biting and crushing. The pharyngeal jaws, so-called because they are positioned within the pharynx, are used to further process the food and move it from the mouth to the stomach.

Cartilaginous fishes, such as sharks and rays, have one set of oral jaws made mainly of cartilage. They do not have pharyngeal jaws. Generally jaws are articulated and oppose vertically, comprising an upper jaw and a lower jaw and can bear numerous ordered teeth. Bony fishes usually develop only one set of teeth *(monophyodont)*. Cartilaginous fishes grow multiple sets *(polyphyodont)* and replace teeth as they wear.

Jaws probably originated in the pharyngeal arches supporting the gills of jawless fish. The earliest jaws appeared in the now extinct placoderms and spiny sharks during the Silurian, about 430 million years ago. The original selective advantage offered by the jaw was probably not related to feeding, but to increased respiration efficiency — the jaws were used in the buccal pump to pump water across the gills. The familiar use of jaws for feeding would then have developed as a secondary function before becoming the primary function in many vertebrates. All vertebrate jaws, including the human jaw, evolved from early fish jaws. The appearance of the early vertebrate jaw has been described as "perhaps the most profound and radical evolutionary step in the vertebrate history". Fish without jaws had more difficulty surviving than fish with jaws, and most jawless fish became extinct.

Jaws use linkage mechanisms. These linkages can be especially common and complex in the head of bony fishes, such as wrasses, which have evolved many specialized feeding mechanisms. Especially advanced are the linkage mechanisms of jaw protrusion. For suction feeding a system of linked four-bar linkages is responsible for the coordinated opening of the mouth and the three-dimensional expansion of the buccal cavity. Other linkages are responsible for protrusion of the premaxilla. Linkage systems are widely distributed in animals. The most thorough overview of the different types of linkages in animals has been provided by M. Muller, who also designed a new classification system, which is especially well suited for biological systems.

Skull

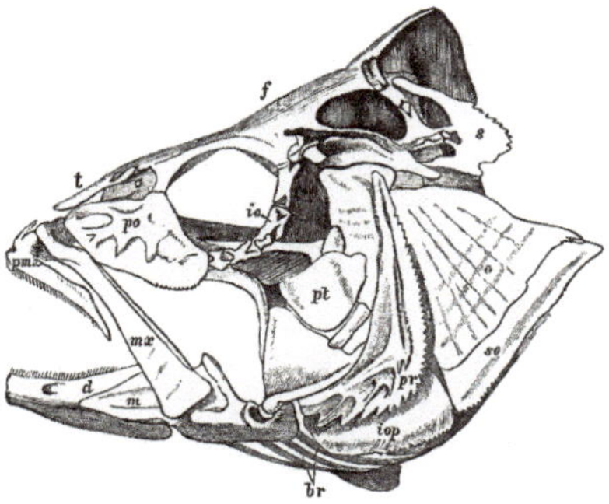

Skeleton of Head of a Perch.

f, frontal.
t, turbinal.
po, preorbital.
io, infraorbital ring.
mx, maxillary.
pmx, premaxillary.
m, mandible.
d, dentary bone.
pt, posttympanic.
s, suprascapula.
o, opercle.
so, subopercle.
pr, preopercle.
iop, interopercle.
br, branchiostegal rays.

The skull of fishes is formed from a series of loosely connected bones. Lampreys and sharks only possess a cartilaginous endocranium, with both the upper and lower jaws being separate elements. Bony fishes have additional dermal bone, forming a more or less coherent skull roof in lungfish and holost fish. The lower jaw defines a chin.

The simpler structure is found in jawless fish, in which the cranium is represented by a trough-like basket of cartilaginous elements only partially enclosing the brain, and associated with the capsules for the inner ears and the single nostril. Distinctively, these fish have no jaws.

Cartilaginous fish, such as sharks, also have simple skulls. The cranium is a single structure forming a case around the brain, enclosing the lower surface and the sides, but always at least partially open at the top as a large fontanelle. The most anterior part of the cranium includes a forward plate of cartilage, the rostrum, and capsules to enclose the olfactory organs. Behind these are the orbits, and then an additional pair of capsules enclosing the structure of the inner ear. Finally, the skull tapers towards the rear, where the foramen magnum lies immediately above a single condyle, articulating with the first vertebra. There are, in addition, at various points throughout the cranium, smaller foramina for the cranial nerves. The jaws consist of separate hoops of cartilage, almost always distinct from the cranium proper.

In ray-finned fishes, there has also been considerable modification from the primitive pattern. The roof of the skull is generally well formed, and although the exact relationship of its bones to those of tetrapods is unclear, they are usually given similar names for convenience. Other elements of the skull, however, may be reduced; there is little cheek region behind the enlarged orbits, and little, if any bone in between them. The upper jaw is often formed largely from the premaxilla, with the maxilla itself located further back, and an additional bone, the symplectic, linking the jaw to the rest of the cranium.

Although the skulls of fossil lobe-finned fish resemble those of the early tetrapods, the same cannot be said of those of the living lungfishes. The skull roof is not fully formed, and consists of multiple, somewhat irregularly shaped bones with no direct relationship to those of tetrapods. The upper jaw is formed from the pterygoids and vomers alone, all of which bear teeth. Much of the skull is formed from cartilage, and its overall structure is reduced.

Oral Jaws

Lower

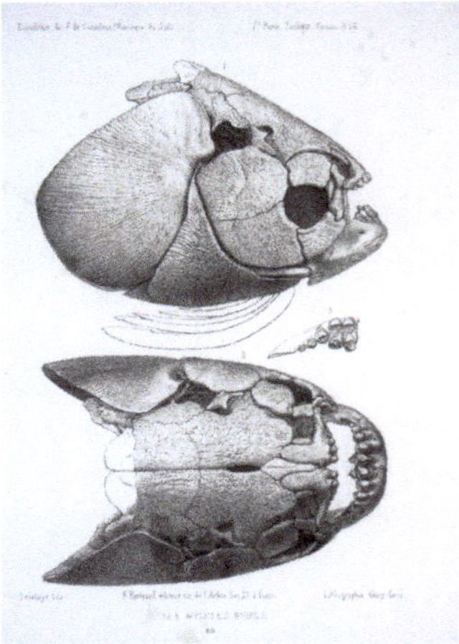

Oral jaw from side and above of Piaractus brachypomus, a close relative of piranhas

In vertebrates, the lower jaw (mandible or jawbone) is a bone forming the skull with the cranium. In lobe-finned fishes and the early fossil tetrapods, the bone homologous to the mandible of mammals is merely the largest of several bones in the lower jaw. It is referred to as the '*dentary bone*, and forms the body of the outer surface of the jaw. It is bordered below by a number of splenial bones, while the angle of the jaw is formed by a lower angular bone and a suprangular bone just above it. The inner surface of the jaw is lined by a *prearticular* bone, while the articular bone forms the articulation with the skull proper. Finally a set of three narrow *coronoid bones* lie above the prearticular bone. As the name implies, the majority of the teeth are attached to the dentary, but there are commonly also teeth on the coronoid bones, and sometimes on the prearticular as well.

This complex primitive pattern has, however, been simplified to various degrees in the great majority of vertebrates, as bones have either fused or vanished entirely. In teleosts, only the dentary, articular, and angular bones remain. Cartilagenous fish, such as sharks, do not have any of the bones found in the lower jaw of other vertebrates. Instead, their lower jaw is composed of a cartilagenous structure homologous with the Meckel's cartilage of other groups. This also remains a significant element of the jaw in some primitive bony fish, such as sturgeons.

Upper

The upper jaw, or maxilla is a fusion of two bones along the palatal fissure that form the upper jaw. This is similar to the mandible (lower jaw), which is also a fusion of two halves at the mandibular symphysis. In bony fish, the maxilla is called the "upper maxilla," with the mandible being the "lower maxilla". The alveolar process of the maxilla holds the upper teeth, and is referred to as the maxillary arch. In most vertebrates, the foremost part of the upper jaw, to which the incisors are attached in mammals consists of a separate pair of bones, the premaxillae. In bony fish, both maxilla and premaxilla are relatively plate-like bones, forming only the sides of the upper jaw, and part of the face, with the premaxilla also forming the lower boundary of the nostrils. Cartilaginous fish, such as sharks and rays also lack a true maxilla. Their upper jaw is instead formed from a cartilagenous bar that is not homologous with the bone found in other vertebrates.

Some fish have permanently protruding upper jawbones called rostrums. Billfish (marlin, swordfish and sailfish) use rostrums (bills) to slash and stun prey. Paddlefish, goblin sharks and hammerhead sharks have rostrums packed with electroreceptors which signal the presence of prey by detecting weak electrical fields. Sawsharks and the critically endangered sawfish have rostrums (saws) which are both electro-sensitive and used for slashing. The rostrums extend ventrally in front of the fish. In the case of hammerheads the rostrum (hammer) extends both ventrally and laterally (sideways).

- Fish with rostrums (extended upper jawbones)

Sailfish, like all billfish, have a rostrum (bill) which evolved from the upper jawbone

The paddlefish has a rostrum packed with electroreceptors

Sawfish have an electro-sensitive rostrum (saw) which is also used to slash at prey

Hammerhead sharks use their rostrum (hammer) to detect and pin rays buried in sand

Pharyngeal Jaws

Pharyngeal jaws are a second set of jaws distinct from the primary (oral) jaws. They are contained within the throat, or pharynx, of most bony fish. They are believed to have originated, in a similar way to oral jaws, as a modification of the fifth gill arch which no longer has a respiratory function. The first four arches still function as gills. Unlike the oral jaw, the pharyngeal jaw has no jaw joint, but is supported instead by a sling of muscles.

Pharyngeal jaw of an asp carrying some pharyngeal teeth

A notable example occurs with the moray eel. The pharyngeal jaws of most fishes are not mobile. The pharyngeal jaws of the moray are highly mobile, perhaps as an adaptation to the constricted nature of the burrows they inhabit which inhibits their ability to swallow as other fishes do by creating a negative pressure in the mouth. Instead, when the moray bites prey, it first bites normally with its oral jaws, capturing the prey. Immediately thereafter, the pharyngeal jaws are brought forward and bite down on the prey to grip it; they then retract, pulling the prey down the moray eel's gullet, allowing it to be swallowed.

All vertebrates have a pharynx, used in both feeding and respiration. The pharynx arises during development through a series of six or more outpocketings called pharyngeal arches on the lateral sides of the head. The pharyngeal arches give rise to a number of different structures in the skeletal, muscular and circulatory systems in a manner which varies across the vertebrates. Pharyngeal arches trace back through chordates to basal deuterostomes who also share endodermal outpocketings of the pharyngeal apparatus. Similar patterns of gene expression can be detected in the developing pharynx of amphioxus and hemichordates. However, the vertebrate pharynx is unique in that it gives rise to endoskeletal support through the contribution of neural crest cells.

Cartilaginous Jaws

Cartilaginous fishes (sharks, rays and skates) have cartilaginous jaws. The jaw's surface (in comparison to the vertebrae and gill arches) needs extra strength due to its heavy exposure to physical stress. It has a layer of tiny hexagonal plates called "tesserae", which are crystal blocks of calcium salts arranged as a mosaic. This gives these areas much of the same strength found in the bony tissue found in other animals.

Generally sharks have only one layer of tesserae, but the jaws of large specimens, such as the bull shark, tiger shark, and the great white shark, have two to three layers or more, depending on body size. The jaws of a large great white shark may have up to five layers. In the rostrum (snout), the cartilage can be spongy and flexible to absorb the power of impacts.

In sharks and other extant elasmobranchs the upper jaw is not fused to the cranium, and the lower jaw is articulated with the upper. The arrangement of soft tissue and any additional articulations connecting these elements is collectively known as the jaw suspension. There are several arche-typal jaw suspensions: amphistyly, orbitostyly, hyostyly, and euhyostyly. In amphistyly, the pala-toquadrate has a postorbital articulation with the chondrocranium from which ligaments primarily suspend it anteriorly. The hyoid articulates with the mandibular arch posteriorly, but it appears to provide little support to the upper and lower jaws. In orbitostyly, the orbital process hinges with the orbital wall and the hyoid provides the majority of suspensory support. In contrast, hyostyly involves an ethmoid articulation between the upper jaw and the cranium, while the hyoid most likely provides vastly more jaw support compared to the anterior ligaments. Finally, in euhyostyly, also known as true hyostyly, the mandibular cartilages lack a ligamentous connection to the cranium. Instead, the hyomandibular cartilages provide the only means of jaw support, while the ceratohyal and basihyal elements articulate with the lower jaw, but are disconnected from the rest of the hyoid.

Teeth

Inside of a shark jaw where new teeth move forward as though on a conveyor belt

Jaws provide a platform in most fishes for simple pointed teeth. Lungfish and chimaera have teeth modified into broad enamel plates with jagged ridges for crushing or grinding. Carp and loach have pharyngeal teeth only. Sea horses, pipefish and adult sturgeon have no teeth of any type.

Shark teeth are embedded in the gums rather than directly affixed to the jaw, and are constantly replaced throughout life. Multiple rows of replacement teeth grow in a groove on the inside of the jaw and steadily moving forward as though on a conveyor belt. Some sharks lose 30,000 or more teeth in their lifetime. The rate of tooth replacement varies from once every 8 to 10 days to several months. In most species, teeth are replaced one at a time as opposed to the simultaneous replacement of an entire row, which is observed in the cookiecutter shark.

Tooth shape depends on the shark's diet: those that feed on mollusks and crustaceans have dense and flattened teeth used for crushing, those that feed on fish have needle-like teeth for gripping, and those that feed on larger prey such as mammals have pointed lower teeth for gripping and triangular upper teeth with serrated edges for cutting. The teeth of plankton-feeders such as the basking shark are small and non-functional.

- Cartilaginous jaws and their teeth

Jaw reconstruction of the extinct Carcharodon megalodon, 1909

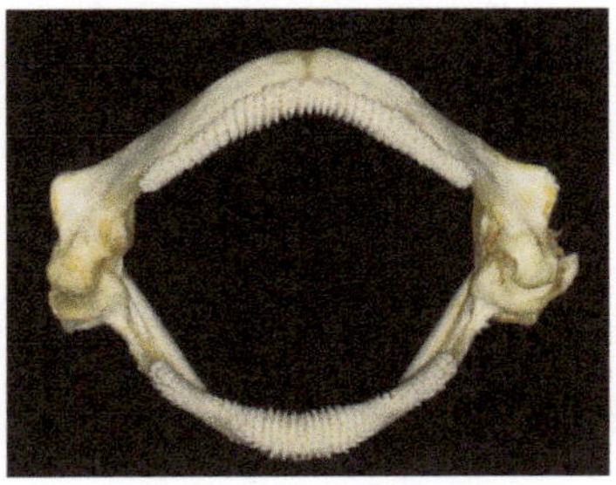

The thornback ray has teeth adapted to feed on crabs, shrimps and small fish.

The shortfin mako shark lunges vertically and tears flesh from prey

Tiger shark teeth are oblique and serrated to saw through flesh

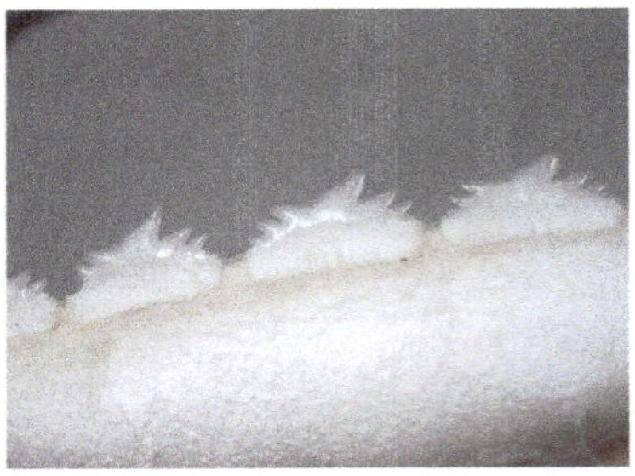

The prickly shark has knife-like teeth with main cusps flanked by lateral cusplets

Examples

Salmon

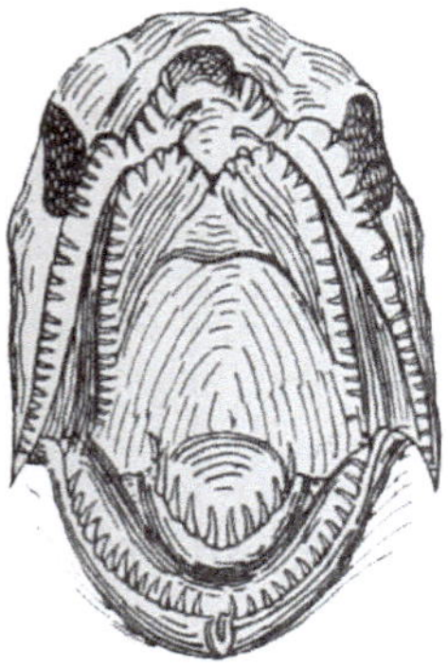

Open mouth of a salmon showing the second set of pharyngeal jaws positioned at the back of the throat

Kype of a spawning male salmon

Male salmon often remodel their jaws during spawning runs so they have a pronounced curvature. These hooked jaws are called kypes. The purpose of the kype is not altogether clear, though they can be used to establish dominance by clamping them around the base of the tail (caudal peduncle) of an opponent.

Cichlids

Dorsal view of right-bending (left) and left-bending (right) jaw morphs

Fish jaws, like vertebrates in general, normally show bilateral symmetry. An exception occurs with the parasitic scale-eating cichlid *Perissodus microlepis*. The jaws of this fish occur in two distinct morphological forms. One morph has its jaw twisted to the left, allowing it to eat scales more readily on its victim's right flank. The other morph has its jaw twisted to the right, which makes it easier to eat scales on its victim's left flank. The relative abundance of the two morphs in populations is regulated by frequency-dependent selection.

Wrasses

Lips of a humphead wrasse

Wrasses have become a primary study species in fish-feeding biomechanics due to their jaw structure. They have protractile mouths, usually with separate jaw teeth that jut outwards. Many species can be readily recognized by their thick lips, the inside of which is sometimes curiously folded, a peculiarity which gave rise the German name of "lip-fishes" (*Lippfische.*)

The nasal and mandibular bones are connected at their posterior ends to the rigid neurocranium, and the superior and inferior articulations of the maxilla are joined to the anterior tips of these two bones, respectively, creating a loop of 4 rigid bones connected by moving joints. This "four-bar linkage" has the property of allowing numerous arrangements to achieve a given mechanical result (fast jaw protrusion or a forceful bite), thus decoupling morphology from function. The actual morphology of wrasses reflects this, with many lineages displaying different jaw morphology that results in the same functional output in a similar or identical ecological niche.

Other

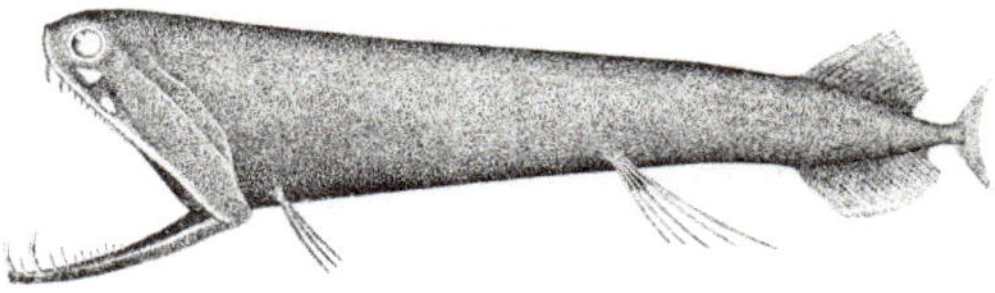

Relative to its size the stoplight loosejaw has one of the widest gapes of any fish.

Stoplight loosejaws are small fish found worldwide in the deep sea. Relative to their size they have one of the widest gapes of any fish. The lower jaw has no ethmoid membrane (floor) and is attached only by the hinge and a modified tongue bone. There are several large, fang-like teeth in the front of the jaws, followed by many small barbed teeth. There are several groups of pharyngeal teeth that serve to direct food down the esophagus.

Another deep sea fish, the pelican eel, has jaws larger than its body. The jaws are lined with small teeth and are loosely hinged. They open wide enough to swallow a fish larger than the eel itself.

Distichodontidae are a family of fresh water fishes which can be divided into genera with protractile upper jaws which are carnivores, and genera with nonprotractile upper jaws which are herbivores or predators of very small organisms.

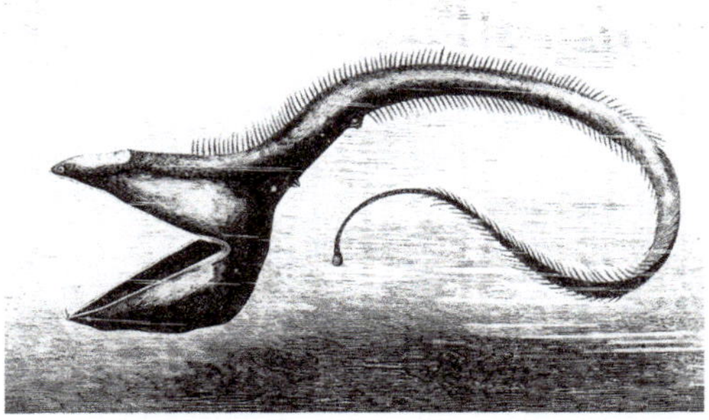

The pelican eel jaws are larger than its body.

Evolution

Vertebrate Classes

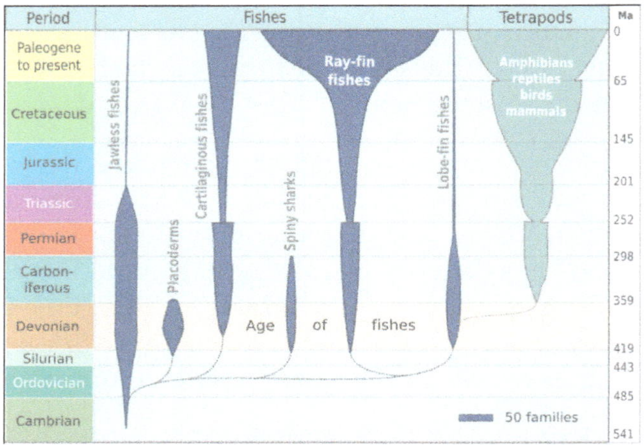

Spindle diagram for the evolution of fish and other vertebrate classes. The earliest classes that developed jaws were the now extinct placoderms and the spiny sharks.

The appearance of the early vertebrate jaw has been described as "a crucial innovation" and "perhaps the most profound and radical evolutionary step in the vertebrate history". Fish without jaws had more difficulty surviving than fish with jaws, and most jawless fish became extinct. However studies of the cyclostomes, the jawless hagfishes and lampreys that did survive, have yielded little insight into the deep remodelling of the vertebrate skull that must have taken place as early jaws evolved.

The customary view is that jaws are homologous to the gill arches. In jawless fishes a series of gills opened behind the mouth, and these gills became supported by cartilaginous elements. The first set of these elements surrounded the mouth to form the jaw. The upper portion of the second embryonic arch supporting the gill became the hyomandibular bone of jawed fishes, which supports the skull and therefore links the jaw to the cranium. The hyomandibula is a set of bones found in the hyoid region in most fishes. It usually plays a role in suspending the jaws or the operculum in the case of teleosts.

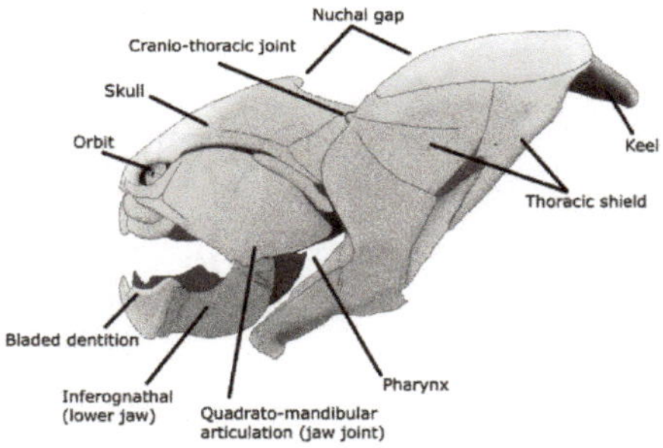

↑ Skull diagram of the huge predatory placoderm fish Dunkleosteus terrelli, which lived about 380–360 million years ago

↑ Reconstruction of Dunkleosteus terrelli

↑ Spiny shark

It is now accepted that the precursors of the jawed vertebrates are the long extinct bony (armoured) jawless fish, the so-called ostracoderms. The earliest known fish with jaws are the now extinct placoderms and spiny sharks.

Placoderms were a class of fish, heavily armoured at the front of their body, which first appeared in the fossil records during the Silurian about 430 million years ago. Initially they very successful, diversifying remarkably during the Devonian. They became extinct by the end of that period, about 360 million years ago. Their largest species, *Dunkleosteus terrelli*, measured up to 10 m (33 ft) and weighed 3.6 t (4.0 short tons). It possessed a four bar linkage mechanism for jaw opening that incorporated connections between the skull, the thoracic shield, the lower jaw and the jaw muscles joined together by movable joints. This mechanism allowed *Dunkleosteus terrelli* to achieve a high speed of jaw opening, opening their jaws in 20 milliseconds and completing the whole process in 50-60 milliseconds, comparable to modern fishes that use suction feeding to assist in prey capture. They could also produce high bite forces when closing the jaw, estimated at 6,000 N (1,350 lb_f) at the tip and 7,400 N (1,660 lb_f) at the blade edge in the largest individuals. The pressures generated in those regions were high enough to puncture or cut through cuticle or dermal armour suggesting that *Dunkleosteus terrelli* was perfectly adapted to prey on free-swimming, armoured prey like arthropods, ammonites, and other placoderms.

Spiny sharks were another class of fish which appeared also in the fossil records during the Silurian at about the same time as the placoderms. They were smaller than most placoderms, usually under 20 centimetres. Spiny sharks did not diversify as much as placoderms, but survived much longer into the Early Permian about 290 million years ago.

The original selective advantage offered by the jaw was not related to feeding, but to increased respiration efficiency. The jaws were used in the buccal pump still observable in modern fish and amphibians, that uses "breathing with the cheeks" to pump water across the gills of fish or air into the lungs in the case of amphibians. Over evolutionary time the more familiar use of jaws (to humans) for feeding was selected for and became a very important function in vertebrates. Many teleost fish have substantially modified jaws for suction feeding and jaw protrusion, resulting in highly complex jaws with dozens of bones involved.

Jaws are thought to derive from the pharyngeal arches that support the gills in fish. The two most anterior of these arches are thought to have become the jaw itself and the hyoid arch, which braces the jaw against the braincase and increases mechanical efficiency. While there is no fossil evidence directly to support this theory, it makes sense in light of the numbers of pharyngeal arches that are visible in extant jawed (the Gnathostomes), which have seven arches, and primitive jawless vertebrates (the Agnatha), which have nine.

Meckel's cartilage is a piece of cartilage from which the mandibles (lower jaws) of vertebrates evolved. Originally it was the lower of two cartilages which supported the first gill arch (nearest the front) in early fish. Then it grew longer and stronger, and acquired muscles capable of closing the developing jaw. In early fish and in chondrichthyans (cartilaginous fish such as sharks), Meckel's cartilage continued to be the main component of the lower jaw. But in the adult forms of osteichthyans (bony fish) and their descendants (amphibians, reptiles, birds and mammals) the cartilage was covered in bone - although in their embryos the jaw initially develops as the Meckel's cartilage. In tetrapods the cartilage partially ossifies (changes to bone) at the rear end of the jaw and becomes the articular bone, which forms part of the jaw joint in all tetrapods except mammals.

Swim Bladder

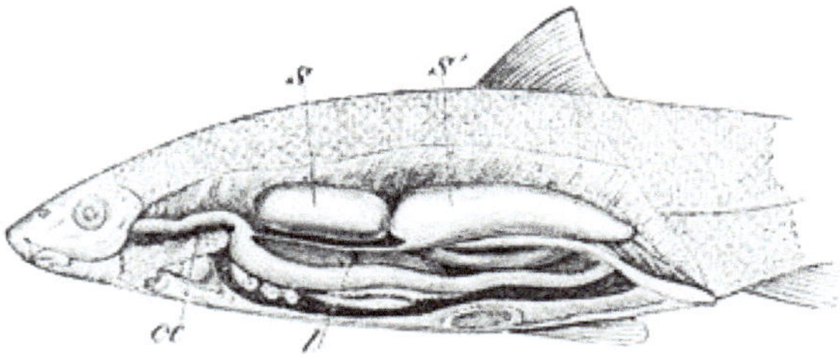

Internal positioning of the swim bladder of a bleakS: anterior, S': posterior portion of the air bladderœ: œsophagus; l: air passage of the air bladder

The swim bladder, gas bladder, fish maw or air bladder is an internal gas-filled organ that contributes to the ability of many bony fish (but not cartilaginous fish) to control their buoyancy, and thus to stay at their current water depth without having to waste energy in swimming. Also, the dorsal position of the swim bladder means the center of mass is below the center of volume, allowing it to act as a stabilizing agent. Additionally, the swim bladder functions as a resonating chamber, to produce or receive sound.

The swim bladder is evolutionarily homologous to the lungs. Charles Darwin remarked upon this in *On the Origin of Species*. Darwin reasoned that the lung in air-breathing vertebrates had derived from a more primitive swim bladder, but scientists now believe that the swim bladder derived from a more primitive lung.

In the embryonic stages some species, such as redlip blenny, have lost the swim bladder again, mostly bottom dwellers like the weather fish. Other fish like the Opah and the Pomfret use their pectoral fins to swim and balance the weight of the head to keep a horizontal position. The normally bottom dwelling sea robin can use their pectoral fins to produce lift while swimming.

The gas/tissue interface at the swim bladder produces a strong reflection of sound, which is used in sonar equipment to find fish.

The cartilaginous fish (e.g. sharks and rays) do not have swim bladders. Some of them can control their depth only by swimming (using dynamic lift); others store fats or oils with density less than that of seawater to produce a neutral or near neutral buoyancy, which does not change with depth.

Structure and Function

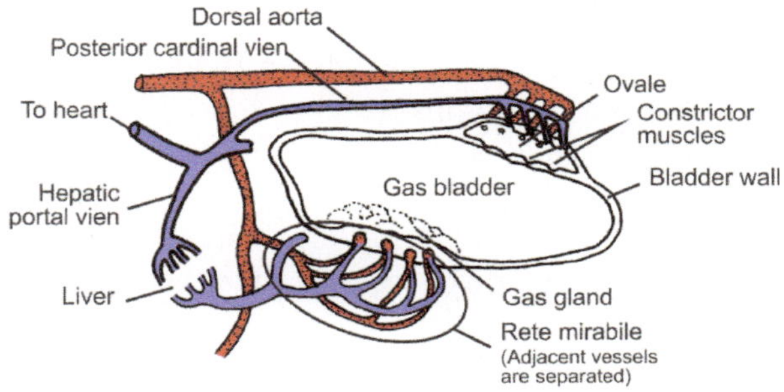

Swim bladder from a bony (teleost) fish

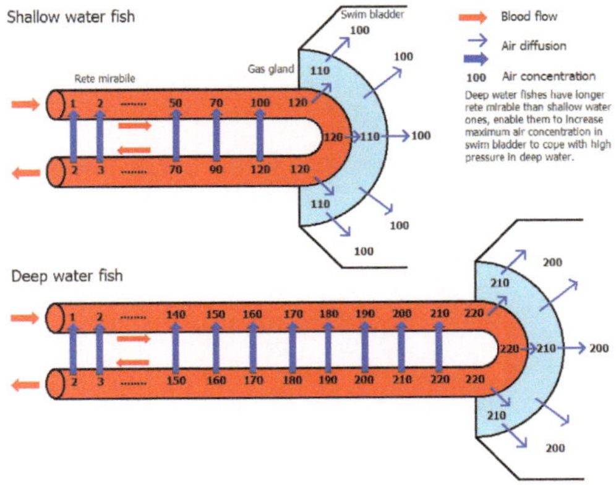

How gas is pumped into the swim bladder using counter-current exchange.

The swim bladder normally consists of two gas-filled sacs located in the dorsal portion of the fish, although in a few primitive species, there is only a single sac. It has flexible walls that contract or expand according to the ambient pressure. The walls of the bladder contain very few blood vessels and are lined with guanine crystals, which make them impermeable to gases. By adjusting the gas pressurising organ using the gas gland or oval window the fish can obtain neutral buoyancy and ascend and descend to a large range of depths. Due to the dorsal position it gives the fish lateral stability.

In physostomous swim bladders, a connection is retained between the swim bladder and the gut, the pneumatic duct, allowing the fish to fill up the swim bladder by "gulping" air. Excess gas can be removed in a similar manner.

In more derived varieties of fish (the physoclisti) the connection to the digestive tract is lost. In early life stages, these fish must rise to the surface to fill up their swim bladders; in later stages, the pneumatic duct disappears, and the gas gland has to introduce gas (usually oxygen) to the bladder to increase its volume and thus increase buoyancy. In order to introduce gas into the bladder, the gas gland excretes lactic acid and produces carbon dioxide. The resulting acidity causes the hemoglobin of the blood to lose its oxygen (Root effect) which then diffuses partly into the swim bladder. The blood flowing back to the body first enters a rete mirabile where virtually all the excess carbon dioxide and oxygen produced in the gas gland diffuses back to the arteries supplying the gas gland. Thus a very high gas pressure of oxygen can be obtained, which can even account for the presence of gas in the swim bladders of deep sea fish like the eel, requiring a pressure of hundreds of bars. Elsewhere, at a similar structure known as the *oval window*, the bladder is in contact with blood and the oxygen can diffuse back out again. Together with oxygen, other gases are salted outin the swim bladder which accounts for the high pressures of other gases as well.

The combination of gases in the bladder varies. In shallow water fish, the ratios closely approximate that of the atmosphere, while deep sea fish tend to have higher percentages of oxygen. For instance, the eel *Synaphobranchus* has been observed to have 75.1% oxygen, 20.5% nitrogen, 3.1% carbon dioxide, and 0.4% argon in its swim bladder.

Physoclist swim bladders have one important disadvantage: they prohibit fast rising, as the bladder would burst. Physostomes can "burp" out gas, though this complicates the process of re-submergence.

In some fish, mainly freshwater species (e.g. common carp, wels catfish), the swim bladder is connected to the labyrinth of the inner ear by the Weberian apparatus, a bony structure derived from the vertebrae, which provides a precise sense of water pressure (and thus depth), and improves hearing.

In red-bellied piranha, the swimbladder may play an important role in sound production as a resonator. The sounds created by piranhas are generated through rapid contractions of the sonic muscles and is associated with the swimbladder.

Evolution

The West African lungfish possesses a lung homologous to swim bladders

The illustration of the swim bladder in fishes ... shows us clearly the highly important fact that an organ originally constructed for one purpose, namely, flotation, may be converted into one for a widely different purpose, namely, respiration. The swim bladder has, also, been worked in as an accessory to the auditory organs of certain fishes. All physiologists admit that the swimbladder is homologous, or "ideally similar" in position and structure with the lungs of the higher vertebrate animals: hence there is no reason to doubt that the swim bladder has actually been converted into lungs, or an organ used exclusively for respiration. According to this view it may be inferred that all vertebrate animals with true lungs are descended by ordinary generation from an ancient and unknown prototype, which was furnished with a floating apparatus or swim bladder.

Charles Darwin, 1859

Swim bladders are evolutionarily closely related (i.e., homologous) to lungs. It is believed that the first lungs, simple sacs connected to the gut that allowed the organism to gulp air under oxygen-poor conditions, evolved into the lungs of today's terrestrial vertebrates and some fish (e.g., lungfish, gar, and bichir) and into the swim bladders of the ray-finned fish. In embryonal development, both lung and swim bladder originate as an outpocketing from the gut; in the case of swim bladders, this connection to the gut continues to exist as the pneumatic duct in the more "primitive" ray-finned fish, and is lost in some of the more derived teleost orders. There are no animals which have both lungs and a swim bladder.

The cartilaginous fish (e.g., sharks and rays) split from the other fishes about 420 million years ago, and lack both lungs and swim bladders, suggesting that these structures evolved after that split. Correspondingly, these fish also have both heterocercal and pectoral fins which provide the necessary lift needed due to the lack of swim bladders. Teleost fish with swim bladders have neutral buoyancy, and have no need for this lift.

Deep Scattering Layer

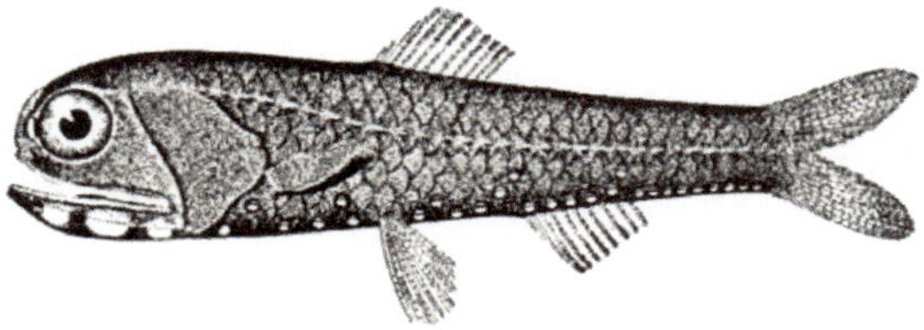

Most mesopelagic fishes are small filter feeders which ascend at night using their swimbladders to feed in the nutrient rich waters of the epipelagic zone. During the day, they return to the dark, cold, oxygen deficient waters of the mesopelagic where they are relatively safe from predators. Lanternfish account for as much as 65 percent of all deep sea fish biomass and are largely responsible for the deep scattering layer of the world's oceans.

Sonar operators, using the newly developed sonar technology during World War II, were puzzled by what appeared to be a false sea floor 300–500 metres deep at day, and less deep at night. This turned out to be due to millions of marine organisms, most particularly small mesopelagic fish, with swimbladders that reflected the sonar. These organisms migrate up into shallower water at dusk to feed on plankton. The layer is deeper when the moon is out, and can become shallower when clouds obscure the moon.

Most mesopelagic fish make daily vertical migrations, moving at night into the epipelagic zone, often following similar migrations of zooplankton, and returning to the depths for safety during the day. These vertical migrations often occur over large vertical distances, and are undertaken with the assistance of a swim bladder. The swim bladder is inflated when the fish wants to move up, and, given the high pressures in the mesoplegic zone, this requires significant energy. As the fish ascends, the pressure in the swimbladder must adjust to prevent it from bursting. When the fish wants to return to the depths, the swimbladder is deflated. Some mesopelagic fishes make daily migrations through the thermocline, where the temperature changes between 10 and 20 °C, thus displaying considerable tolerance for temperature change.

Sampling via deep trawling indicates that lanternfish account for as much as 65% of all deep sea fish biomass. Indeed, lanternfish are among the most widely distributed, populous, and diverse of all vertebrates, playing an important ecological role as prey for larger organisms. The estimated global biomass of lanternfish is 550–660 million metric tonnes, several times the entire world fisheries catch. Lanternfish also account for much of the biomass responsible for the deep scattering layer of the world's oceans. Sonar reflects off the millions of lanternfish swim bladders, giving the appearance of a false bottom.

Human Uses

In some Asian cultures, the swim bladders of certain large fishes are considered a food delicacy. In China they are known as *fish maw*, and are served in soups or stews. Swim bladders are also used in the food industry as a source of collagen. They can be made into a strong, water-resistant glue, or used to make isinglass for the clarification of beer. In earlier times they were used to make condoms.

Swim Bladder Disease

Swim bladder disease is a common ailment in aquarium fish. A fish with swim bladder disorder can float nose down tail up, or can float to the top or sink to the bottom of the aquarium.

Risk of Injury

Fish with a swim bladder are at greater risk of injury after exposure to loud impulsive sounds than those without.

Similar Structures in other Organisms

Siphonophores have a special swim bladder that allows the jellyfish-like colonies to float along the surface of the water while their tentacles trail below. This organ is unrelated to the one in fish.

Fish Fin

Fins are usually the most distinctive features of a fish. They are composed of bony spines or rays protruding from the body with skin covering them and joining them together, either in a webbed fashion, as seen in most bony fish, or similar to a flipper, as seen in sharks. Apart from the tail or caudal fin, fish fins have no direct connection with the spine and are supported only by muscles. Their principal function is to help the fish swim. Fins located in different places on the fish serve different purposes such as moving forward, turning, keeping an upright position or stopping. Most fish use fins when swimming, flying fish use pectoral fins for gliding, and frogfish use them for crawling. Fins can also be used for other purposes; male sharks and mosquitofish use a modified fin to deliver sperm, thresher sharks use their caudal fin to stun prey, reef stonefish have spines in their dorsal fins that inject venom, anglerfish use the first spine of their dorsal fin like a fishing rod to lure prey, and triggerfish avoid predators by squeezing into coral crevices and using spines in their fins to lock themselves in place.

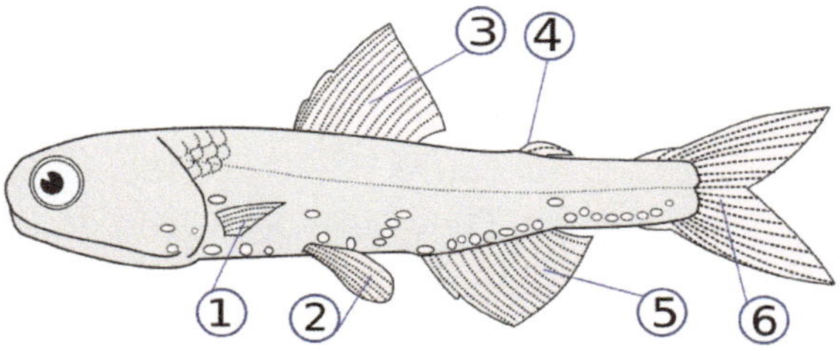

Ray fins on a teleost fish, Hector's lanternfish(1) pectoral fins (paired), (2) pelvic fins (paired), (3) dorsal fin,(4) adipose fin, (5) anal fin, (6) caudal (tail) fin

Types

For every type of fin, there are a number of fish species in which this particular fin has been lost during evolution.

Bony Fishes

Bony fishes form a taxonomic group called Osteichthyes. They have skeletons made of bone, and can be contrasted with cartilaginous fishes which have skeletons made of cartilage. Bony fishes are divided into ray-finned and lobe-finned fish. Most fish are ray-finned, an extremely diverse and abundant group consisting of over 30,000 species. It is the largest class of vertebrates in existence today. In the distant past, lobe-finned fish were abundant. Nowadays they are mainly extinct, with only eight living species. Bony fish have fin spines and rays called lepidotrichia. They typically have swim bladders, which allows the fish to create a neutral balance between sinking and floating without having to use its fins. However, these are absent in many species, and have developed into primitive lungs in the lungfishes. Bony fishes also have an operculum, which helps them breathe without having to use fins to swim.

Lobe-finned

Lobe-finned fishes are a class of bony fishes called Sarcopterygii. They have fleshy, lobed, paired fins, which are joined to the body by a single bone. The fins of lobe-finned fish differ from those of all other fish in that each is borne on a fleshy, lobelike, scaly stalk extending from the body. Pectoral and pelvic fins have articulations resembling those of tetrapod limbs. These fins evolved into legs of the first tetrapod land vertebrates, amphibians. They also possess two dorsal fins with separate bases, as opposed to the single dorsal fin of ray-finned fish.

Lobe-finned fishes, like this coelacanth, have fins that are borne on a fleshy, lobelike, scaly stalk extending from the body. Due to the high number of fins it possesses, the coelacanth has high maneuverability and can orient their bodies in almost any direction in the water.

The coelacanth is another lobe-finned fish which is still extant. It is thought to have evolved into roughly its current form about 408 million years ago, during the early Devonian. Locomotion of the coelacanths is unique to their kind. To move around, coelacanths most commonly take advantage of up or downwellings of the current and drift. They use their paired fins to stabilize their movement through the water. While on the ocean floor their paired fins are not used for any kind of movement. Coelacanths can create thrust for quick starts by using their caudal fins. Due to the high number of fins they possess, coelacanths have high maneuverability and can orient their bodies in almost any direction in the water. They have been seen doing headstands and swimming belly up. It is thought that their rostral organ helps give the coelacanth electroperception, which aids in their movement around obstacles.

Ray-finned

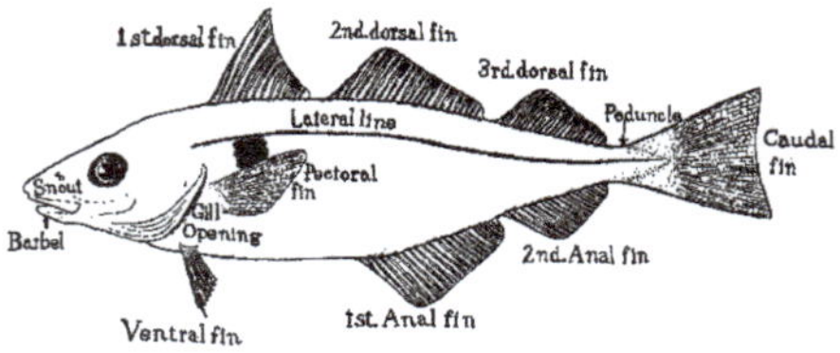

The haddock, a type of cod, is ray-finned. It has three dorsal and two anal fins

Ray-finned fishes are a class of bony fishes called Actinopterygii. Their fins contain spines or rays. A fin may contain only spiny rays, only soft rays, or a combination of both. If both are present, the spiny rays are always anterior. Spines are generally stiff and sharp. Rays are generally soft, flexible, segmented, and may be branched. This segmentation of rays is the main difference that separates them from spines; spines may be flexible in certain species, but they will never be segmented.

Spines have a variety of uses. In catfish, they are used as a form of defense; many catfish have the ability to lock their spines outwards. Triggerfish also use spines to lock themselves in crevices to prevent them being pulled out.

Lepidotrichia are bony, bilaterally paired, segmented fin rays found in bony fishes. They develop around actinotrichia as part of the dermal exoskeleton. Lepidotrichia are usually composed of bone, but in early osteichthyans such as *Cheirolepis*, there was also dentine and enamel. They are segmented and appear as a series of disks stacked one on top of another. The genetic basis for the formation of the fin rays is thought to be genes coded for the production of certain proteins. It has been suggested that the evolution of the tetrapod limb from lobe-finned fishes is related to the loss of these proteins.

Cartilaginous Fishes

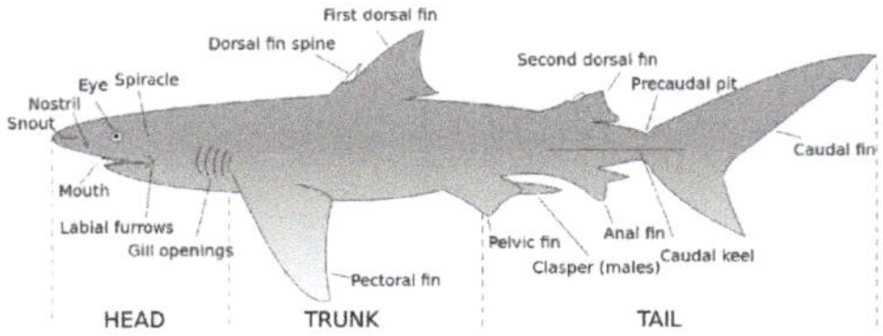

Cartilaginous fishes, like this shark, have fins that are elongated and supported with soft and unsegmented rays named ceratotrichia, filaments of elastic protein resembling the horny keratin in hair and feathers

Caudal fin of a grey reef shark

Cartilaginous fishes are a class of fishes called Chondrichthyes. They have skeletons made of cartilage rather than bone. The class includes sharks, rays and chimaeras. Shark fin skeletons are elongated and supported with soft and unsegmented rays named ceratotrichia, filaments of elastic protein resembling the horny keratin in hair and feathers. Originally the pectoral and pelvic girdles, which do not contain any dermal elements, did not connect. In later forms, each pair of fins became ventrally connected in the middle when scapulocoracoid and pubioischiadic bars evolved. In rays, the pectoral fins have connected to the head and are very flexible. One of the primary characteristics present in most sharks is the heterocercal tail, which aids in locomotion. Most sharks have eight fins. Sharks can only drift away from objects directly in front of them because their fins do not allow them to move in the tail-first direction.

As with most fish, the tails of sharks provide thrust, making speed and acceleration dependent on tail shape. Caudal fin shapes vary considerably between shark species, due to their evolution in separate environments. Sharks possess a heterocercal caudal fin in which the dorsal portion is usually noticeably larger than the ventral portion. This is because the shark's vertebral column extends into that dorsal portion, providing a greater surface area for muscle attachment. This allows more efficient locomotion among these negatively buoyant cartilaginous fish. By contrast, most bony fish possess a homocercal caudal fin.

Tiger sharks have a large upper lobe, which allows for slow cruising and sudden bursts of speed. The tiger shark must be able to twist and turn in the water easily when hunting to support its varied diet, whereas the porbeagle shark, which hunts schooling fish such as mackerel and herring, has a large lower lobe to help it keep pace with its fast-swimming prey. Other tail adaptations help sharks catch prey more directly, such as the thresher shark's usage of its powerful, elongated upper lobe to stun fish and squid.

Generating Thrust

Foil shaped fins generate thrust when moved, the lift of the fin sets water or air in motion and pushes the fin in the opposite direction. Aquatic animals get significant thrust by moving fins back and forth in water. Often the tail fin is used, but some aquatic animals generate thrust from pectoral fins.

Moving Fins can Provide Thrust

Finlets may influence the way a vortex develops around the tail fin.

Cavitation occurs when negative pressure causes bubbles (cavities) to form in a liquid, which then promptly and violently collapse. It can cause significant damage and wear. Cavitation damage can occur to the tail fins of powerful swimming marine animals, such as dolphins and tuna. Cavitation is more likely to occur near the surface of the ocean, where the ambient water pressure is relatively low. Even if they have the power to swim faster, dolphins may have to restrict their speed because collapsing cavitation bubbles on their tail are too painful. Cavitation also slows tuna, but for a different reason. Unlike dolphins, these fish do not feel the bubbles, because they have bony fins without nerve endings. Nevertheless, they cannot swim faster because the cavitation bubbles create a vapor film around their fins that limits their speed. Lesions have been found on tuna that are consistent with cavitation damage.

Fish get thrust moving vertical tail fins from side to side

Stingrays get thrust from large pectoral fins

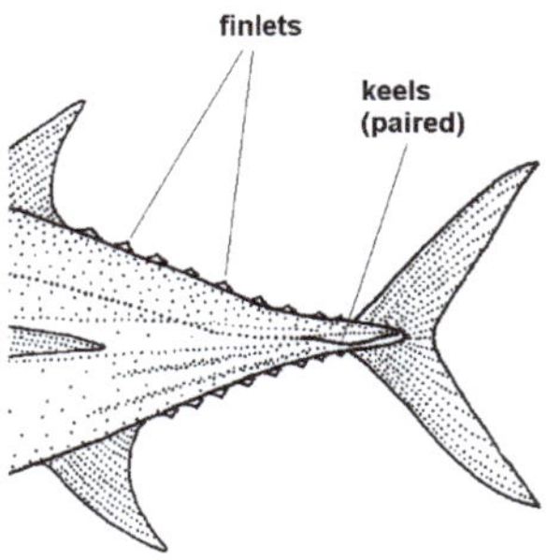

Drawing by Dr Tony Ayling

Scombrid fishes (tuna, mackerel and bonito) are particularly high-performance swimmers. Along the margin at the rear of their bodies is a line of small rayless, non-retractable fins, known as finlets. There has been much speculation about the function of these finlets. Research done in 2000 and 2001 by Nauen and Lauder indicated that "the finlets have a hydrodynamic effect on local flow during steady swimming" and that "the most posterior finlet is oriented to redirect flow into the developing tail vortex, which may increase thrust produced by the tail of swimming mackerel".

Fish use multiple fins, so it is possible that a given fin can have a hydrodynamic interaction with another fin. In particular, the fins immediately upstream of the caudal (tail) fin may be proximate fins that can directly affect the flow dynamics at the caudal fin. In 2011, researchers using volumetric imaging techniques were able to generate "the first instantaneous three-dimensional views of wake structures as they are produced by freely swimming fishes". They found that "continuous tail beats resulted in the formation of a linked chain of vortex rings" and that "the dorsal and anal fin wakes are rapidly entrained by the caudal fin wake, approximately within the timeframe of a subsequent tail beat".

Controlling Motion

Once motion has been established, the motion itself can be controlled with the use of other fins.

Specialised Fins are used to Control Motion

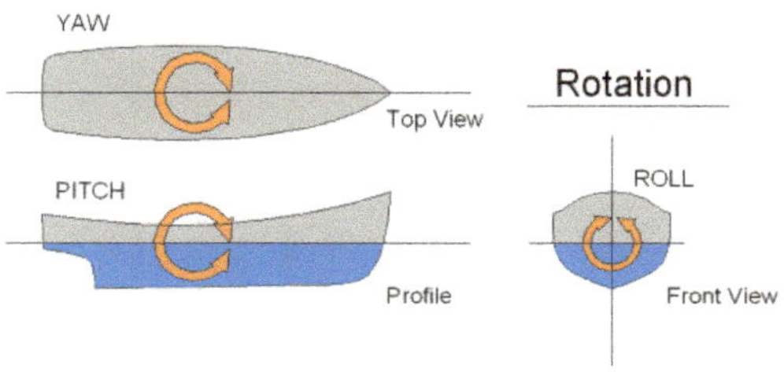

Like boats and airplanes, fish need some control over six degrees of freedom, three translational (heaving, swaying and surging) and three rotational (pitching, yawing and rolling)

Many reef fish have pectoral and pelvic fins optimised for flattened bodies

The dorsal fin of a white shark contain dermal fibers that work "like riggings that stabilize a ship's mast", and stiffen dynamically as the shark swims faster to control roll and yaw.

The bodies of reef fishes are often shaped differently from open water fishes. Open water fishes are usually built for speed, streamlined like torpedoes to minimise friction as they move through the water. Reef fish operate in the relatively confined spaces and complex underwater landscapes of coral reefs. For this manoeuvrability is more important than straight line speed, so coral reef fish have developed bodies which optimize their ability to dart and change direction. They outwit predators by dodging into fissures in the reef or playing hide and seek around coral heads. The pectoral and pelvic fins of many reef fish, such as butterflyfish, damselfish and angelfish, have evolved so they can act as brakes and allow complex manoeuvres. Many reef fish, such as butterflyfish, damselfish and angelfish, have evolved bodies which are deep and laterally compressed like a pancake, and will fit into fissures in rocks. Their pelvic and pectoral fins are designed differently, so they act together with the flattened body to optimise manoeuvrability. Some fishes, such as puffer fish, filefish and trunkfish, rely on pectoral fins for swimming and hardly use tail fins at all.

Reproduction

This male mosquitofish has a gonopodium, an anal fin which functions as an intromittent organ

Male cartilaginous fishes (sharks and rays), as well as the males of some live-bearing ray finned fishes, have fins that have been modified to function as intromittent organs, reproductive appendages which allow internal fertilization. In ray finned fish they are called *gonopodia* or *andropodia*, and in cartilaginous fish they are called *claspers*.

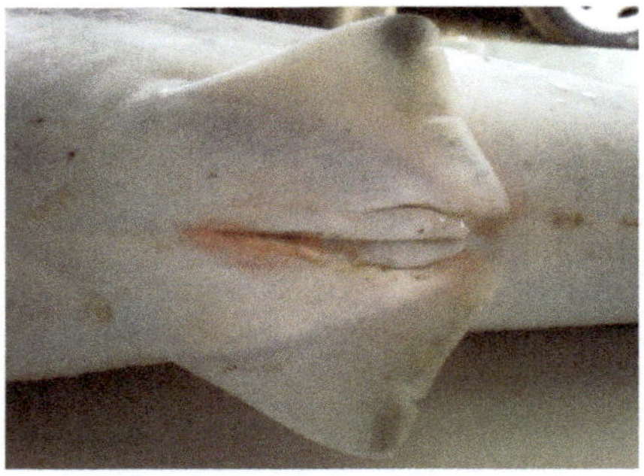

This young male spinner shark has claspers, a modification to the pelvic fins which also function as intromittent organs

Gonopodia are found on the males of some species in the Anablepidae and Poeciliidae families. They are anal fins that have been modified to function as movable intromittent organs and are used to impregnate females with milt during mating. The third, fourth and fifth rays of the male's anal fin are formed into a tube-like structure in which the sperm of the fish is ejected. When ready for mating, the gonopodium becomes erect and points forward towards the female. The male shortly inserts the organ into the sex opening of the female, with hook-like adaptations that allow the fish to grip onto the female to ensure impregnation. If a female remains stationary and her partner contacts her vent with his gonopodium, she is fertilized. The sperm is preserved in the female's oviduct. This allows females to fertilize themselves at any time without further assistance from males. In some species, the gonopodium may be half the total body length. Occasionally the fin is too long to be used, as in the "lyretail" breeds of *Xiphophorus helleri*. Hormone treated females may develop gonopodia. These are useless for breeding.

Similar organs with similar characteristics are found in other fishes, for example the *andropodium* in the *Hemirhamphodon* or in the Goodeidae.

Claspers are found on the males of cartilaginous fishes. They are the posterior part of the pelvic fins that have also been modified to function as intromittent organs, and are used to channel semen into the female's cloaca during copulation. The act of mating in sharks usually includes raising one of the claspers to allow water into a siphon through a specific orifice. The clasper is then inserted into the cloaca, where it opens like an umbrella to anchor its position. The siphon then begins to contract expelling water and sperm.

Other Uses

The Indo-Pacific sailfish has a prominent dorsal fin. Like scombroids and other billfish, they streamline themselves by retracting their dorsal fins into a grove in their body when they swim. The huge dorsal fin, or sail, of the sailfish is kept retracted most of the time. Sailfish raise them if they want to herd a school of small fish, and also after periods of high activity, presumably to cool down.

Frogfish use their pectoral and pelvic fins to walk along the ocean bottom

Flying fish achieve sufficient lift to glide above the surface of the water thanks to their enlarged pectoral fins

Large retractable dorsal fin of the Indo-Pacific sailfish

The thresher shark uses its caudal fin to stun prey

The oriental flying gurnard has large pectoral fins which it normally holds against its body, and expands when threatened to scare predators. Despite its name, it is a demersal fish, not a flying fish, and uses its pelvic fins to walk along the bottom of the ocean.

Fins can have an adaptive significance as sexual ornaments. During courtship, the female cichlid, *Pelvicachromis taeniatus*, displays a large and visually arresting purple pelvic fin. "The researchers found that males clearly preferred females with a larger pelvic fin and that pelvic fins grew in a more disproportionate way than other fins on female fish."

Other Uses of Fins

The Oriental flying gurnard has large pectoral fins with eye spots which it displays to scare predators

During courtship, the female cichlid, *Pelvicachromis taeniatus*, displays her visually arresting purple pelvic fin

Triggerfish squeeze into coral crevices to avoid predators, and lock themselves in place with the first spine of their dorsal fin

The first spine of the dorsal fin of the anglerfish is modified so it functions like a fishing rod with a lure

In some Asian countries shark fins are a culinary delicacy

Evolution

Evolution of Paired Fins

There are two prevailing hypotheses that have been historically debated as models for the evolution of paired fins in fish: the gill arch theory and the lateral fin-fold theory. The former, commonly referred to as the "Gegenbaur hypothesis," was posited in 1870 and proposes that the "paired fins are derived from gill structures". This fell out of popularity in favor of the lateral fin-fold theory, first suggested in 1877, which proposes that paired fins budded from longitudinal, lateral folds along the epidermis just behind the gills. There is weak support for both hypotheses in the fossil record and in embryology. However, recent insights from developmental patterning have prompted reconsideration of both theories in order to better elucidate the origins of paired fins.

Classical Theories

Karl Gegenbaur's concept of the "Archipterygium" was introduced in 1876. It was described as a gill ray, or "joined cartilaginous stem," that extended from the gill arch. Additional rays arose from along the arch and from the central gill ray. Gegenbaur suggested a model of transformative homology – that all vertebrate paired fins and limbs were transformations of the Archipterygium. Based on this theory, paired appendages such as pectoral and pelvic fins would have differentiated from the branchial arches and migrated posteriorly. However, there has been limited support for this hypothesis in the fossil record both morphologically and phylogenically. In addition, there was little to no evidence of an anterior-posterior migration of pelvic fins. Such shortcomings of the gill-arch theory led to its early demise in favor of the lateral fin-fold theory proposed by St. George Jackson Mivart, Francis Balfour, and James Kingsley Thacher.

The lateral fin-fold theory hypothesized that paired fins developed from lateral folds along the body wall of the fish. Just as segmentation and budding of the median fin fold gave rise to the median fins, a similar mechanism of fin bud segmentation and elongation from a lateral fin fold was proposed to have given rise to the paired pectoral and pelvic fins. However, there was little evidence of a lateral fold-to-fin transition in the fossil record. In addition, it was later demonstrated phylogenically that pectoral and pelvic fins arise from distinct evolutionary and mechanistic origins.

Evolutionary Developmental Biology

Recent studies in the ontogeny and evolution of paired appendages have compared finless vertebrates – such as lampreys – with chondricthyes, the most basal living vertebrate with paired fins. In 2006, researchers found that the same genetic programming involved in the segmentation and development of median fins was found in the development of paired appendages in catsharks. Although these findings do not directly support the lateral fin-fold hypothesis, the original concept of a shared median-paired fin evolutionary developmental mechanism remains relevant.

We find a similar renovation of an old theory in the developmental programming of chondricthyan gill arches and paired appendages. In 2009, researchers at the University of Chicago demonstrated that there are shared molecular patterning mechanisms in the early development of the chondricthyan gill arch and paired fins. Findings such as these have prompted reconsideration of the once-debunked gill-arch theory.

From Fins to Limbs

Fish are the ancestors of all mammals, reptiles, birds and amphibians. In particular, terrestrial tetrapods (four-legged animals) evolved from fish and made their first forays onto land 400 million years ago. They used paired pectoral and pelvic fins for locomotion. The pectoral fins developed into forelegs (arms in the case of humans) and the pelvic fins developed into hind legs. Much of the genetic machinery that builds a walking limb in a tetrapod is already present in the swimming fin of a fish.

Aristotle recognised the distinction between analogous and homologous structures, and made the following prophetic comparison: *"Birds in a way resemble fishes. For birds have their wings in the upper part of their bodies and fishes have two fins in the front part of their bodies. Birds have feet on their underpart and most fishes have a second pair of fins in their under-part and near their front fins."*

– Aristotle, De incessu animalium

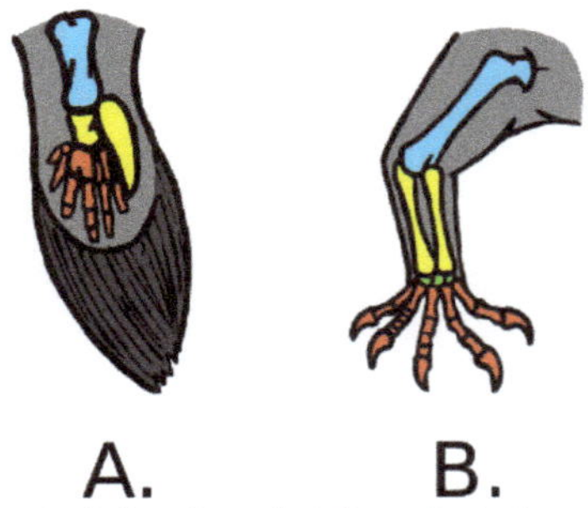

Comparison between A) the swimming fin of a lobe-finned fish and B) the walking leg of a tetrapod. Bones considered to correspond with each other have the same color.

In a parallel but independent evolution, the ancient reptile Ichthyosaurus communis developed fins (or flippers) very similar to fish (or dolphins)

In 2011, researchers at Monash University in Australia used primitive but still living lungfish "to trace the evolution of pelvic fin muscles to find out how the load-bearing hind limbs of the tetrapods evolved." Further research at the University of Chicago found bottom-walking lungfishes had already evolved characteristics of the walking gaits of terrestrial tetrapods.

In a classic example of convergent evolution, the pectoral limbs of pterosaurs, birds and bats further evolved along independent paths into flying wings. Even with flying wings there are many similarities with walking legs, and core aspects of the genetic blueprint of the pectoral fin have been retained.

The first mammals appeared during the Permian period (between 298.9 and 252.17 million years ago). Several groups of these mammals started returning to the sea, including the cetaceans (whales, dolphins and porpoises). Recent DNA analysis suggests that cetaceans evolved from within the even-toed ungulates, and that they share a common ancestor with the hippopotamus. About 23 million years ago another group of bearlike land mammals started returning to the sea. These were the seals. What had become walking limbs in cetaceans and seals evolved independently into new forms of swimming fins. The forelimbs became flippers, while the hindlimbs were either lost (cetaceans) or also modified into flipper (pinnipeds). In cetaceans, the tail gained two fins at the end, called a fluke. Fish tails are usually vertical and move from side to side. Cetacean flukes are horizontal and move up and down, because cetacean spines bend the same way as in other mammals.

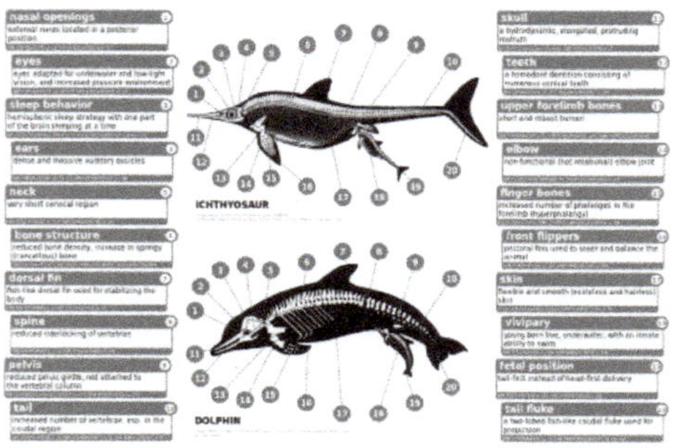

Similar adaptations for fully aquatic lifestyle are found both in dolphins and ichthyosaurs

Ichthyosaurs are ancient reptiles that resembled dolphins. They first appeared about 245 million years ago and disappeared about 90 million years ago.

"This sea-going reptile with terrestrial ancestors converged so strongly on fishes that it actually evolved a dorsal fin and tail fin for improved aquatic locomotion. These structures are all the more remarkable because they evolved from nothing — the ancestral terrestrial reptile had no hump on its back or blade on its tail to serve as a precursor."

The biologist Stephen Jay Gould said the ichthyosaur was his favorite example of convergent evolution.

Fins or flippers of varying forms and at varying locations (limbs, body, tail) have also evolved in a number of other tetrapod groups, including diving birds such as penguins (modified from wings), sea turtles (forelimbs modified into flippers), mosasaurs (limbs modified into flippers), and sea snakes (vertically expanded, flattened tail fin).

Robotic Fins

In the 1990s, the CIA built a robotic catfish called *Charlie*, designed to collect underwater intelligence undetected

The use of fins for the propulsion of aquatic animals can be remarkably effective. It has been calculated that some fish can achieve a propulsive efficiency greater than 90%. Fish can accelerate and maneuver much more effectively than boats or submarine, and produce less water disturbance and noise. This has led to biomimetic studies of underwater robots which attempt to emulate the locomotion of aquatic animals. An example is the Robot Tuna built by the Institute of Field Robotics, to analyze and mathematically model thunniform motion. In 2005, the Sea Life London Aquarium displayed three robotic fish created by the computer science department at the University of Essex. The fish were designed to be autonomous, swimming around and avoiding obstacles like real fish. Their creator claimed that he was trying to combine "the speed of tuna, acceleration of a pike, and the navigating skills of an eel."

The *AquaPenguin*, developed by Festo of Germany, copies the streamlined shape and propulsion by front flippers of penguins. Festo also developed *AquaRay*, *AquaJelly* and *AiraCuda*, respectively emulating the locomotion of manta rays, jellyfish and barracuda.

In 2004, Hugh Herr at MIT prototyped a biomechatronic robotic fish with a living actuator by surgically transplanting muscles from frog legs to the robot and then making the robot swim by pulsing the muscle fibers with electricity.

Robotic fish offer some research advantages, such as the ability to examine an individual part of a fish design in isolation from the rest of the fish. However, this risks oversimplifying the biology so key aspects of the animal design are overlooked. Robotic fish also allow researchers to vary a single parameter, such as flexibility or a specific motion control. Researchers can directly measure forces, which is not easy to do in live fish. "Robotic devices also facilitate three-dimensional kinematic studies and correlated hydrodynamic analyses, as the location of the locomotor surface can be known accurately. And, individual components of a natural motion (such as outstroke vs. instroke of a flapping appendage) can be programmed separately, which is certainly difficult to achieve when working with a live animal."

Diversity of Fins

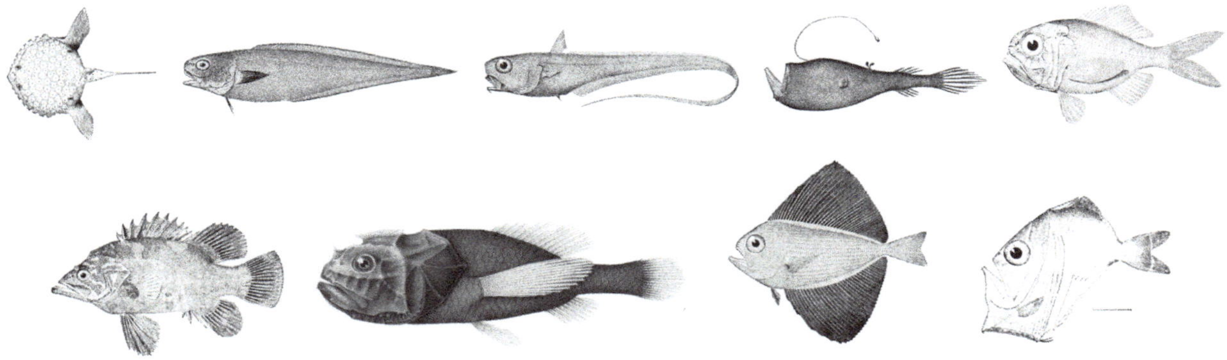

Actinopterygii

Actinopterygii or the ray-finned fishes, constitute a class or subclass of the bony fishes.

The ray-finned fishes are so called because they possess lepidotrichia or "fin rays", their fins being webs of skin supported by bony or horny spines ("rays"), as opposed to the fleshy, lobed fins that characterize the class Sarcopterygii which also, however, possess lepidotrichia. These actinopterygian fin rays attach directly to the proximal or basal skeletal elements, the radials, which represent the link or connection between these fins and the internal skeleton (e.g., pelvic and pectoral girdles).

Numerically, actinopterygians are the dominant class of vertebrates, comprising nearly 99% of the over 30,000 species of fish. They are ubiquitous throughout freshwater and marine environments from the deep sea to the highest mountain streams. Extant species can range in size from *Paedocypris*, at 8 mm (0.3 in), to the massive ocean sunfish, at 2,300 kg (5,070 lb), and the long-bodied oarfish, at 11 m (36 ft).

Characteristics

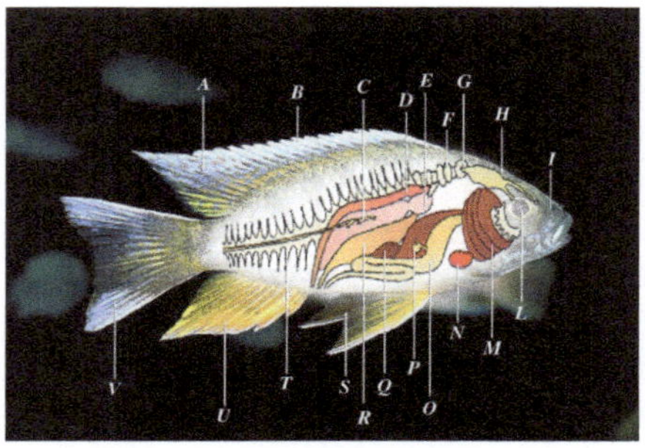

Anatomy of a typical ray-finned fish

A – dorsal fin: B – fin rays: C – lateral line: D – kidney: E – swim bladder: F – Weberian apparatus: G – inner ear: H – brain: I – nostrils: L – eye: M – gills: N – heart O – stomach: P – gall bladder: Q – spleen: R – internal sex organs (ovaries or testes): S – ventral fins: T – spine: U – anal fin: V – tail (caudal fin). Possible other parts not shown: barbels, adipose fin, external genitalia (gonopodium)

Ray-finned fishes occur in many variant forms. The main features of a typical ray-finned fish are shown in the diagram at the left.

Fin arrangements

Ray-finned fish are varied in size, shape and the arrangement and number of their ray-fins.

Tuna are streamlined for straight line speed with a deeply forked tail

The swordfish is even faster and more streamlined than the tuna

Salmon generate enough thrust with their powerful tail fin to jump obstacles during river migrations

Cod have three dorsal and two anal fins, which give them great maneuverability

Catfish

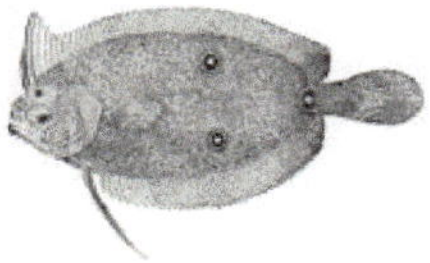

Flatfish have developed partially symmetric dorsal and pelvic fins

Sturgeon

Lanternfish

Elongated bristlemouth

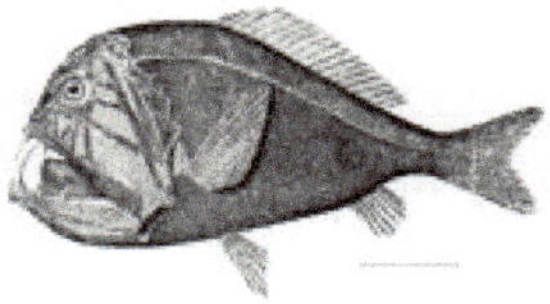

Fangtooth are indifferent swimmer who try to ambush their prey

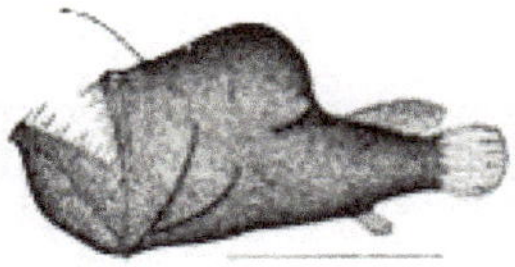

The first spine of the dorsal fin of anglerfish is modified like a fishing rod with a lure

Alfonsino

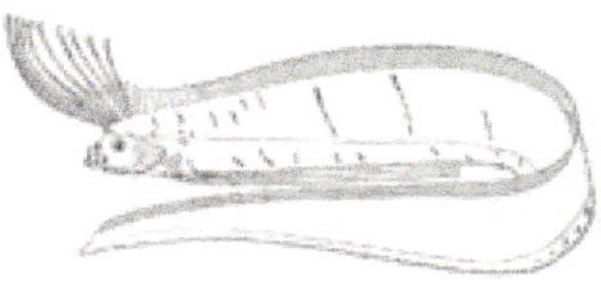

King of herrings

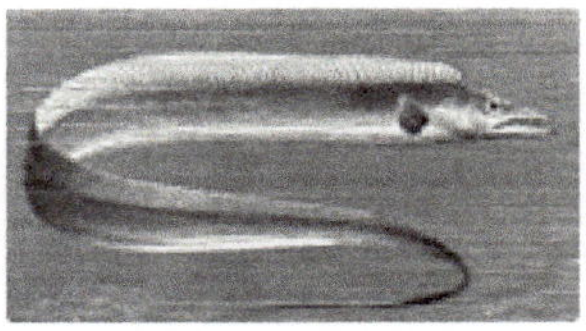

Cutlassfish

European conger are ray-finned fish

So are sea horses

Reproduction

Three-spined stickleback males (red belly) build nests and compete to attract females to lay eggs in them. Males then defend and fan the eggs. Painting by Alexander Francis Lydon, 1879

In nearly all ray-finned fish, the sexes are separate, and in most species the females spawn eggs that are fertilized externally, typically with the male inseminating the eggs after they are laid. Development then proceeds with a free-swimming larval stage. However other patterns of ontogeny exist, with one of the commonest being sequential hermaphroditism. In most cases this involves protogyny, fish starting life as females and converting to males at some stage, triggered by some internal or external factor. This may be advantageous as females become less prolific as they age while male fecundity increases with age. Protandry, where a fish converts from male to female, is much less common than protogyny. Most families use external rather than internal fertilization. Of the oviparous teleosts, most (79%) do not provide parental care. Viviparity, ovoviviparity, or some form of parental care for eggs, whether by the male, the female, or both parents is seen in a significant fraction (21%) of the 422 teleost families; no care is likely the ancestral condition. Viviparity is relatively rare and is found in about 6% of teleost species; male care is far more common than female care. Male territoriality "preadapts" a species for evolving male parental care.

There are a few examples of fish that self-fertilise. The mangrove rivulus is an amphibious, simultaneous hermaphrodite, producing both eggs and spawn and having internal fertilisation. This mode of reproduction may be related to the fish's habit of spending long periods out of water in the mangrove forests it inhabits. Males are occasionally produced at temperatures below 19 °C (66 °F) and can fertilise eggs that are then spawned by the female. This maintains genetic variability in a species that is otherwise highly inbred.

Fossil Record

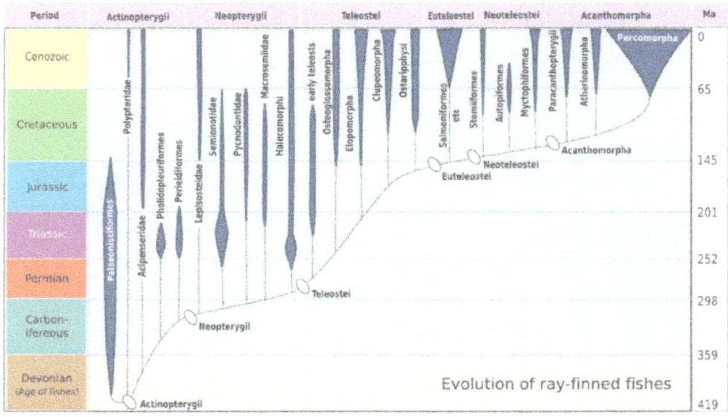

The earliest known fossil actinopterygiian is Andreolepis hedei, dating back 420 million years (Late Silurian). Remains have been found in Russia, Sweden, and Estonia.

Sarcopterygii

The Sarcopterygii or lobe-finned fish – sometimes considered synonymous with Crossopterygii – constitute a clade (traditionally a class or subclass) of the bony fish, though a strict cladistic view includes the terrestrial vertebrates. The living sarcopterygians are the coelacanths and lungfish; additionally, all tetrapods are sarcopterygians or descendants of them.

Characteristics

Early lobe-finned fishes are bony fish with fleshy, lobed, paired fins, which are joined to the body by a single bone. The fins of lobe-finned fishes differ from those of all other fish in that each is borne on a fleshy, lobelike, scaly stalk extending from the body. The scales of sarcopterygians are true scaloids, consisting of lamellar bone surrounded by layers of vascular bone, dentine-like cosmine, and external keratin. Pectoral and pelvic fins have articulations resembling those of tetrapod limbs. These fins evolved into the legs of the first tetrapod land vertebrates, amphibians. They also possess two dorsal fins with separate bases, as opposed to the single dorsal fin of actinopterygians (ray-finned fish). The braincase of sarcopterygians primitively has a hinge line, but this is lost in tetrapods and lungfish. Many early sarcopterygians have a symmetrical tail. All sarcopterygians possess teeth covered with true enamel.

Guiyu oneiros, the earliest known bony fish, lived during the Late Silurian, 419 million years ago). It has the combination of both ray-finned and lobe-finned features, although analysis of the totality of its features place it closer to lobe-finned fish.

Most species of lobe-finned fishes are extinct. The largest known lobe-finned fish was *Rhizodus hibberti* from the Carboniferous period of Scotland which may have exceeded 7 meters in length. Among the two groups of extant (living) species, the coelacanths and the lungfishes, the largest species is the West Indian Ocean coelacanth, reaching 2 m (6.5 ft) in length and weighing up 110 kg (240 lb). The largest lungfish is the African lungfish which can reach 2 m (6.6 ft) in length and weigh up to 50 kg (110 lb).

Classification

Taxonomists who subscribe to the cladistic approach include the grouping Tetrapoda within this group, which in turn consists of all species of four-limbed vertebrates. The fin-limbs of lobe-finned fishes such as the coelacanths show a strong similarity to the expected ancestral form of tetrapod limbs. The lobe-finned fishes apparently followed two different lines of development and are accordingly separated into two subclasses, the Rhipidistia (including the Dipnoi, the lungfish, and the Tetrapodomorpha which include the Tetrapoda) and the Actinistia (coelacanths).

Taxonomy

The classification below follows Benton 2004, and uses a synthesis of rank-based Linnaean taxonomy and also reflects evolutionary relationships. Benton included the Superclass Tetrapoda in the Subclass Sarcopterygii in order to reflect the direct descent of tetrapods from lobe-finned fish, despite the former being assigned a higher taxonomic rank.

Actinistia	 West Indian Ocean coelacanth	Actinistia, coelacanths, are a subclass of mostly fossil lobe-finned fishes. This subclass contains the coelacanths, including the two living coelacanths, the West Indian Ocean coelacanth and the Indonesian coelacanth.
Dipnoi	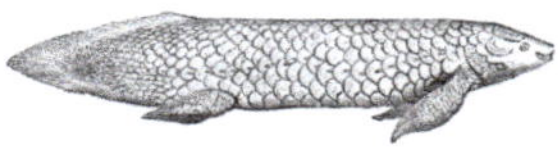 Queensland lungfish	Dipnoi, lungfish, also known as salamanderfish, are a subclass of freshwater fish. Lungfish are best known for retaining characteristics primitive within the bony fishes, including the ability to breathe air, and structures primitive within the lobe-finned fishes, including the presence of lobed fins with a well-developed internal skeleton. Today, lungfish live only in Africa, South America, and Australia. While vicariance would suggest this represents an ancient distribution limited to the Mesozoic supercontinent Gondwana, the fossil record suggests advanced lungfish had a widespread freshwater distribution and the current distribution of modern lungfish species reflects extinction of many lineages following the breakup of Pangaea, Gondwana, and Laurasia.
Tetrapodomorpha	 Advanced tetrapodomorph *Tiktaalik*	Tetrapodomorpha, tetrapods and their extinct relatives, are a clade of vertebrates consisting of tetrapods (four-limbed vertebrates) and their closest sarcopterygian relatives that are more closely related to living tetrapods than to living lungfish. Advanced forms transitional between fish and the early labyrinthodonts, like *Tiktaalik*, have been referred to as "fishapods" by their discoverers, being half-fish, half-tetrapods, in appearance and limb morphology. The Tetrapodomorpha contain the crown group tetrapods (the last common ancestor of living tetrapods and all of its descendants) and several groups of early stem tetrapods, and several groups of related lobe-finned fishes, collectively known as the osteolepiforms. The Tetrapodamorpha minus the crown group Tetrapoda are the Stem Tetrapoda, a paraphyletic unit encompassing the fish to tetrapod transition. Among the characters defining tetrapodomorphs are modifications to the fins, notably a humerus with convex head articulating with the glenoid fossa (the socket of the shoulder joint). Tetrapodomorph fossils are known from the early Devonian onwards, and include *Osteolepis*, *Panderichthys*, *Kenichthys*, and *Tungsenia*.

A modern coelacanth, Latimeria chalumnae

Queensland lungfish

- Subclass Sarcopterygii
    - †Order Onychodontida
    - Order Actinistia
    - Infraclass Dipnomorpha
        - †Order Porolepiformes
        - Subclass Dipnoi
            - Order Ceratodontiformes
            - Order Lepidosireniformes
    - Infraclass Tetrapodomorpha
        - †Order Rhizodontida
        - Superorder Osteolepidida
            - †Order Osteolepiformes
                - †Family Tristichopteridae
            - †Order Panderichthyida
            - Superclass Tetrapoda

Evolution

Evolution of Lobe-finned Fishes

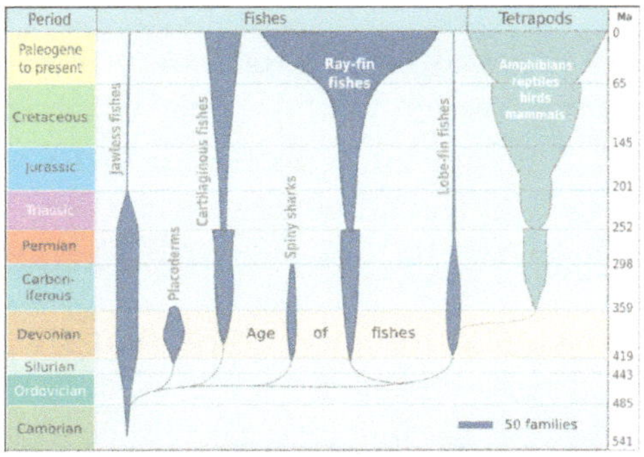

Spindle diagram for the evolution of lobe-finned fishes, tetrapods and other vertebrate classes.

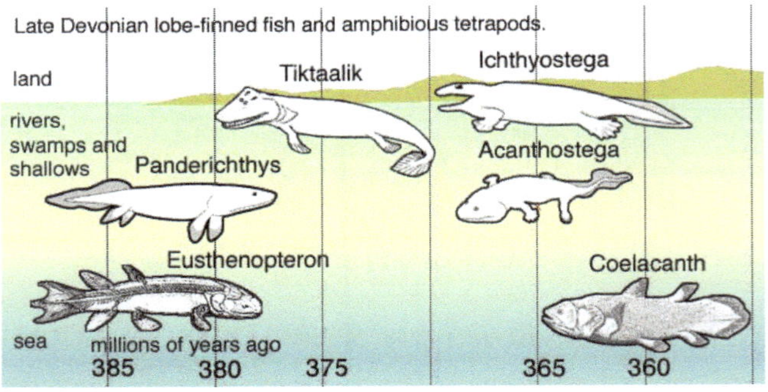

In Late Devonian vertebrate speciation, descendants of pelagic lobe-finned fish — like *Eusthenopteron* — exhibited a sequence of adaptations:

- *Panderichthys*, suited to muddy shallows;
- *Tiktaalik* with limb-like fins that could take it onto land;
- Early tetrapods in weed-filled swamps, such as:
    - *Acanthostega*, which had feet with eight digits,
    - *Ichthyostega* with limbs.

Descendants also included pelagic lobe-finned fish such as coelacanth species.

Lobe-finned fishes (sarcopterygians) and their relatives the ray-finned fishes (actinopterygians) comprise the superclass of bony fishes (Osteichthyes) characterized by their bony skeleton rather than cartilage. There are otherwise vast differences in fin, respiratory, and circulatory structures between the Sarcopterygii and the Actinopterygii, such as the presence of cosmoid layers in the scales of sarcopterygians. The earliest fossils of sarcopterygians, found in the uppermost Silurian (ca 418 Ma), closely resembled the acanthodians (the "spiny fish", a taxon that became extinct at the end of the Paleozoic). In the early–middle Devonian (416 - 385 Ma), while the predatory placoderms dominated the seas, some sarcopterygians came into freshwater habitats.

In the Early Devonian (416 - 397 Mya), the sarcopterygians split into two main lineages — the coelacanths and the rhipidistians. Coelacanths never left the oceans and their heyday was the late Devonian and Carboniferous, from 385 to 299 Ma, as they were more common during those periods than in any other period in the Phanerozoic; coelacanths still live today in the oceans (genus *Latimeria*).

The Rhipidistians, whose ancestors probably lived in the oceans near the river mouths (estuaries), left the ocean world and migrated into freshwater habitats. They in turn split into two major groups: the lungfish and the tetrapodomorphs. The lungfish's greatest diversity was in the Triassic period; today there are fewer than a dozen genera left. The lungfish evolved the first proto-lungs and proto-limbs; developing the ability to live outside a water environment in the middle Devonian (397 - 385 Ma).

There are three major hypotheses as to how they evolved their stubby fins (proto-limbs). The traditional explanation is the "shrinking waterhole hypothesis" or "desert hypothesis" posited by the American paleontologist Alfred Romer. He believed that limbs and lungs may have evolved from the necessity of having to find new bodies of water as old waterholes dried up.

The second hypothesis is the "inter-tidal hypothesis" put forward in 2010 by a team of Polish paleontologists (Grzegorz Niedźwiedzki, Piotr Szrek, Katarzyna Narkiewicz, Marek Narkiewicz, & Per Ahlberg). They argued that sarcopterygians may have first emerged unto land from intertidal zones rather than inland bodies of water. Their hypothesis is based on the discovery of the 395 million-year-old Zachełmie tracks in Zachełmie, Świętokrzyskie Voivodeship, Poland, the oldest ever discovered fossil evidence of tetrapods.

The third hypothesis is dubbed the "woodland hypothesis" and was proposed by the American paleontologist Greg Retallack in 2011. He argues that limbs may have developed in shallow bodies of water in woodlands as a means of navigating in environments filled with roots and vegetation. He based his conclusions on the evidence that transitional tetrapod fossils are consistently found in habitats that were formerly humid and wooded floodplains.

The first tetrapodomorphs, which included the gigantic rhizodonts, had the same general anatomy as the lungfish, who were their closest kin, but they appear not to have left their water habitat until the late Devonian epoch (385 - 359 Ma), with the appearance of tetrapods (four-legged vertebrates). Tetrapods are the only tetrapodomorphs which survived after the Devonian.

Non-tetrapod sarcopterygians continued until towards the end of Paleozoic era, suffering heavy losses during the Permian-Triassic extinction event (251 Ma).

Vision in Fishes

An oscar, Astronotus ocellatus, surveys its world

Vision is an important sensory system for most species of fish. Fish eyes are similar to terrestrial vertebrates like birds and mammals, but have a more spherical lens. Their retinas generally have both rod cells and cone cells (for scotopic and photopic vision), and most species have colour vision. Some fish can see ultraviolet and some are sensitive to polarized light. Amongst jawless fish, the lamprey has well-developed eyes, while the hagfish has only primitive eyespots. The ancestors of modern hagfish, thought to be the protovertebrate were evidently pushed to very deep, dark waters, where they were less vulnerable to sighted predators, and where it is advantageous to have a convex eye-spot, which gathers more light than a flat or concave one. Unlike humans, fish normally adjust focus by moving the lens closer to or further from the retina.

Fish vision shows evolutionary adaptation to their visual environment, for example deep sea fish have eyes suited to the dark environment.

Water as a Visual Environment

Fish and other aquatic animals live in a different light environment than terrestrial species. Water absorbs light so that with increasing depth the amount of light available decreases quickly. The optical properties of water also lead to different wavelengths of light being absorbed to different degrees. For example, visible light of long wavelengths (e.g. red, orange) is absorbed quicker than light of shorter wavelengths (green, blue). Ultraviolet light (even shorter wavelength than violet) is absorbed quicker yet. Besides these universal qualities of water, different bodies of water may absorb light of different wavelengths due to varying salt and/or chemical presence in the water.

Structure and Function

Fish eyes are broadly similar to those of other vertebrates – notably the tetrapods (amphibians, reptiles, birds and mammals – all of which evolved from a fish ancestor). Light enters the eye at the cornea, passing through the pupil to reach the lens. Most fish species seem to have a fixed pupil size, but elasmobranches (like sharks and rays) have a muscular iris which allows pupil diameter to be adjusted. Pupil shape varies, and may be e.g. circular or slit-like.

Lenses are normally spherical but can be slightly elliptical in some species. Compared to terrestrial vertebrates, fish lenses are generally more dense and spherical. In the aquatic environment there is not a major difference in the refractive index of the cornea and the surrounding water (compared to air on land) so the lens has to do the majority of the refraction. Due to "a refractive index gradient within the lens — exactly as one would expect from optical theory" the spherical lenses of fish are able to form sharp images free from spherical aberration.

Once light passes through the lens it is transmitted through a transparent liquid medium until it reaches the retina, containing the photoreceptors. Like other vertebrates, the photoreceptors are on the inside layer so light must pass through layers of other neurons before it reaches them. The retina contains rod cells and cone cells.

The Retina

Within the retina, rod cells provide high visual sensitivity (at the cost of acuity), being used in low light conditions. Cone cells provide higher spatial and temporal resolution than rods can, and allow for the possibility of colour vision by comparing absorbances across different types of cones which are more sensitive to different wavelengths. The ratio of rods to cones depends on the ecology of the fish species concerned, *e.g.*, those mainly active during the day in clear waters will have more cones than those living in low light environments. Colour vision is more useful in environments with a broader range of wavelengths available, *e.g.*, near the surface in clear waters rather than in deeper water where only a narrow band of wavelengths persist.

The distribution of photoreceptors across the retina is not uniform. Some areas have higher densities of cone cells, for example. Fish may have two or three areas specialized for high acuity (e.g. for prey capture) or sensitivity (e.g. from dim light coming from below). The distribution of photoreceptors may also change over time during development of the individual. This is especially the case when the species typically moves between different light environments during its life cycle (e.g. shallow to deep waters, or fresh water to ocean).

Some species have a tapetum, a reflective layer which bounces light that passes through the retina back through it again. This enhances sensitivity in low light conditions, such as nocturnal and deep sea species, by giving photons a second chance to be captured by photoreceptors. However this comes at a cost of reduced resolution. Some species are able to effectively turn their tapetum off in bright conditions, with a dark pigment layer covering it as needed.

The retina uses a lot of oxygen compared to most other tissues, and is supplied with plentiful oxygenated blood to ensure optimal performance.

Humans have a vestibulo-ocular reflex, which is a reflex eye movement that stabilizes images on the retina during head movement by producing an eye movement in the direction opposite to head movement, thus preserving the image on the center of the visual field. In a similar manner, fish have a vestibulo-ocular reflex which stabilizes visual images on the retina when it moves its tail.

Accommodation

Accommodation is the process by which the vertebrate eye adjusts focus on an object as it moves closer or further away. Whereas birds and mammals achieve accommodation by deforming the lens of their eyes, fish and amphibians normally adjust focus by moving the lens closer or further from the retina. They use a special muscle which changes the distance of the lens from the retina. In bony fishes the muscle is called the *retractor lentis*, and is relaxed for near vision, whereas for cartilaginous fishes the muscle is called the *protractor lentis*, and is relaxed for far vision. Thus bony fishes accommodate for distance vision by moving the lens further from the retina, while cartilaginous fishes accommodate for near vision by moving the lens closer to the retina.

Stabilising Images

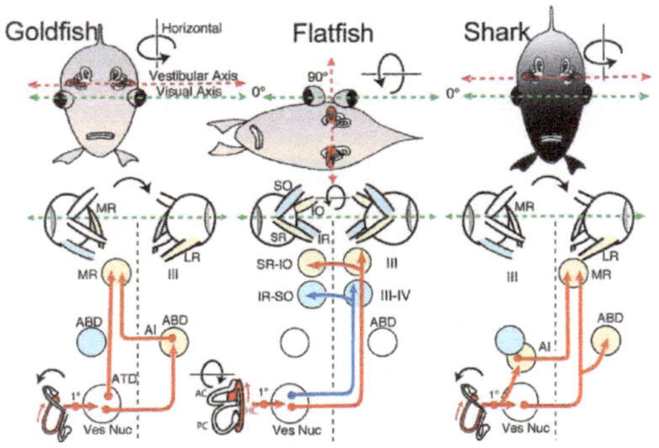

Horizontal vestibulo-ocular reflex in goldfish, flatfish and sharks

There is a need for some mechanism that stabilises images during rapid head movements. This is achieved by the vestibulo-ocular reflex, which is a reflex eye movement that stabilises images on the retina by producing eye movements in the direction opposite to head movements, thus preserving the image on the centre of the visual field. For example, when the head moves to the right, the eyes move to the left, and vice versa. In many animals, including human beings, the inner ear functions as the biological analogue of an accelerometer in camera image stabilization systems, to stabilize the image by moving the eyes. When a rotation of the head is detected, an inhibitory signal is sent to the extraocular muscles on one side and an excitatory signal to the muscles on the other side. The result is a compensatory movement of the eyes. Typical human eye movements lag head movements by less than 10 ms.

The diagram on the right shows the horizontal vestibulo-ocular reflex circuitry in bony and cartilaginous fish.

- "Goldfish" shows the principal three-neuronal vestibulo-ocular reflex linking the horizontal semicircular canal with contralateral abducens (ABD) and ipsilateral MR motoneurons.
- "Flatfish" shows that after 90° displacement of the vestibular relative to visual axis (metamorphosis) compensatory eye movements are produced by redirecting horizontal canal signals to vertical and oblique motoneurons.
- In "Shark" horizontal canal/second order neurons project to contralateral ABD and MR motoneurons including ipsilateral AI neurons. 1° , first order vestibular neuron; ATD, Ascending Tract of Deiter's.

Ultraviolet

Fish vision is mediated by four visual pigments that absorb various wavelengths of light. Each pigment is constructed from a chromophore and the transmembrane protein, known as opsin. Mutations in opsin have allowed for visual diversity, including variation in wavelength absorption. A mutation of the opsin on the SWS-1 pigment allows some vertebrates to absorb UV light (≈360 nm), so they can see objects to reflect UV light. A wide range of fish species has developed and maintained this visual trait throughout evolution, suggesting it is advantageous. UV vision may be related to foraging, communication, and mate selection.

The leading theory regarding the evolutionary selection of UV vision in select fish species is due to its strong role in mate selection. Behavioral experiments show that African cichlids utilize visual cues when choosing a mate. Their breeding sites are typically in shallow waters with high clarity and UV light penetration. Male African cichlids are largely a blue color that is reflective in UV light. Females are able to correctly choose a mate of their species when these reflective visual cues are present. This suggests that UV light detection is crucial for correct mate selection. UV reflective color patterns also enhance male attractiveness in guppies and three-spined sticklebacks. In experimental settings, female guppies spent significantly more time inspecting males with UV-reflective coloring than those with UV reflection blocked. Similarly, female three-spined sticklebacks preferred males viewed in full spectrum over those viewed in UV blocking filters. These results strongly suggest the role of UV detection in sexual selection and, thus, reproductive fitness. The prominent role of UV light detection in fish mate choice has allowed the trait to be maintained over time. UV vision may also be related to foraging and other communication behaviors.

Many species of fish can see the ultraviolet end of the spectrum, beyond the violet.

Ultraviolet vision is sometimes used during only part of the life cycle of a fish. For example, juvenile brown trout live in shallow water where they use ultraviolet vision to enhance their ability to detect zooplankton. As they get older, they move to deeper waters where there is little ultraviolet light.

The two stripe damselfish, *Dascyllus reticulatus*, has ultraviolet-reflecting colouration which they appear to use as an alarm signal to other fish of their species. Predatory species cannot see this if their vision is not sensitive to ultraviolet. There is further evidence for this view that some fish use ultraviolet as a "high-fidelity secret communication channel hidden from predators", while yet other species use ultraviolet to make social or sexual signals.

Polarized Light

It is not easy to establish whether a fish is sensitive to polarized light, though it appears likely in a number of taxa. It has been unambiguously demonstrated in anchovies. The ability to detect polarized light may provide better contrast and/or directional information for migrating species. Polarized light is most abundant at dawn and dusk. Polarized light reflected from the scales of a fish may enable other fish to better detect it against a diffuse background, and may provide useful information to schooling fish about their proximity and orientation relative to neighbouring fish.

Double Cones

Most fish have double cones, a pair of cone cells joined to each other. Each member of the double cone may have a different peak absorbance, and behavioural evidence supports the idea that each type of individual cone in a double cone can provide separate information (i.e. the signal from individual members of the double cone are not necessarily summed together).

Adaptation to Habitat

Four-eyed fishThe four-eyed fish feeds at the surface of the water with eyes that allow it to see both above and below the surface at the same time.

Fishes that live in surface waters down to about 200 metres, epipelagic fishes, live in a sunlit zone where visual predators use visual systems which are designed pretty much as might be expected. But even so, there can be unusual adaptations. Four-eyed fish have eyes raised above the top of the head and divided in two different parts, so that they can see below and above the water surface at the same time. Four-eyed fish actually have only two eyes, but their eyes are specially adapted for their surface-dwelling lifestyle. The eyes are positioned on the top of the head, and the fish floats at the water surface with only the lower half of each eye underwater. The two halves are divided by a band of tissue and the eye has two pupils, connected by part of the iris. The upper half of the eye is adapted for vision in air, the lower half for vision in water. The lens of the eye changes in thickness top to bottom to account for the difference in the refractive indices of air versus water. These fish spend most of their time at the surface of the water. Their diet mostly consists of the terrestrial insects which are available at the surface.

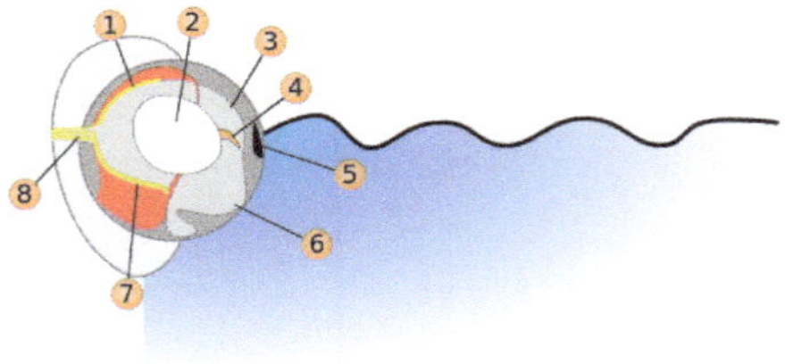

Eye of a four-eyed fish1) Underwater retina 2) Lens 3) Air pupil4) Tissue band 5) Iris 6) Underwater pupil 7) Air retina 8) Optic nerve

Deepwater fishes, like this Antarctic toothfish, often have large, upward looking eyes, adapted to detect prey silhouetted against the gloom above.

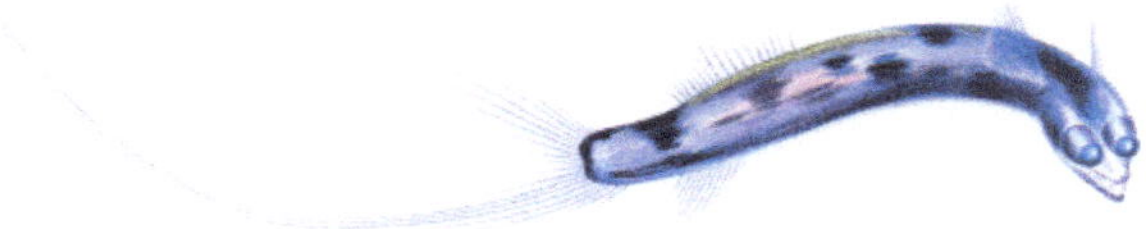

The telescopefish has large, forward-pointing telescoping eyes with large lenses.

Mesopelagic fishes live in deeper waters, in the twilight zone down to depths of 1000 metres, where the amount of sunlight available is not sufficient to support photosynthesis. These fish are adapted for an active life under low light conditions. Most of them are visual predators with large eyes. Some of the deeper water fish have tubular eyes with big lenses and only rod cells that look upwards. These give binocular vision and great sensitivity to small light signals. This adaptation gives improved terminal vision at the expense of lateral vision, and allows the predator to pick out squid, cuttlefish, and smaller fish that are silhouetted against the gloom above them. For more sensitive vision in low light, some fish have a retroreflector behind the retina. Flashlight fish have this plus photophores, which they use in combination to detect eyeshine in other fish.

The mesopelagic sabertooth is an ambush predator with telescopic, upward-pointing eyes.

Still deeper down the water column, below 1000 metres, are found the bathypelagic fishes. At this depth the ocean is pitch black, and the fish are sedentary, adapted to outputting minimum energy in a habitat with very little food and no sunlight. Bioluminescence is the only light available at these depths. This lack of light means the organisms have to rely on senses other than vision. Their eyes are small and may not function at all.

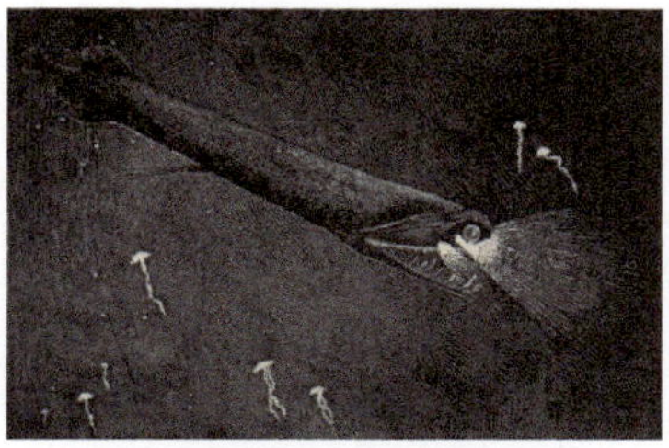

Most deep-sea fish cannot see red light. The deepwater stoplight loosejaw produces red bioluminescence so it can hunt with an effectively invisible beam of light.

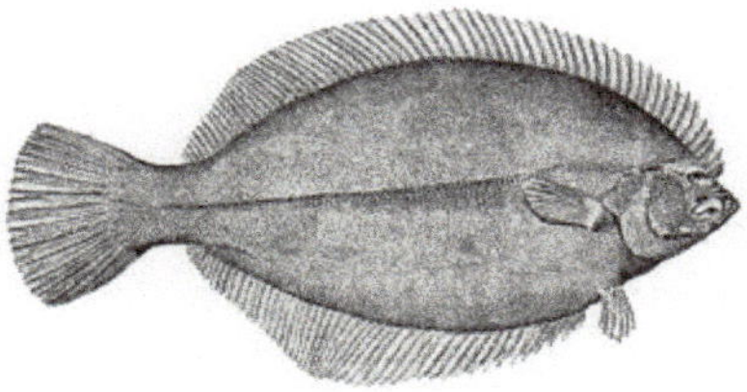

When the larvae of a flatfish grows, the eye on one side rotates to the other side so the fish can rest on the seafloor

At the very bottom of the ocean flatfish can be found. Flatfish are benthic fish with a negative buoyancy so they can rest on the seafloor. Although flatfish are bottom dwellers, they are not usually deep sea fish, but are found mainly in estuaries and on the continental shelf. When flatfish larvae hatch they have the elongated and symmetric shape of a typical bony fish. The larvae do not dwell on the bottom, but float in the sea as plankton. Eventually they start metamorphosing into the adult form. One of the eyes migrates across the top of the head and onto the other side of the body, leaving the fish blind on one side. The larva loses its swim bladder and spines, and sinks to the bottom, laying its blind side on the underlying surface. Richard Dawkins explains this as an example of evolutionary adaptation

The European plaice is a flatfish with raised eyes, so when it buries itself in sand for camouflage it can still see

...bony fish as a rule have a marked tendency to be flattened in a vertical direction.... It was natural, therefore, that when the ancestors of [flatfish] took to the sea bottom, they should have lain on one *side*.... But this raised the problem that one eye was always looking down into the sand and was effectively useless. In evolution this problem was solved by the lower eye 'moving' round to the upper side.

Prey usually have eyes on the sides of their head so they have a large field of view, from which to avoid predators. Predators usually have eyes in front of their head so they have better depth perception. Benthic predators, like flatfish, have eyes arranged so they have a binocular view of what is above them as they lie on the bottom.

Coloration

Fish have evolved sophisticated ways of using colouration. For example, prey fish have ways of using colouration to make it more difficult for visual predators to see them. In pelagic fish, these adaptations are mainly concerned with a reduction in silhouette, a form of camouflage. One method of achieving this is to reduce the area of their shadow by lateral compression of the body. Another method, also a form of camouflage, is by countershading in the case of epipelagic fish and by counter-illumination in the case of mesopelagic fish. Countershading is achieved by colouring the fish with darker pigments at the top and lighter pigments at the bottom in such a way that the colouring matches the background. When seen from the top, the darker dorsal area of the animal blends into the darkness of the water below, and when seen from below, the lighter ventral area blends into the sunlight from the surface. Counter illumination is achieved via bioluminescence by the production of light from ventral photophores, aimed at matching the light intensity from the underside of the fish with the light intensity from the background.

Benthic fish, which rest on the seafloor, physically hide themselves by burrowing into sand or retreating into nooks and crannies, or camouflage themselves by blending into the background or by looking like a rock or piece of seaweed.

While these tools may be effective as predator avoidance mechanisms, they also serve as equally effective tools for the predators themselves. For example, the deepwater velvet belly lantern shark uses counter-illumination to hide from its prey.

Epipelagic fish, like this Atlantic bluefin tuna, are typically countershaded with silvery colours

The foureye butterflyfish has false eyes on its back end, confusing predators about which is the front end of the fish

John Dory

Some fish species also display false eyespots. The foureye butterflyfish gets its name from a large dark spot on the rear portion of each side of the body. This spot is surrounded by a brilliant white ring, resembling an eyespot. A black vertical bar on the head runs through the true eye, making it hard to see. This can result in a predator thinking the fish is bigger than it is, and confusing the back end with the front end. The butterflyfish's first instinct when threatened is to flee, putting the false eyespot closer to the predator than the head. Most predators aim for the eyes, and this false eyespot tricks the predator into believing that the fish will flee tail first.

The John Dory is a benthopelagic coastal fish with a high laterally compressed body. Its body is so thin that it can hardly be seen from the front. It also has a large dark spot on both sides, which is used to flash an "evil eye" if danger approaches. The large eyes at the front of the head provide it with the bifocal vision and depth perception it needs to catch prey. The John Dory's eye spot on the side of its body also confuses prey, which is then scooped up in its mouth.

Barreleyes

Barreleyes are a family of small, unusual-looking mesopelagic fishes, named for their barrel-shaped, tubular eyes which are generally directed upwards to detect the silhouettes of available prey. Barreleyes have large, telescoping eyes which dominate and protrude from the skull. These eyes generally gaze upwards, but can also be swivelled forwards in some species. Their eyes have a large lens and a retina with an exceptional number of rod cells and a high density of rhodopsin (the "visual purple" pigment); there are no cone cells.

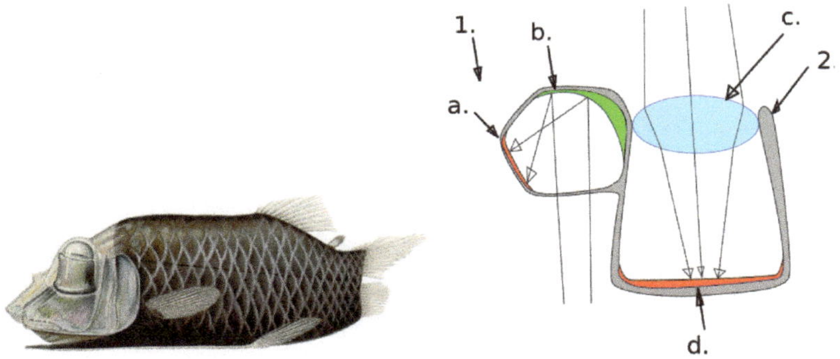

Left: The barreleye has barrel-shaped, telescopic eyes which aregenerally directed upwards, but can also be swivelled forward Right: The brownsnout spookfish is the only vertebrate knownto employ a mirror eye (as well as a lens):(1) diverticulum (2) main eye(a) retina (b) reflective crystals (c) lens (d) retina

The barreleye species, *Macropinna microstoma*, has a transparent protective dome over the top of its head, somewhat like the dome over an airplane cockpit, through which the lenses of its eyes can be seen. The dome is tough and flexible, and presumably protects the eyes from the nematocysts (stinging cells) of the siphonophores from which it is believed the barreleye steals food.

Another barreleye species, the brownsnout spookfish, is the only vertebrate known to employ a mirror, as opposed to a lens, to focus an image in its eyes. It is unusual in that it utilizes both refractive and reflective optics to see. The main tubular eye contains a lateral ovoid swelling called a diverticulum, largely separated from the eye by a septum. The retina lines most of the interior of the eye, and there are two corneal openings, one directed up and the other down, that allow light into the main eye and the diverticulum respectively. The main eye employs a lens to focus its image, as in other fishes. However, inside the diverticulum the light is reflected and focused onto the retina by a curved composite mirror derived from the retinal tapetum, composed of many layers of small reflective plates possibly made of guanine crystals. The split structure of the brownsnout spookfish eye allows the fish to see both up and down at the same time. In addition, the mirror system is superior to a lens in gathering light. It is likely that the main eye serves to detect objects silhouetted against the sunlight, while the diverticulum serves to detect bioluminescent flashes from the sides and below.

Sharks

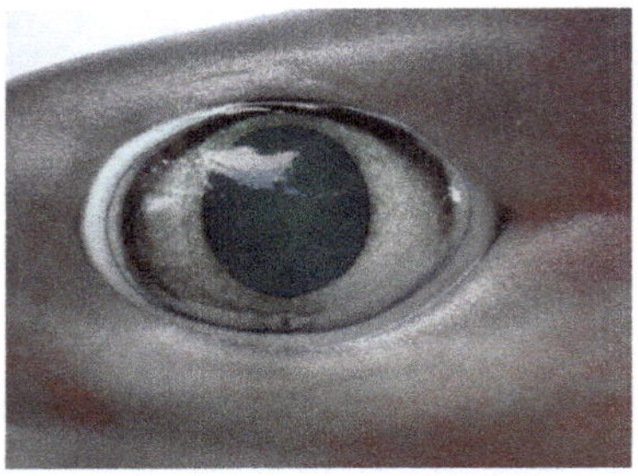

Eye of a bigeyed sixgill shark

Shark eyes are similar to the eyes of other vertebrates, including similar lenses, corneas and retinas, though their eyesight is well adapted to the marine environment with the help of a tissue called tapetum lucidum. This tissue is behind the retina and reflects light back to it, thereby increasing visibility in the dark waters. The effectiveness of the tissue varies, with some sharks having stronger nocturnal adaptations. Many sharks can contract and dilate their pupils, like humans, something no teleost fish can do. Sharks have eyelids, but they do not blink because the surrounding water cleans their eyes. To protect their eyes some species have nictitating membranes. This membrane covers the eyes while hunting and when the shark is being attacked. However, some species, including the great white shark (*Carcharodon carcharias*), do not have this membrane, but instead roll their eyes backwards to protect them when striking prey. The importance of sight in shark hunting behavior is debated. Some believe that electro- and chemoreception are more significant, while others point to the nictating membrane as evidence that sight is important. Presumably, the shark would not protect its eyes were they unimportant. The use of sight probably varies with species and water conditions. The shark's field of vision can swap between monocular and stereoscopic at any time. A micro-spectrophotometry study of 17 species of shark found 10 had only rod photoreceptors and no cone cells in their retinas giving them good night vision while making them colorblind. The remaining seven species had in addition to rods a single type of cone photoreceptor sensitive to green and, seeing only in shades of grey and green, are believed to be effectively colorblind. The study indicates that an object's contrast against the background, rather than colour, may be more important for object detection.

Other Examples

Small fish often school together for safety. This can have visual advantages, both by visually confusing predator fishes, and by providing many eyes for the school regarded as a body. The "predator confusion effect" is based on the idea that it becomes difficult for predators to pick out individual prey from groups because the many moving targets create a sensory overload of the predator's visual channel. "Shoaling fish are the same size and silvery, so it is difficult for a visually oriented predator to pick an individual out of a mass of twisting, flashing fish and then have enough time to grab its prey before it disappears into the shoal." The "many eyes effect" is based on the idea that as the size of the group increases, the task of scanning the environment for predators can be spread out over many individuals, a mass collaboration presumably providing a higher level of vigilance.

By schooling together, small fish provide many eyes as a precaution against ambush, and can visually confuse predators if they do attack

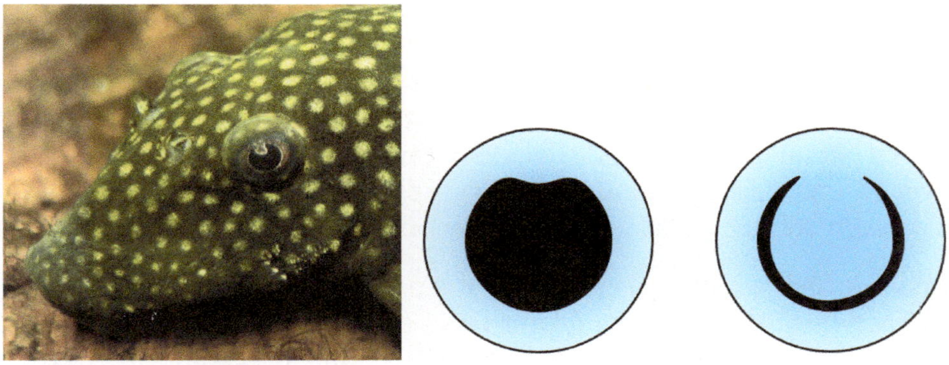

The omega iris allows Loricariids to adjust the amount of light that enters their eye

Fish are normally cold-blooded, with body temperatures the same as the surrounding water. However, some oceanic predatory fish, such as swordfish and some shark and tuna species, can warm parts of their body when they hunt for prey in deep and cold water. The highly visual swordfish uses a heating system involving its muscles which raises the temperature in its eyes and brain by up to 15 °C. The warming of the retina improves the rate at which the eyes respond to changes in rapid motion made by its prey by as much as ten times.

Some fish have eyeshine. Eyeshine is the result of a light-gathering layer in the eyes called the tapetum lucidum, which reflects white light. It does not occur in humans, but can be seen in other species, such as deer in a headlight. Eyeshine allows fish to see well in low-light conditions as well as in turbid (stained or rough, breaking) waters, giving them an advantage over their prey. This enhanced vision allows fish to populate the deeper regions in the ocean or a lake. In particular, freshwater walleye are so named because their eyeshine.

Many species of Loricariidae, a family of catfish, have a modified iris called an *omega iris*. The top part of the iris descends to form a loop which can expand and contract called an iris operculum; when light levels are high, the pupil reduces in diameter and the loop expands to cover the center of the pupil giving rise to a crescent shaped light transmitting portion. This feature gets its name from its similarity to an upside-down Greek letter omega (Ω). The origins of this structure are unknown, but it has been suggested that breaking up the outline of the highly visible eye aids camouflage in what are often highly mottled animals.

Distance Sensory Systems

Blind cavefish find their way around by means of lateral lines, which are highly sensitive to changes in water pressure.

Visual systems are distance sensory systems which provide fish with data about location or objects at a distance without a need for the fish to directly touch them. Such distance sensing systems are important, because they allow communication with other fish, and provide information about the location of food and predators, and about avoiding obstacles or maintaining position in fish schools. For example, some schooling species have “schooling marks” on their sides, such as visually prominent stripes which provide reference marks and help adjacent fish judge their relative positions. But the visual system is not the only one that can perform such functions. Some schooling fish also have a lateral line running the length of their bodies. This lateral line enables the fish to sense changes in water pressure and turbulence adjacent to its body. Using this information, schooling fish can adjust their distance from adjacent fish if they come too close or stray too far.

The visual system in fish is augmented by other sensing systems with comparable or complimentary functions. Some fish are blind, and must rely entirely on alternate sensing systems. Other senses which can also provide data about location or distant objects include hearing and echolocation, electroreception, magnetoception and chemoreception (smell and taste). For example, catfish have chemoreceptors across their entire bodies, which means they “taste” anything they touch and “smell” any chemicals in the water. “In catfish, gustation plays a primary role in the orientation and location of food”.

Cartilaginous fish (sharks, stingrays and chimaeras) use magnetoception. They possess special electroreceptors called the *ampullae of Lorenzini* which detect a slight variation in electric potential. These receptors, located along the mouth and nose of the fish, operate according to the principle that a time-varying magnetic field moving through a conductor induces an electric potential across the ends of the conductor. The ampullae may also allow the fish to detect changes in water temperature. As in birds, magnetoception may provide information which help the fish map migration routes.

Sensory Systems in Fish

Most fish possess highly developed sense organs. Nearly all daylight fish have color vision that is at least as good as a human’s Many fish also have chemoreceptors that are responsible for extraordinary senses of taste and smell. Although they have ears, many fish may not hear very well. Most fish have sensitive receptors that form the lateral line system, which detects gentle currents and vibrations, and senses the motion of nearby fish and prey. Sharks can sense frequencies in the range of 25 to 50 Hz through their lateral line.

Fish orient themselves using landmarks and may use mental maps based on multiple landmarks or symbols. Fish behavior in mazes reveals that they possess spatial memory and visual discrimination.

Vision

Vision is an important sensory system for most species of fish. Fish eyes are similar to those of terrestrial vertebrates like birds and mammals, but have a more spherical lens. Their retinas generally have both rod cells and cone cells (for scotopic and photopic vision), and most species have colour vision. Some fish can see ultraviolet and some can see polarized light. Amongst jawless fish, the lamprey has well-developed eyes, while the hagfish has only primitive eyespots. Fish vision shows adaptation to their visual environment, for example deep sea fishes have eyes suited to the dark environment.

Fish and other aquatic animals live in a different light environment than terrestrial species. Water absorbs light so that with increasing depth the amount of light available decreases quickly. The optic properties of water also lead to different wavelengths of light being absorbed to different degrees, for example light of long wavelengths (e.g. red, orange) is absorbed quite quickly compared to light of short wavelengths (blue, violet), though ultraviolet light (even shorter wavelength than blue) is absorbed quite quickly as well. Besides these universal qualities of water, different bodies of water may absorb light of different wavelengths because of salts and other chemicals in the water.

Hearing

Hearing is an important sensory system for most species of fish. Hearing threshold and the ability to localize sound sources are reduced underwater, in which the speed of sound is faster than in air. Underwater hearing is by bone conduction, and localization of sound appears to depend on differences in amplitude detected by bone conduction. Aquatic animals such as fish, however, have a more specialized hearing apparatus that is effective underwater.

Fish can sense sound through their lateral lines and their otoliths (ears). Some fishes, such as some species of carp and herring, hear through their swim bladders, which function rather like a hearing aid.

Hearing is well-developed in carp, which have the Weberian organ, three specialized vertebral processes that transfer vibrations in the swim bladder to the inner ear.

Although it is hard to test sharks' hearing, they may have a sharp sense of hearing and can possibly hear prey many miles away. A small opening on each side of their heads (not the spiracle) leads directly into the inner ear through a thin channel. The lateral line shows a similar arrangement, and is open to the environment via a series of openings called lateral line pores. This is a reminder of the common origin of these two vibration- and sound-detecting organs that are grouped together as the acoustico-lateralis system. In bony fish and tetrapods the external opening into the inner ear has been lost.

Current Detection

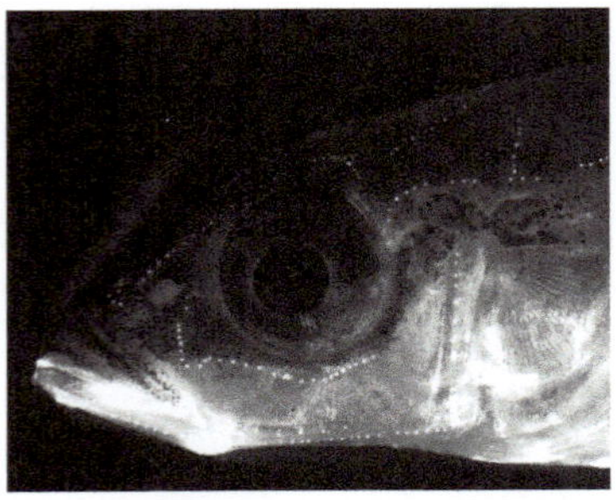

A three-spined stickleback with stained neuromasts

Hair cells in fish are used to detect water movements around their bodies. These hair cells are embedded in a jelly-like protrusion called cupula. The hair cells therefore can not be seen and do not appear on the surface of skin.

The lateral line in fish and aquatic forms of amphibians is a detection system of water currents, consisting mostly of vortices. The lateral line is also sensitive to low-frequency vibrations. The mechanoreceptors are hair cells, the same mechanoreceptors for vestibular sense and hearing. It is used primarily for navigation, hunting, and schooling. The receptors of the electrical sense are modified hair cells of the lateral line system.

Fish and some aquatic amphibians detect hydrodynamic stimuli via a lateral line. This system consists of an array of sensors called neuromasts along the length of the fish's body. Neuromasts can be free-standing (superficial neuromasts) or within fluid-filled canals (canal neuromasts). The sensory cells within neuromasts are polarized hair cells contained within a gelatinous cupula. The cupula, and the stereocilia within, are moved by a certain amount depending on the movement of the surrounding water. Afferent nerve fibers are excited or inhibited depending on whether the hair cells they arise from are deflected in the preferred or opposite direction. Lateral line receptors form somatotopic maps within the brain informing the fish of amplitude and direction of flow at different points along the body. These maps are located in the medial octavolateral nucleus (MON) of the medulla and in higher areas such as the torus semicircularis.

Pressure Detection

Pressure detection uses the organ of Weber, a system consisting of three appendages of vertebrae transferring changes in shape of the gas bladder to the middle ear. It can be used to regulate the buoyancy of the fish. Fish like the weather fish and other loaches are also known to respond to low pressure areas but they lack a swim bladder.

Chemoreception

Sharks have keen olfactory senses, located in the short duct (which is not fused, unlike bony fish) between the anterior and posterior nasal openings, with some species able to detect as little as one part per million of blood in seawater.

Sharks have the ability to determine the direction of a given scent based on the timing of scent detection in each nostril. This is similar to the method mammals use to determine direction of sound.

The shape of the hammerhead shark's head may enhance olfaction by spacing the nostrils further apart.

They are more attracted to the chemicals found in the intestines of many species, and as a result often linger near or in sewage outfalls. Some species, such as nurse sharks, have external barbels that greatly increase their ability to sense prey.

The MHC genes are a group of genes present in many animals and important for the immune system; in general, offspring from parents with differing MHC genes have a stronger immune system. Fish are able to smell some aspect of the MHC genes of potential sex partners and prefer partners with MHC genes different from their own.

Salmon have a strong sense of smell. Speculation about whether odours provide homing cues, go back to the 19th century. In 1951, Hasler hypothesised that, once in vicinity of the estuary or entrance to its birth river, salmon may use chemical cues which they can smell, and which are unique to their natal stream, as a mechanism to home onto the entrance of the stream. In 1978, Hasler and his students convincingly showed that the way salmon locate their home rivers with such precision was indeed because they could recognise its characteristic smell. They further demonstrated that the smell of their river becomes imprinted in salmon when they transform into smolts, just before they migrate out to sea. Homecoming salmon can also recognise characteristic smells in tributary streams as they move up the main river. They may also be sensitive to characteristic pheromones given off by juvenile conspecifics. There is evidence that they can "discriminate between two populations of their own species".

Electroreception and Magnetoreception

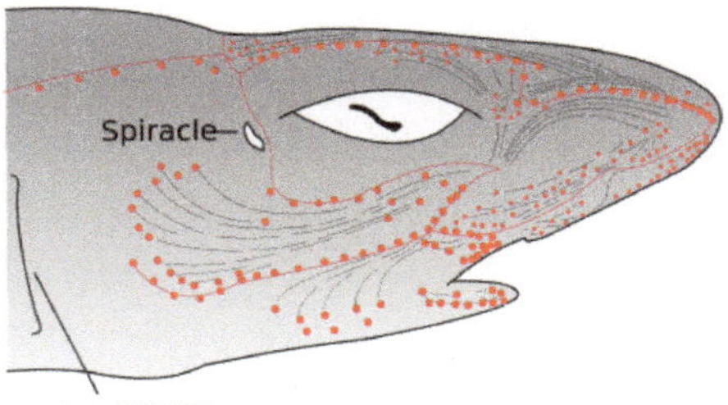

Electromagnetic field receptors (ampullae of Lorenzini) and motion detecting canals in the head of a shark

Electroreception, or electroception, is the ability to detect electric fields or currents. Some fish, such as catfish and sharks, have organs that detect weak electric potentials on the order of millivolts. Other fish, like the South American electric fishes Gymnotiformes, can produce weak electric currents, which they use in navigation and social communication. In sharks, the ampullae of Lorenzini are electroreceptor organs. They number in the hundreds to thousands. Sharks use the ampullae of Lorenzini to detect the electromagnetic fields that all living things produce. This helps sharks (particularly the hammerhead shark) find prey. The shark has the greatest electrical sensitivity of any animal. Sharks find prey hidden in sand by detecting the electric fields they produce. Ocean currents moving in the magnetic field of the Earth also generate electric fields that sharks can use for orientation and possibly navigation.

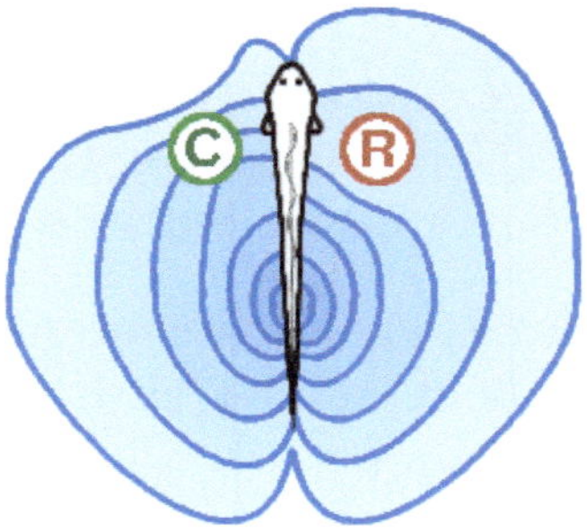

Active electrolocation. Conductive objects concentrate the field and resistive objects spread the field.

Electric field proximity sensing is used by the electric catfish to navigate through muddy waters. These fish make use of spectral changes and amplitude modulation to determine factors such shape, size, distance, velocity, and conductivity. The abilities of the electric fish to communicate and identify sex, age, and hierarchy within the species are also made possible through electric fields. EF gradients as low as 5nV/cm can be found in some saltwater weakly electric fish.

The paddlefish (*Polyodon spathula*) hunts plankton using thousands of tiny passive electroreceptors located on its extended snout, or rostrum. The paddlefish is able to detect electric fields that oscillate at 0.5–20 Hz, and large groups of plankton generate this type of signal.

Electric fishes use an active sensory system to probe the environment and create active electrodynamic imaging.

In 1973, it was shown that Atlantic salmon have conditioned cardiac responses to electric fields with strengths similar to those found in oceans. "This sensitivity might allow a migrating fish to align itself upstream or downstream in an ocean current in the absence of fixed references."

Magnetoception, or magnetoreception, is the ability to detect the direction one is facing based on the Earth's magnetic field. In 1988, researchers found iron, in the form of single domain magnetite, resides in the skulls of sockeye salmon. The quantities present are sufficient for magnetoception.

Fish Navigation

Salmon regularly migrate thousands of miles to and from their breeding grounds.

Salmon spend their early life in rivers, and then swim out to sea where they live their adult lives and gain most of their body mass. When they have matured, they return to the rivers to spawn. Usually they return with uncanny precision to the natal river where they were born, and even to the very spawning ground of their birth. It is thought that, when they are in the ocean, they use magnetoception to locate the general position of their natal river, and once close to the river, that they use their sense of smell to home in on the river entrance and even their natal spawning ground.

After several years wandering huge distances in the ocean, most surviving salmon return to the same natal rivers where they were spawned. Then most of them swim up the rivers until they reach the very spawning ground that was their original birthplace.

There are various theories about how this happens. One theory is that there are geomagnetic and chemical cues which the salmon use to guide them back to their birthplace. The fish may be sensitive to the Earth's magnetic field, which could allow the fish to orient itself in the ocean, so it can navigate back to the estuary of its natal stream.

Pain

Experiments done by William Tavolga provide evidence that fish have pain and fear responses. For instance, in Tavolga's experiments, toadfish grunted when electrically shocked and over time they came to grunt at the mere sight of an electrode.

In 2003, Scottish scientists at the University of Edinburgh and the Roslin Institute concluded that rainbow trout exhibit behaviors often associated with pain in other animals. Bee venom and acetic acid injected into the lips resulted in fish rocking their bodies and rubbing their lips along the sides and floors of their tanks, which the researchers concluded were attempts to relieve pain, similar to what mammals would do. Neurons fired in a pattern resembling human neuronal patterns.

Hooked sailfish

Professor James D. Rose of the University of Wyoming claimed the study was flawed since it did not provide proof that fish possess "conscious awareness, particularly a kind of awareness that is meaningfully like ours". Rose argues that since fish brains are so different from human brains, fish are probably not conscious in the manner humans are, so that reactions similar to human reactions to pain instead have other causes. Rose had published a study a year earlier arguing that fish cannot feel pain because their brains lack a neocortex. However, animal behaviorist Temple Grandin argues that fish could still have consciousness without a neocortex because "different species can use different brain structures and systems to handle the same functions."

Animal welfare advocates raise concerns about the possible suffering of fish caused by angling. Some countries, such as Germany have banned specific types of fishing, and the British RSPCA now prosecutes individuals who are cruel to fish.

References

- Magnuson JJ (1978) "Locomotion by scombrid fishes: Hydromechanics, morphology and behavior" in Fish Physiology, Volume 7: Locomotion, WS Hoar and DJ Randall (Eds) Academic Press. Page 240–308. ISBN

9780123504074.

- Kapoor BG and Khanna B (2004) Ichthyology Handbook pp. 497–498, Springer Science & Business Media. ISBN 9783540428541.
- Helfman G, Collette BB, Facey DH and Bowen BW (2009) The Diversity of Fishes: Biology, Evolution, and Ecology p. 35, Wiley-Blackwell. ISBN 978-1-4051-2494-2.
- Heinicke, Matthew P.; Naylor, Gavin J. P.; Hedges, S. Blair (2009). The Timetree of Life: Cartilaginous Fishes (Chondrichthyes). Oxford University Press. p. 320. ISBN 0191560154.
- Laurin, M. (2010). How Vertebrates Left the Water. Berkeley, California, USA.: University of California Press. ISBN 978-0-520-26647-6.
- Hall, Brian K (2007) Fins into Limbs: Evolution, Development, and Transformation University of Chicago Press. ISBN 9780226313375.
- Clack, Jennifer A (2012) "From fins to feet" Chapter 6, pages 187–260, in: Gaining Ground, Second Edition: The Origin and Evolution of Tetrapods, Indiana University Press. ISBN 9780253356758.
- Gould,Stephen Jay (1993) "Bent Out of Shape" in Eight Little Piggies: Reflections in Natural History. Norton, 179–94. ISBN 9780393311396.
- Kardong, KV (1998) Vertebrates: Comparative Anatomy, Function, Evolution2nd edition, illustrated, revised. Published by WCB/McGraw-Hill, p. 12 ISBN 0-697-28654-1
- Hulley, P. Alexander (1998). Paxton, J.R.; Eschmeyer, W.N., eds. Encyclopedia of Fishes. San Diego: Academic Press. pp. 127–128. ISBN 0-12-547665-5.
- Johnson, Erik L. and Richard E. Hess (2006) Fancy Goldfish: A Complete Guide to Care and Collecting, Weatherhill, Shambhala Publications, Inc. ISBN 0-8348-0448-4
- Kardong, Kenneth (2015). Vertebrates: Comparative Anatomy, Function, Evolution. New York: McGraw-Hill Education. pp. 99–100. ISBN 978-0-07-802302-6.
- Baylis, Jeffrey (1981). "The Evolution of Parental Care in Fishes, with reference to Darwin's rule of male sexual selection". Environmental Biology of Fishes. 6 (2). doi:10.1007/BF00002788. Retrieved 16 September 2013.

4

Insect Physiology

Insect physiology studies the physiology of an insect's biological system. Insects are usually made up of three main body regions, the head, thorax and abdomen. Insect flight, insect olfaction, pupa and insect reproductive system are also explained in the following section.

Insect Physiology

Insect physiology includes the physiology and biochemistry of insect organ systems.

Although diverse, insects are quite indifferent in overall design, internally and externally. The insect is made up of three main body regions (tagmata), the head, thorax and abdomen. The head comprises six fused segments with compound eyes, ocelli, antennae and mouthparts, which differ according to the insect's particular diet, e.g. grinding, sucking, lapping and chewing. The thorax is made up of three segments: the pro, meso and meta thorax, each supporting a pair of legs which may also differ, depending on function, e.g. jumping, digging, swimming and running. Usually the middle and the last segment of the thorax have paired wings. The abdomen generally comprises eleven segments and contains the digestive and reproductive organs. A general overview of the internal structure and physiology of the insect is presented, including digestive, circulatory, respiratory, muscular, endocrine and nervous systems, as well as sensory organs, temperature control, flight and molting.

Digestive System

An insect uses its digestive system to extract nutrients and other substances from the food it consumes. Most of this food is ingested in the form of macromolecules and other complex substances (such as proteins, polysaccharides, fats, and nucleic acids) which must be broken down by catabolic reactions into smaller molecules (i.e. amino acids, simple sugars, etc.) before being used by cells of the body for energy, growth, or reproduction. This break-down process is known as digestion.

The insect's digestive system is a closed system, with one long enclosed coiled tube called the alimentary canal which runs lengthwise through the body. The alimentary canal only allows food to enter the mouth, and then gets processed as it travels toward the anus. The insects alimentary canal has specific sections for grinding and food storage, enzyme production and nutrient absorption. Sphincters control the food and fluid movement between three regions. The three regions include the foregut (stomatodeum)(27,) the midgut (mesenteron)(13), and the hindgut (proctodeum)(16).

In addition to the alimentary canal, insects also have paired salivary glands and salivary reservoirs. These structures usually reside in the thorax (adjacent to the fore-gut). The salivary glands (30) produce saliva, the salivary ducts lead from the glands to the reservoirs and then forward through the head to an opening called the salivarium behind the hypopharynx; which movements of the mouthparts help mix saliva with food in the buccal cavity. Saliva mixes with food which travels through salivary tubes into the mouth, beginning the process of breaking it down.

The stomatedeum and proctodeum are invaginations of the epidermis and are lined with cuticle (intima). The mesenteron is not lined with cuticle but with rapidly dividing and therefore constantly replaced, epithelial cells. The cuticle sheds with every moult along with the exoskeleton. Food is moved down the gut by muscular contractions called peristalsis.

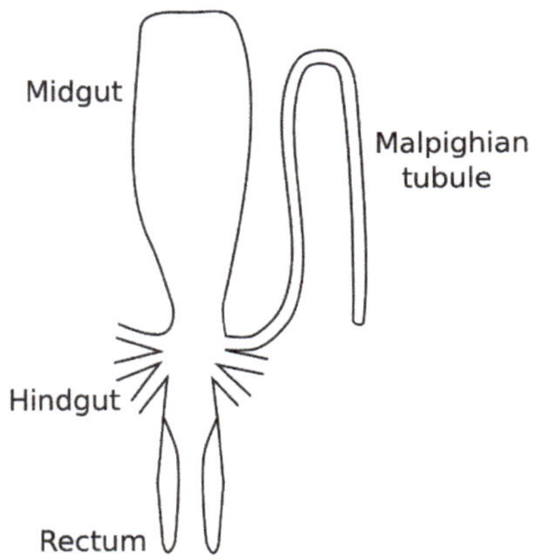

Stylised diagram of insect digestive tract showing Malpighian tubule (Orthopteran type)

1. Stomatodeum (foregut): This region stores, grinds and transports food to the next region. Included in this are the buccal cavity, the pharynx, the oesophagus, the crop (stores food), and proventriculus or gizzard (grinds food). Salivary secretions from the labial glands dilute the ingested food. In mosquitoes (Diptera), which are blood-feeding insects, anticoagulants and blood thinners are also released here.

2. Mesenteron (midgut): Digestive enzymes in this region are produced and secreted into the lumen and here nutrients are absorbed into the insect's body. Food is enveloped by this part of the gut as it arrives from the foregut by the peritrophic membrane which is a mucopolysaccharide layer secreted from the midgut's epithelial cells. It is thought that this membrane prevents food pathogens from contacting the epithelium and attacking the insects' body. It also acts as a filter allowing small molecules through, but preventing large molecules and particles of food from reaching the midgut cells. After the large substances are broken down into smaller ones, digestion and consequent nutrient absorption takes place at the surface of the epithelium. Microscopic projections from the mid-gut wall, called microvilli, increase surface area and allow for maximum absorption of nutrients.

3. Proctodeum (hindgut): This is divided into three sections; the anterior is the ileum, the middle portion, the colon, and the wider, posterior section is the rectum. This extends from the pyloric valve which is located between the mid and the hindgut to the anus. Here absorption of water, salts and other beneficial substances take place before excretion. Like other animals, the removal of toxic metabolic waste requires water. However, for very small animals like insects, water conservation is a priority. Because of this, blind-ended ducts called Malpighian tubules come into play. These ducts emerge as evaginations at the anterior end of the hindgut and are the main organs of

osmoregulation and excretion. These extract the waste products from the haemolymph, in which all the internal organs are bathed). These tubules continually produce the insect's uric acid, which is transported to the hindgut, where important salts and water are re-absorbed by both the hindgut and rectum. Excrement is then voided as insoluble and non-toxic uric acid granules. Excretion and osmoregulation in insects are not orchestrated by the Malpighian tubules alone, but require a joint function of the ileum and/or rectum.

Circulatory System

The main function of insect blood, hemolymph, is that of transport and it bathes the insect's body organs. Making up usually less than 25% of an insect's body weight, it transports hormones, nutrients and wastes and has a role in osmoregulation, temperature control, immunity, storage (water, carbohydrates and fats) and skeletal function. It also plays an essential part in the moulting process. An additional role of the hemolymph in some orders, can be that of predatory defence. It can contain unpalatable and malodourous chemicals that will act as a deterrent to predators.

Hemolymph contains molecules, ions and cells. Regulating chemical exchanges between tissues, hemolymph is encased in the insect body cavity or haemocoel. It is transported around the body by combined heart (posterior) and aorta (anterior) pulsations which are located dorsally just under the surface of the body. It differs from vertebrate blood in that it doesn't contain any red blood cells and therefore is without high oxygen carrying capacity, and is more similar to lymph found in vertebrates.

Body fluids enter through one way valved ostia which are openings situated along the length of the combined aorta and heart organ. Pumping of the hemolymph occurs by waves of peristaltic contraction, originating at the body's posterior end, pumping forwards into the dorsal vessel, out via the aorta and then into the head where it flows out into the haemocoel. The hemolymph is circulated to the appendages unidirectionally with the aid of muscular pumps or accessory pulsatile organs which are usually found at the base of the antennae or wings and sometimes in the legs. Pumping rate accelerates due to periods of increased activity. Movement of hemolymph is particularly important for thermoregulation in orders such as Odonata, Lepidoptera, Hymenoptera and Diptera.

Respiratory System

Insect respiration is accomplished without lungs using a system of internal tubes and sacs through which gases either diffuse or are actively pumped, delivering oxygen directly to tissues that need oxygen and eliminate carbon dioxide via their cells. Since oxygen is delivered directly, the circulatory system is not used to carry oxygen, and is therefore greatly reduced; it has no closed vessels (i.e., no veins or arteries), consisting of little more than a single, perforated dorsal tube which pulses peristaltically, and in doing so helps circulate the hemolymph inside the body cavity.

Air is taken in through spiracles, openings which are positioned laterally in the pleural wall, usually a pair on the anterior margin of the meso and meta thorax, and pairs on each of the eight or less abdominal segments, Numbers of spiracles vary from 1 to 10 pairs. The oxygen passes through the tracheae to the tracheoles, and enters the body by the process of diffusion. Carbon dioxide leaves the body by the same process.

The major tracheae are thickened spirally like a flexible vacuum hose to prevent them from collapsing and often swell into air sacs. Larger insects can augment the flow of air through their tracheal system, with body movement and rhythmic flattening of the tracheal air sacs. Spiracles are closed and opened by means of valves and can remain partly or completely closed for extended periods in some insects, which minimises water loss.

There are many different patterns of gas exchange demonstrated by different groups of insects. Gas exchange patterns in insects can range from continuous, diffusive ventilation, to discontinuous gas exchange.

Terrestrial and a large proportion of aquatic insects perform gaseous exchange as previously mentioned under an open system. Other smaller numbers of aquatic insects have a closed tracheal system, for example, Odonata, Tricoptera, Ephemeroptera, which have tracheal gills and no functional spiracles. Endoparasitic larvae are without spiracles and also operate under a closed system. Here the tracheae separate peripherally, covering the general body surface which results in a cutaneous form of gaseous exchange. This peripheral tracheal division may also lie within the tracheal gills where gaseous exchange may also take place.

Muscular System

Many insects are able to lift twenty times their own body weight and may jump distances that are many times greater than their own length. This is not because they are strong but because their energy output is high in relation to their body mass.

The muscular system of insects ranges from a few hundred muscles to a few thousand. Unlike vertebrates that have both smooth and striated muscles, insects have only striated muscles. Muscle cells are amassed into muscle fibers and then into the functional unit, the muscle. Muscles are attached to the body wall, with attachment fibers running through the cuticle and to the epicuticle, where they can move different parts of the body including appendages such as wings. The muscle fiber has many cells with a plasma membrane and outer sheath or sarcolemma. The sarcolemma is invaginated and can make contact with the tracheole carrying oxygen to the muscle fiber. Arranged in sheets or cylindrically, contractile myofibrils run the length of the muscle fiber. Myofibrils comprising a fine actin filament enclosed between a thick pair of myosin filaments slide past each other instigated by nerve impulses.

Muscles can be divided into four categories:

1. Visceral: these muscles surround the tubes and ducts and produce peristalsis as demonstrated in the digestive system.

2. Segmental: causing telescoping of muscle segments required for moulting, increase in body pressure and locomotion in legless larvae.

3. Appendicular: originating from either the sternum or the tergum and inserted on the coxae these muscles move appendages as one unit. These are arranged segmentally and usually in antagonistic pairs. Appendage parts of some insects, e.g. the galea and the lacinia of the maxillae, only have flexor muscles. Extension of these structures is by haemolymph pressure and cuticle elasticity.

4. Flight: Flight muscles are the most specialised category of muscle and are capable of rapid contractions. Nerve impulses are required to initiate muscle contractions and therefore flight. These muscles are also known as neurogenic or synchronous muscles. This is because there is a one-to-one correspondence between action potentials and muscle contractions. In insects with higher wing stroke frequencies the muscles contract more frequently than at the rate that the nerve impulse reaches them and are known as asynchronous muscles.

Flight has allowed the insect to disperse, escape from enemies and environmental harm, and colonise new habitats. One of the insect's key adaptations is flight, the mechanics of which differ from those of other flying animals because their wings are not modified appendages. Fully developed and functional wings occur only in adult insects. To fly, gravity and drag (air resistance to movement) have to be overcome. Most insects fly by beating their wings and to power their flight they have either direct flight muscles attached to the wings, or an indirect system where there is no muscle-to-wing connection and instead they are attached to a highly flexible box-like thorax.

Direct flight muscles generate the upward stroke by the contraction of the muscles attached to the base of the wing inside the pivotal point. Outside the pivotal point the downward stroke is generated through contraction of muscles that extend from the sternum to the wing. Indirect flight muscles are attached to the tergum and sternum. Contraction makes the tergum and base of the wing pull down. In turn this movement lever the outer or main part of the wing in strokes upward. Contraction of the second set of muscles, which run from the back to the front of the thorax, powers the downbeat. This deforms the box and lifts the tergum.

Endocrine System

Hormones are the chemical substances that are transported in the insect's body fluids (haemolymph) that carry messages away from their point of synthesis to sites where physiological processes are influenced. These hormones are produced by glandular, neuroglandular and neuronal centres. Insects have several organs that produce hormones, controlling reproduction, metamorphosis and moulting. It has been suggested that a brain hormone is responsible for caste dermination in termites and diapause interruption in some insects.

Four endocrine centers have been identified:

1. Neurosecretory cells in the brain can produce one or more hormones that affect growth, reproduction, homeostasis and metamorphosis.

2. Corpora cardiaca are a pair of neuroglandular bodies that are found behind the brain and on either sides of the aorta. These not only produce their own neurohormones but they store and release other neurohormones including PTTH prothoracicotropic hormone (brain hormone), which stimulates the secretory activity of the prothoracic glands, playing an integral role in moulting.

3. Prothoracic glands are diffuse, paired glands located at the back of the head or in the thorax. These glands secrete an ecdysteroid called ecdysone, or the moulting hormone, which initiates the epidermal moulting process. Additionally it plays a role in accessory reproductive glands in the female, differentiation of ovarioles and in the process of egg production.

4. Corpora allata are small, paired glandular bodies originating from the epithelium located on either side of the foregut. They secrete the juvenile hormone, which regulate reproduction and metamorphosis.

Nervous System

Insects have a complex nervous system which incorporates a variety of internal physiological information as well as external sensory information. As in the case of vertebrates, the basic component is the neuron or nerve cell. This is made up of a dendrite with two projections that receive stimuli and an axon, which transmits information to another neuron or organ, like a muscle. As for vertebrates, chemicals (neurotransmitters such as acetylcholine and dopamine) are released at synapses.

Central Nervous System

An insect's sensory, motor and physiological processes are controlled by the central nervous system along with the endocrine system. Being the principal division of the nervous system, it consists of a brain, a ventral nerve cord and a subesophageal ganglion. This is connected to the brain by two nerves, extending around each side of the oesophagus.

The brain has three lobes:

- Protocerebrum, innervating the compound eyes and the ocelli
- Deutocerebrum, innervating the antennae
- Tritocerebrum, innervating the foregut and the labrum.

The ventral nerve cord extends from the suboesophageal ganglion posteriorly. A layer of connective tissue called the neurolemma covers the brain, ganglia, major peripheral nerves and ventral nerve cords.

The head capsule (made up of six fused segments) has six pairs of ganglia. The first three pairs are fused into the brain, while the three following pairs are fused into the subesophageal ganglion. The thoracic segments have one ganglion on each side, which are connected into a pair, one pair per segment. This arrangement is also seen in the abdomen but only in the first eight segments. Many species of insects have reduced numbers of ganglia due to fusion or reduction. Some cockroaches have just six ganglia in the abdomen, whereas the wasp *Vespa crabro* has only two in the thorax and three in the abdomen. And some, like the house fly *Musca domestica*, have all the body ganglia fused into a single large thoracic ganglion. The ganglia of the central nervous system act as the coordinating centres with their own specific autonomy where each may coordinate impulses in specified regions of the insect's body.

Peripheral Nervous System

This consists of motor neuron axons that branch out to the muscles from the ganglia of the central nervous system, parts of the sympathetic nervous system and the sensory neurons of the cuticular sense organs that receive chemical, thermal, mechanical or visual stimuli from the insects environment. The sympathetic nervous system includes nerves and the ganglia that innervate the gut both posteriorly and anteriorly, some endocrine organs, the spiracles of the tracheal system and the reproductive organs.

Sensory Organs

Chemical senses include the use of chemoreceptors, related to taste and smell, affecting mating, habitat selection, feeding and parasite-host relationships. Taste is usually located on the mouthparts of the insect but in some insects, such as bees, wasps and ants, taste organs can also be found on the antennae. Taste organs can also be found on the tarsi of moths, butterflies and flies. Olfactory sensilla enable insects to smell and are usually found in the antennae. Chemoreceptor sensitivity related to smell in some substances, is very high and some insects can detect particular odours that are at low concentrations miles from their original source.

Mechanical senses provide the insect with information that may direct orientation, general movement, flight from enemies, reproduction and feeding and are elicited from the sense organs that are sensitive to mechanical stimuli such as pressure, touch and vibration. Hairs (setae) on the cuticle are responsible for this as they are sensitive to vibration touch and sound.

Hearing structures or tympanal organs are located on different body parts such as, wings, abdomen, legs and antennae. These can respond to various frequencies ranging from 100 to 240 kHz depending on insect species. Many of the joints of the insect have tactile setae that register movement. Hair beds and groups of small hair like sensilla, determine proprioreception or information about the position of a limb, and are found on the cuticle at the joints of segments and legs. Pressure on the body wall or strain gauges are detected by the campiniform sensilla and internal stretch receptors sense muscle distension and digestive system stretching.

The compound eye and the ocelli supply insect vision. The compound eye consists of individual light receptive units called ommatidia. Some ants may have only one or two, however dragonflies may have over 10,000. The more ommatidia the greater the visual acuity. These units have a clear lens system and light sensitive retina cells. By day, the image flying insects receive is made up of a mosaic of specks of differing light intensity from all the different ommatidia. At night or dusk, visual acuity is sacrificed for light sensitivity. The ocelli are unable to form focussed images but are sensitive mainly, to differences in light intensity. Colour vision occurs in all orders of insects. Generally insects see better at the blue end of the spectrum than at the red end. In some orders sensitivity ranges can include ultraviolet.

A number of insects have temperature and humidity sensors and insects being small, cool more quickly than larger animals. Insects are generally considered cold-blooded or ectothermic, their body temperature rising and falling with the environment. However, flying insects raise their body temperature through the action of flight, above environmental temperatures.

The body temperature of butterflies and grasshoppers in flight may be 5 °C or 10 °C above environmental temperature, however moths and bumblebees, insulated by scales and hair, during flight, may raise flight muscle temperature 20–30 °C above the environment temperature. Most flying insects have to maintain their flight muscles above a certain temperature to gain power enough to fly. Shivering, or vibrating the wing muscles allow larger insects to actively increase the temperature of their flight muscles, enabling flight.

Until very recently, no one had ever documented the presence of nociceptors (the cells that detect and transmit sensations of pain) in insects, though recent findings of nociception in larval fruit flies challenges this and raises the possibility that some insects may be capable of feeling pain.

Reproductive System

Most insects have a high reproductive rate. With a short generation time, they evolve faster and can adjust to environmental changes more rapidly than other slower breeding animals. Although there are many forms of reproductive organs in insects, there remains a basic design and function for each reproductive part. These individual parts may vary in shape (gonads), position (accessory gland attachment), and number (testicular and ovarian glands), with different insect groups.

Female

The female insect's main reproductive function is to produce eggs, including the egg's protective coating, and to store the male spermatozoa until egg fertilisation is ready. The female reproductive organs include, paired ovaries which empty their eggs (oocytes) via the calyces into lateral oviducts, joining to form the common oviduct. The opening (gonopore) of the common oviduct is concealed in a cavity called the genital chamber and this serves as a copulatory pouch (bursa copulatrix) when mating. The external opening to this is the vulva. Often in insects the vulva is narrow and the genital chamber becomes pouch or tube like and is called the vagina. Related to the vagina is a saclike structure, the spermatheca, where spermatozoa are stored ready for egg fertilisation. A secretory gland nourishes the contained spermatozoa in the vagina.

Egg development is mostly completed by the insect's adult stage and is controlled by hormones that control the initial stages of oogenesis and yolk deposition. Most insects are oviparous, where the young hatch after the eggs have been laid.

Insect sexual reproduction starts with sperm entry that stimulates oogenesis, meiosis occurs and the egg moves down the genital tract. Accessory glands of the female secrete an adhesive substance to attach eggs to an object and they also supply material that provides the eggs with a protective coating. Oviposition takes place via the female ovipositor.

Male

The male's main reproductive function is to produce and store spermatozoa and provide transport to the reproductive tract of the female. Sperm development is usually completed by the time the insect reaches adulthood. The male has two testes, which contain follicles in which the spermatozoa are produced. These open separately into the sperm duct or vas deferens and this stores the sperm. The vas deferentia then unite posteriorally to form a central ejaculatory duct, this opens to the outside on an aedeagus or a penis. Accessory glands secrete fluids that comprise the spermatophore. This becomes a package that surrounds and carries the spermatozoa, forming a sperm-containing capsule.

Sexual and Asexual Reproduction

Most insects reproduce via sexual reproduction, i.e. the egg is produced by the female, fertilised by the male and oviposited by the female. Eggs are usually deposited in a precise microhabitat on or near the required food. However, some adult females can reproduce without male input. This is known as parthenogenesis and in the most common type of parthenogenesis the offspring are essentially identical to the mother. This is most often seen in aphids and scale insects.

Life Cycle

An insect's life-cycle can be divided into three types:

- Ametabolous, no metamorphosis, these insects are primitively wingless where the only difference between adult and nymph is size, e.g. Order: Thysanura (Silverfish).
- Hemimetabolous, or incomplete metamorphosis. The terrestrial young are called nymphs and aquatic young are called naiads. Insect young are usually similar to the adult. Wings appear as buds on the nymphs or early instars. When the last moult is completed the wings expand to the full adult size, e.g. Order: Odonata (Dragonflies).
- Holometabolus, or complete metamorphosis. These insects have a different form in their immature and adult stages, have different behaviours and live in different habitats. The immature form is called larvae and remains similar in form but increases in size. They usually have chewing mouthparts even if the adult form mouth parts suck. At the last larval instar phase the insect forms into a pupa, it doesn't feed and is inactive, and here wing development is initiated, and the adult emerges e.g. Order: Lepidoptera (Butterflies and Moths).

Moulting

As an insect grows it needs to replace the rigid exoskeleton regularly. Moulting may occur up to three or four times or, in some insects, fifty times or more during its life. A complex process controlled by hormones, it includes the cuticle of the body wall, the cuticular lining of the tracheae, foregut, hindgut and endoskeletal structures.

The stages of molting:

1. Apolysis—moulting hormones are released into the haemolymph and the old cuticle separates from the underlying epidermal cells. The epidermis increases in size due to mitosis and then the new cuticle is produced. Enzymes secreted by the epidermal cells digest the old endocuticle, not affecting the old sclerotised exocuticle.
2. Ecdysis—this begins with the splitting of the old cuticle, usually starting in the midline of the thorax's dorsal side. The rupturing force is mostly from haemolymph pressure that has been forced into thorax by abdominal muscle contractions caused by the insect swallowing air or water. After this the insect wriggles out of the old cuticle.
3. Sclerotisation—after emergence the new cuticle is soft and this a particularly vulnerable time for the insect as its hard protective coating is missing. After an hour or two the exocuticle hardens and darkens. The wings expand by the force of haemolymph into the wing veins.

Insect Flight

Insects are the only group of invertebrates that have evolved wings and flight. Two insect groups, the dragonflies and the mayflies, have flight muscles attached directly to the wings. Other insects have the flight muscles attached to the thorax, making it oscillate and indirectly causing the wings to beat. Some very small insects make use not of steady-state aerodynamics but of the Weis-Fogh clap and fling mechanism, generating large lift forces at the expense of wear and tear on the wings. Many insects can hover, maintaining height and controlling their position. Some insects such as moths have the forewings coupled to the hindwings so these can work in unison.

A tau emerald (Hemicordulia tau) dragonfly has flight muscles attached directly to its wings.

Insects first flew in the Carboniferous, some 350 million years ago. Wings may have evolved from appendages on the sides of existing limbs, which already had nerves, joints, and muscles used for other purposes. These may initially have been used for sailing on water, or to slow the rate of descent when gliding.

Mechanisms

Direct Flight

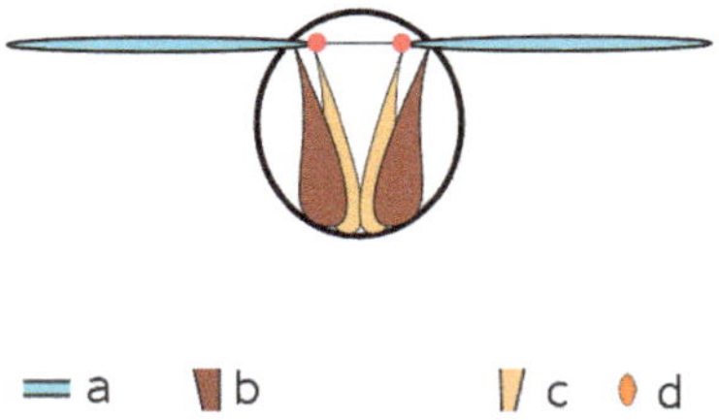

Direct flight: muscles attached to wings. Large insects only

The Odonata (like this Australian emperor dragonfly) have direct flight musculature.

Unlike other insects, the wing muscles of the Ephemeroptera (Mayflies) and Odonata (dragonflies and damselflies) insert directly at the wing bases, which are hinged so that a small movement of the wing base downward, lifts the wing itself upward, very much like rowing through the air.

Dragonflies and damselflies have fore and hind wings similar in shape and size. Each operates independently, which gives a degree of fine control and mobility in terms of the abruptness with which they can change direction and speed, not seen in other flying insects. This is not surprising, given that odonates are all aerial predators, and they have always hunted other airborne insects.

Indirect Flight

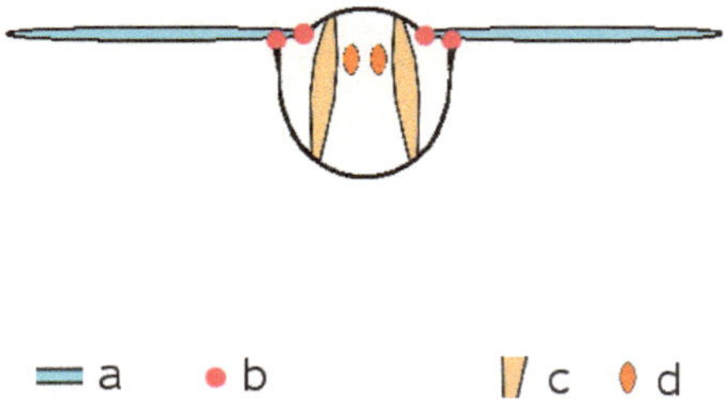

Indirect flight: muscles make thorax oscillate in most insects.

The Neoptera, including butterflies and most other insects, have indirect flight musculature.

Other than the two orders with direct flight muscles, all other living winged insects fly using a different mechanism, involving indirect flight muscles. This mechanism evolved once, and is the defining feature (synapomorphy) for the infraclass Neoptera; it corresponds, probably not coincidentally, with the appearance of a wing-folding mechanism, which allows Neopteran insects to fold the wings back over the abdomen when at rest (though this ability has been lost secondarily in some groups, such as in the butterflies).

In the higher groups with two functional pairs of wings, both pairs are linked together mechanically in various ways, and function as a single wing, although this is not true in the more primitive groups. There are also exceptions to be found among the more advanced Neoptera; the ghost moth is able to unlock its pair of wings and move them independently, allowing them to hover like dragonflies. What all Neoptera share, however, is the way the muscles in the thorax work: these muscles, rather than attaching to the wings, attach to the thorax and deform it; since the wings are extensions of the thoracic exoskeleton, the deformations of the thorax cause the wings to move as well. A set of *dorsal longitudinal muscles* compress the thorax from front to back, causing the dorsal surface of the thorax (notum) to bow upward, making the wings flip down. A set of *tergosternal muscles* pull the notum downward again, causing the wings to flip upward. In a few groups, the downstroke is accomplished solely through the elastic recoil of the thorax when the tergosternal muscles are relaxed. Several small sclerites at the wing base have other, separate, muscles attached and these are used for fine control of the wing base in such a way as to allow various adjustments in the tilt and amplitude of the wing beats. One of the final refinements that has appeared in some of the higher Neoptera (Coleoptera, Diptera, and Hymenoptera) is a type of muscular or neural control system whereby a single nerve impulse causes a muscle fiber to contract multiple times; this allows the frequency of wing beats to exceed the rate at which the nervous system can send impulses. This specialized form of muscle is termed *asynchronous flight muscle*. The overall effect is that many higher Neoptera can beat their wings much faster than insects with direct flight muscles.

Aerodynamics

There are two basic aerodynamic models of insect flight: creating a leading edge vortex, and using fling and clap.

Leading Edge Vortex

Most insects use a method that creates a spiralling leading edge vortex. These flapping wings move through two basic half-strokes. The downstroke starts up and back and is plunged downward and forward. Then the wing is quickly flipped over (supination) so that the leading edge is pointed backward. The upstroke then pushes the wing upward and backward. Then the wing is flipped again (pronation) and another downstroke can occur. The frequency range in insects with synchronous flight muscles typically is 5 to 200 hertz (Hz). In those with asynchronous flight muscles, wing beat frequency may exceed 1000 Hz. When the insect is hovering, the two strokes take the same amount of time. A slower downstroke, however, provides thrust.

Identification of major forces is critical to understanding insect flight. The first attempts to understand flapping wings assumed a quasi-steady state. This means that the air flow over the wing at any given time was assumed to be the same as how the flow would be over a non-flapping, steady-state wing at the same angle of attack. By dividing the flapping wing into a large number of motionless positions and then analyzing each position, it would be possible to create a timeline of the instantaneous forces on the wing at every moment. The calculated lift was found to be too small by a factor of three, so researchers realized that there must be unsteady phenomena providing aerodynamic forces. There were several developing analytical models attempting to approximate flow close to a flapping wing. Some researchers predicted force peaks at supination. With a dynamically scaled model of a fruit fly, these predicted forces later were confirmed. Others argued that the force peaks during supination and pronation are caused by an unknown rotational effect that fundamentally is different from the translational phenomena. There is some disagreement with this argument. Through computational fluid dynamics, some researchers argue that there is no rotational effect. They claim that the high forces are caused by an interaction with the wake shed by the previous stroke.

Similar to the rotational effect mentioned above, the phenomena associated with flapping wings are not completely understood or agreed upon. Because every model is an approximation, different models leave out effects that are presumed to be negligible. For example, the Wagner effect says that circulation rises slowly to its steady-state due to viscosity when an inclined wing is accelerated from rest. This phenomenon would explain a lift value that is less than what is predicted. Typically, the case has been to find sources for the added lift. It has been argued that this effect is negligible for flow with a Reynolds number that is typical of insect flight. The Wagner effect was ignored, consciously, in at least one recent model. One of the most important phenomena that occurs during insect flight is leading edge suction. This force is significant to the calculation of efficiency. The concept of leading edge suction first was put forth to describe vortex lift on sharp-edged delta wings. At high angles of attack, the flow separates over the leading edge, but reattaches before reaching the trailing edge. Within this bubble of separated flow is a vortex. Because the angle of attack is so high, a lot of momentum is transferred downward into the flow. These two features create a large amount of lift force as well as some additional drag. The important feature, however, is the lift. Because the flow has separated, yet it still provides large amounts of lift, this phenomenon is called *stall delay*. This effect was observed in flapping insect flight and it was proven to be capable of providing enough lift to account for the deficiency in the quasi-steady-state models. This effect is used by canoeists in a sculling draw stroke.

All of the effects on a flapping wing may be reduced to three major sources of aerodynamic phenomena: the leading edge vortex, the steady-state aerodynamic forces on the wing, and the wing's contact with its wake from previous strokes. The size of flying insects ranges from about 20 micrograms to about 3 grams. As insect body mass increases, wing area increases and wing beat frequency decreases. For larger insects, the Reynolds number (Re) may be as high as 10000. For smaller insects, it may be as low as 10. This means that viscous effects are much more important to the smaller insects, although the flow is still laminar, even in the largest fliers.

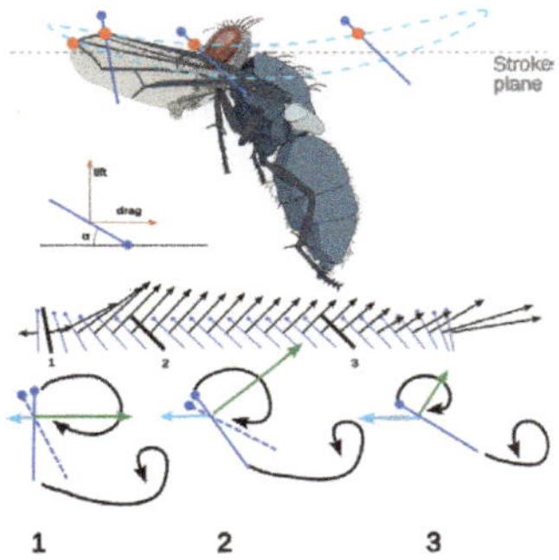

downstroke

Another interesting feature of insect flight is the body tilt. As flight speed increases, the insect body tends to tilt nose-down and become more horizontal. This reduces the frontal area and therefore, the body drag. Since drag also increases as forward velocity increases, the insect is making its flight more efficient as this efficiency becomes more necessary. Additionally, by changing the geometric angle of attack on the downstroke, the insect is able to keep its flight at an optimal efficiency through as many manoeuvres as possible. The development of general thrust is relatively small compared with lift forces. Lift forces may be more than three times the insect's weight, while thrust at even the highest speeds may be as low as 20% of the weight. This force is developed primarily through the less powerful upstroke of the flapping motion.

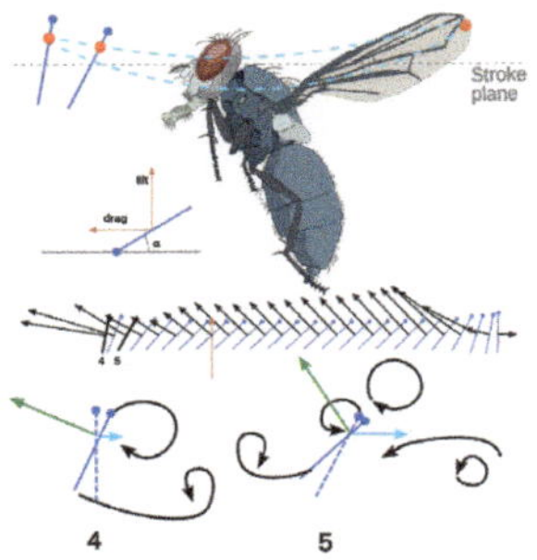

upstroke

Clap and Fling

The second method of flight, clap and fling, or the Weis-Fogh mechanism, functions differently. In this process, found in very small insects such as thrips and chalcid wasps, the wings clap together above the insect's body and then fling apart. As they fling open, the air gets sucked in and creates a vortex over each wing. This bound vortex then moves across the wing and, in the clap, acts as the starting vortex for the other wing. By this effect, circulation and thus, lift are increased to the extent of being higher, in most cases, than the typical leading edge vortex effect. One of the reasons this method is not employed by more insects is the inevitable damage and wear to the wings caused by repeated clapping. It is prevalent, however, among insects that are very small and experience low Reynolds numbers. The mechanism is also employed by the marine mollusc *Limacina helicina*, a sea butterfly.

In addition to the Reynolds number, there are at least two other relevant dimensionless parameters. A wing has three velocity scales: the flapping velocity with respect to the body (u), the forward velocity of the body (U_0), and the pitching velocity (Ωc). The ratios of them form two dimensionless variables, U_0/u and $\Omega c/u$, the former is often referred to as the advance ratio, and it is also related to the reduced frequency, fc/U_0.

If an insect wing is rigid, for example, a *Drosophila* wing is approximately so, its motion relative to a fixed body can be described by three variables: the position of the tip in spherical coordinates, (Θ(t),Φ(t)), and the pitching angle ψ(t), about the axis connecting the root and the tip. To estimate the aerodynamic forces based on blade-element analysis, it is also necessary to determine the angle of attack (α). The typical angle of attack at 70% wingspan ranges from 25° to 45° in hovering insects (15° in hummingbirds). Despite the wealth of data available for many insects, relatively few experiments report the time variation of α during a stroke. They include wind tunnel experiments of a tethered locust and a tethered fly, and free hovering flight of a fruit fly.

Because they are relatively easy to measure, the wing-tip trajectories have been reported more frequently. For example, selecting only flight sequences that produced enough lift to support a weight, will show that the wing tip follows an elliptical shape. Noncrossing shapes were also reported for other insects. Regardless of their exact shapes, the plugging-down motion indicates that insects may use aerodynamic drag in addition to lift to support its weight.

Elasticity

Insects gain kinetic energy, provided by the muscles, when they accelerate. When the wings begin to decelerate toward the end of the stroke, this energy must dissipate. During the downstroke, the kinetic energy is dissipated by the muscles themselves and is converted into heat (this heat is sometimes used to maintain core body temperature). Some insects are able to utilize the kinetic energy in the upward movement of the wings to aid in their flight. The wing joints of these insects contain a pad of elastic, rubber-like protein called resilin. During the upstroke of the wing, the resilin is stretched. The kinetic energy of the wing is converted into potential energy in the stretched resilin, which stores the energy much like a spring. When the wing moves down, this energy is released and aids in the downstroke.

Wing Coupling

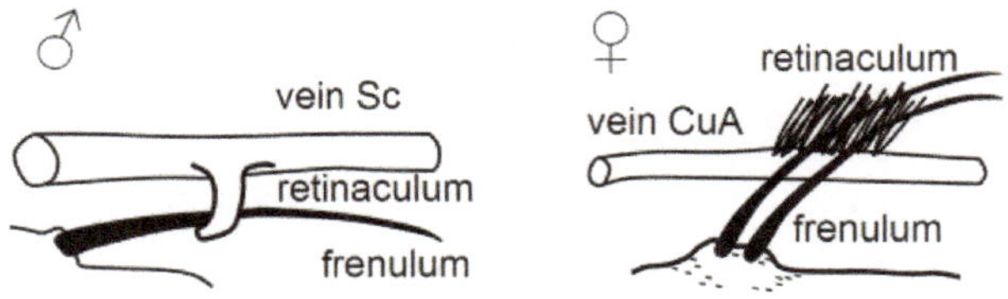

Frenulo-retinacular wing coupling in male and female moths

Some four-winged insect orders, such as the Lepidoptera, have developed a wide variety of morphological wing coupling mechanisms in the imago which render these taxa as "functionally dipterous". All, but the most basal forms, exhibit this wing-coupling.

The mechanisms are of three different types - jugal, frenulo-retinacular and amplexiform.

The more primitive groups have an enlarged lobe-like area near the basal posterior margin, i.e. at the base of the forewing, called *jugum*, that folds under the hindwing in flight.

Other groups have a frenulum on the hindwing that hooks under a retinaculum on the forewing.

In the butterflies (except the male of one species of hesperiid) and in the Bombycoidea (except the Sphingidae), there is no arrangement of frenulum and retinaculum to couple the wings. Instead, an enlarged humeral area of the hindwing is broadly overlapped by the forewing. Despite the absence of a specific mechanical connection, the wings overlap and operate in phase. The power stroke of the forewing pushes down the hindwing in unison. This type of coupling is a variation of frenate type but where the frenulum and retinaculum are completely lost.

Biochemistry

The biochemistry of insect flight has been a focus of considerable study. While many insects use carbohydrates and lipids as the energy source for flight, many beetles and flies use the amino acid proline as their energy source. Some species also use a combination of sources and moths such as *Manduca sexta* use carbohydrates for pre-flight warm-up.

Insects that beat their wings less than one hundred times a second use synchronous muscle. A Synchronous muscle is a type of muscle that contracts once for every nerve impulse, which is more efficient for fast flight. Insects that beat their wings more rapidly, such as the bumblebee, use asynchronous muscle; this is a type of muscle that contracts more than once per nerve impulse. This is achieved by the muscle being stimulated to contract again by a release in tension in the muscle, which can happen more rapidly than through simple nerve stimulation alone.

Evolution and Adaptation

Hoverfly (Xanthogramma pedissequum) has indirect flight musculature.

Sometime in the Carboniferous Period, some 350 million years ago, when there were only two major land masses, insects began flying. How and why insect wings developed is not well understood, largely due to the scarcity of appropriate fossils from the period of their development in the Lower Carboniferous. Three main theories on the origins of insect flight are that wings developed from paranotal lobes, extensions of the thoracic terga; that they are modifications of movable abdominal gills as found on aquatic naiads of mayflies; or that they developed from thoracic protrusions used as radiators.

Paranotal Hypothesis

The paranotal hypothesis suggests that the insect's wings developed from paranotal lobes, a preadaptation found in insect fossils that is believed to have assisted stabilization while hopping or falling. In favor of this hypothesis is the tendency of most insects, when startled while climbing on branches, to escape by dropping to the ground. Such lobes would have served as parachutes and enable the insect to land more softly. The theory suggests that these lobes gradually grew larger and in a later stage developed a joint with the thorax. Even later would appear the muscles to move these crude wings. This model implies a progressive increase in the effectiveness of the wings, starting with parachuting, then gliding and finally active flight. Still, lack of substantial fossil evidence of the development of the wing joints and muscles poses a major difficulty to the theory, as does the seemingly spontaneous development of articulation and venation, and it has been largely rejected by experts in the field.

Epicoxal Hypothesis

Some entomologists have suggested that a possible origin for insect wings might have been movable abdominal gills found in many aquatic insects, such as on naiads of mayflies. According to this theory these tracheal gills, which started their way as exits of the respiratory system and over time were modified into locomotive purposes, eventually developed into wings. The tracheal gills are equipped with little winglets that perpetually vibrate and have their own tiny straight muscles.

Endite-exite Hypothesis

The hypothesis with perhaps the strongest evidence is that which stems from the adaptation of endites and exites, appendages on the respective inner and outer aspects of the primitive arthropod limb. This was advanced by Trueman based on a study by Goldschmidt in 1945 on *Drosophila melanogaster*, in which a *pod* variation displayed a mutation transforming normal wings to what was interpreted as a triple-jointed leg arrangement with some additional appendages but lacking the tarsus, where the wing's costal surface normally would be. This mutation was reinterpreted as strong evidence for a dorsal exite and endite fusion, rather than a leg, with the appendages fitting in much better with this hypothesis. The innervation, articulation and musculature required for the evolution of wings are already present in podomeres.

Other Hypotheses

Suggestions have been made that wings may have evolved initially for sailing on the surface of water as seen in some stoneflies. An alternative idea is that it drives from directed aerial gliding descent—a preflight phenomena found in some apterygote, a wingless sister taxa to the winged insects.

Insect Olfaction

Insect olfaction refers to the function of chemical receptors that enable insects to detect and identify volatile compounds for foraging, predator avoidance, finding mating partners (via pheromones) and locating oviposition habitats. Thus, it is the most important sensation for insects. Most important insect behaviors must be timed perfectly which is dependent on what they smell and when they smell it. For example olfaction is essential for hunting in many species of wasps, including *Polybia sericea*.

The two organs insects primarily use for detecting odors are the antennae and specialized mouth parts called the maxillary palps. Inside of these olfactory organs there are neurons called olfactory receptor neurons which, as the name implies, house receptors for scent molecules in their cell membrane.The majority of olfactory receptor neurons typically reside in the antenna. These neurons can be very abundant, for example Drosophila flies have 2,600 olfactory sensory neurons.

Insects are capable of smelling and differentiating between thousands of volatile compounds both sensitively and selectively. Sensitively is how attune the insect is to very small amounts of an odorant or small changes in the concentration of an odorant. Selectivity refers to the insects ability to tell one odorant apart from another. These compounds are commonly broken into three classes short chain carboxylic acids, aldehydes and low molecular weight nitrogenous compounds.

Insects have been used as a model system to study mammal and especially human olfaction. Yet, unlike vertebrates who use G protein coupled receptors (GPCRs) insects ORs (olfactory receptors), GRs (gustatory receptors) and IRs (ionotropic receptors) are all heteromeric ligand-gated ion channels. Much like in vertebrates, axons from the sensory neurons converge into glomeruli. But they differ in where the glomeruli are housed. In mammals they are located in the olfactory bulbs but are in the antennal lobe of insects.

Details of Olfactory Cascade

Sensory neurons in the antenna generate odor-specific electrical signals called spikes (action potentials) in response to binding of odors to cell surface proteins like the insect pheromone-binding protein. The sensory neurons send this information via their axons to the antennal lobe. Inside the antennal lobe they synapse with other neurons in semidelineated (with membrane boundaries) structures called glomeruli.

Specifically the process is as follows: first the odorant wafts towards an insects antenna or maxillary palp which is covered with hair like projections called sensilla. The odorant then enters through tiny pours in the exoskeleton (or cuticle) of that sensillum and diffuses into the fluid between the cells called extracellular fluids. There the odorant molecule binds to a odorant binding protein which transports it to a receptor and co-recepor (Orco) team on the surface of the olfactory receptor neuron (ORN). This leads to the neuron firing an action potential down the axon. This signal is sent to the antennal lobe of the insects brain where it can then integrate the information with other signals from other sensilla.

These ORNs are bipolar, on one end are the olfactory dendrites with the receptors for the odors and on the other end are the axons that carry the action potential to the antennal lobe of the brain. The antennal lobes have two kinds of neurons, projection neurons (mostly excitatory) and local neurons (inhibitory, with some excitatory). The projection neurons send their axon terminals to a part of the insect brain called the mushroom bodies (important in regulating learned odor responses) and another part of the brain called the lateral horn (important in regulating innate odor responses). Both of these regions are part of the protocerebrum of the insect brain.

How Insect Olfaction is Studied

Action potential recordings are conducted in three different ways electroantenograms, electropalpograms, and single sensillum recordings (SSR). In electroantenograms (EAG) and electropalpograms (EPG) the action potentials from the entire antenna or maxillary palp, respectively, is recorded. EAGs and EPGs provide an overall view of olfaction in the respective organ. During a SSR an electrode is inserted into just one sensillum and the recording is made from only the ORNs which are contained within that sensillum, providing more detailed information·

Any of these methods can be combined with a high resolution gas chromatography to isolate volatile compounds from important animals or habitats. For example this method could be used to determine which compound from a particular flower is the most attractive to a bee. Recordings from projection neurons show that in some insects there is strong specialization and discrimination for the odors presented by the ORNs. This is especially true for the projection neurons of the macroglomeruli, a specialized complex of glomeruli responsible for the pheromones detection.

Repellents and Attractants

Humans exploit the insect olfactory system to control agricultural and disease carrying pests. For some agricultural pests manufactured sex pheromones are placed in traps to capture adults before they can oviposit (lay their eggs) leading to the hatching of their destructive larvae. While there are thousands of chemicals insects can detect there is a limited range that insects use as cues to move towards or away from the source of the odorant.

The art of finding an attractant or repellent for a particular insect of interest is complicated and a long, intensive process. For example, using pharamones only attracts insects in their reproductive stage, a short period in their lives. While scents of food may be attractive to hungry insects they would not be effective in a field full of a crop that is palatable to that insect.

Situationally Dependent Attractants / Repellents

Insects use the same signal for many different uses depending on the situation this is called chemical parsimony. Situations that may change how an insect behaves in reaction to a scent are things like the concentration of the compound, the life stage of the insect, its mating status, other olfactory cues, the insects feeding state (hungry or full), the time of day, or even the insects body position. For example, Drosophila are very attracted to apple cider vinegar but in very high concentrations an additional olfactory receptor (that has low affinity for the vinegar, Or85a) is activated which changes the fly's behavior from attraction to aversion. These different behaviors to the same cue is called behavioral plasticity.

Carbon Dioxide

Many insects are capable of detecting very minute changes in the concentration of CO_2. While CO_2 has been found to be an attractant in every arthropod studied and it is very important in mosquito monitoring and control, even this stereotyped reaction can be plastic. Drosophila avoid CO_2 when walking but move towards it when in flight.

Repellants (DEET)

Many insects (and other arthropods) have been shown to avoid areas containing N,N-diethyl-3-methylbenzamide or DEET. They innately avoid DEET, likely because it is a "confusant" that stimulates gustatory, ionotropic, and olfactory receptors and "distorts" other odorants interaction with those receptors.

Pupa

Pupa of the rose chafer beetle, Cetonia aurata

A pupa is the life stage of some insects undergoing transformation. The pupal stage is found only in holometabolous insects, those that undergo a complete metamorphosis, going through four life stages: embryo, larva, pupa and imago.

Tumbler (pupa) of a mosquito. Unlike most pupae, tumblers can swim around actively.

The pupae of different groups of insects have different names such as *chrysalis* for the pupae of butterflies and *tumbler* for those of the mosquito family. Pupae may further be enclosed in other structures such as cocoons, nests or shells.

Position in Life Cycle

In the life of an insect the pupal stage follows the larval stage and precedes adulthood (*imago*). It is during the time of pupation that the adult structures of the insect are formed while the larval structures are broken down. Pupae are inactive, and usually sessile (not able to move about). However, the pupae may be exarate (have movable legs, wings, antennae, etc.) or obtect (covered in a hard case with the legs and wings attached immovably against the body). They have a hard protective coating and often use camouflage to evade potential predators.

Duration

Pupation may last weeks, months or even years. For example, it is two weeks in monarch butterflies. The pupa may enter dormancy or diapause until the appropriate season for the adult insect. In temperate climates pupae usually stay dormant during winter, while in the tropics pupae usually do so during the dry season. Anise swallowtails sometimes emerge after years as a chrysalis.

Emergence

Eclosion of Papilio dardanus

Insects emerge (eclose) from pupae by splitting the pupal case, and the whole process of pupation is controlled by the insect's hormones. Most butterflies emerge in the morning. In mosquitoes the emergence is in the evening or night. In fleas the process is triggered by vibrations that indicate the possible presence of a suitable host. Prior to emergence, the adult inside the pupal exoskeleton is termed *pharate*. Once the pharate adult has eclosed from the pupa, the empty pupal exoskeleton is called an *exuvia*; in most hymenopterans (ants, bees and wasps) the exuvia is so thin and membranous that it becomes "crumpled" as it is shed.

Pupal Mating

Mating in pierid *Catopsilia pyranthe* of male with newly emerged female.

In a few taxa of the Lepidoptera, especially *Heliconius*, pupal mating is an extreme form of reproductive strategy where adult males mate with female pupa about to emerge or with the newly moulted female; this is accompanied by other actions such as capping of the reproductive system of the female with the sphragis, denying access to other males, or by exuding an anti-aphrodisiac pheromone.

Defense

Pupae are usually immobile and are largely defenseless. To overcome this, a common feature is concealed placement. There are some species of Lycaenid butterflies who are protected in their pupal stage by ants. Another means of defense by pupae of other species is the capability of making sounds or vibrations to scare potential predators. A few species use chemical defenses including toxic secretions. The pupae of social hymenopterans are protected by adult members of the hive.

Types

Based on the presence or absence of articulated mandibles that are employed in emerging from a cocoon or pupal case, the pupae can be classified in to two types:

- Decticous pupa - pupae with articulated mandibles. Examples are pupae of the orders Neuroptera, Mecoptera, Trichoptera and few Lepidoptera families.
- Adecticous pupa - pupae without articulated mandibles. Examples include orders Strepsiptera, Coleoptera, Hymenoptera, Diptera and Siphonaptera.

Based on whether the pupal appendages are free or attached to the body, the pupae can be classified in three types:

- Exarate pupa - appendages are free and are not usually encapsulated within a cocoon. All decticous pupa and some adecticous pupa are always exarate. (Neuroptera, Trichoptera, Cyclorrhapha of Dipterans, Siphonaptera, most Coleoptera, Hymenoptera, and few Lepidoptera).
- Obtect pupa - appendages are attached closely to the body and are commonly encapsulated within a cocoon. Some adecticous pupa are obtect forms. (Most Lepidoptera, Nematocera and Brachycera of Dipterans, Staphylinidae and Chrysomelidae Coleopterans, many Chalcidoidea Hymenopterans)
- Coarctate pupa - enclosed in a hardened cuticle of the penultimate larval instar called puparium. However, the pupa itself is of exarate adecticous pupa forms. (Cyclorrhapha of Dipterans).

Chrysalis

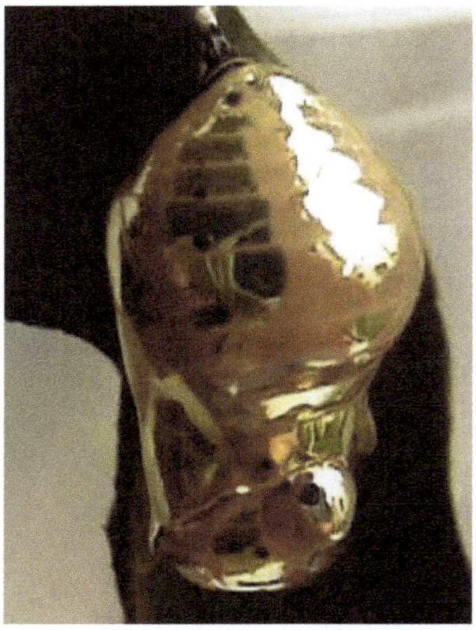

Common crow butterfly (Euploea core) chrysalis illustrating the Greek origin of the term: (*chrysós*) for gold

A chrysalis or nympha is the pupal stage of butterflies. The term is derived from the metallic gold-coloration found in the pupae of many butterflies.

When the caterpillar is fully grown, it makes a button of silk which it uses to fasten its body to a leaf or a twig. Then the caterpillar's skin comes off for the final time. Under this old skin is a hard skin called a chrysalis.

Because chrysalides are often showy and are formed in the open, they are the most familiar examples of pupae. Most chrysalides are attached to a surface by a Velcro-like arrangement of a silken pad spun by the caterpillar, usually cemented to the underside of a perch, and the *cremastral* hook or hooks protruding from the rear of the chrysalis or *cremaster* at the tip of the pupal abdomen by which the caterpillar fixes itself to the pad of silk. (Gr. *kremastos* 'suspended')

Like other types of pupae, the chrysalis stage in most butterflies is one in which there is little movement. However, some butterfly pupae are capable of moving the abdominal segments to produce sounds or to scare away potential predators. Within the chrysalis, growth and differentiation occur. The adult butterfly emerges (ecloses) from this and expands its wings by pumping haemolymph into the wing veins. Although this sudden and rapid change from pupa to imago is often called metamorphosis, metamorphosis is really the whole series of changes that an insect undergoes from egg to adult.

When emerging, the butterfly uses a liquid, sometimes called cocoonase, which softens the shell of the chrysalis. Additionally, it uses two sharp claws located on the thick joints at the base of the forewings to help make its way out. Having emerged from the chrysalis, the butterfly will usually sit on the empty shell in order to expand and harden its wings. However, if the chrysalis was near the ground (such as if it fell off from its silk pad), the butterfly would find another vertical surface to rest upon and harden its wings (such as a wall or fence).

Moth pupae are usually dark in color and either formed in underground cells, loose in the soil, or their pupa is contained in a protective silk case called a *cocoon*.

It is important to differentiate between pupa, chrysalis and cocoon. The pupa is the stage between the larva and adult stages. The chrysalis generally refers to a butterfly pupa although the term maybe misleading as there are some moths whose pupae resembles a chrysalis, eg: the plume winged moths of the family Pterophoridae and some geometrid moths. A cocoon is a silk case that moths, and sometimes other insects, spin around the pupa.

Cocoon

The tough brown cocoon of an emperor gum moth

A cocoon is a casing spun of silk by many moth caterpillars, and numerous other holometabolous insect larvae as a protective covering for the pupa.

Cocoons may be tough or soft, opaque or translucent, solid or meshlike, of various colors, or composed of multiple layers, depending on the type of insect larva producing it. Many moth caterpillars shed the larval hairs (setae) and incorporate them into the cocoon; if these are urticating hairs then the cocoon is also irritating to the touch. Some larvae attach small twigs, fecal pellets or pieces of vegetation to the outside of their cocoon in an attempt to disguise it from predators. Others spin their cocoon in a concealed location—on the *underside* of a leaf, in a crevice, down near the base of a tree trunk, suspended from a twig or concealed in the leaf litter.

The silk in the cocoon of the silk moth can be unravelled to get silk fibre which makes this moth the most economically important of all Lepidopterans. The silk moth is the only completely domesticated Lepidopteran and does not exist in the wild.

Insects that pupate in a cocoon must escape from it, and they do this either by the pupa cutting its way out, or by secreting fluids, sometimes called cocoonase, that soften the cocoon. Some cocoons are constructed with built-in lines of weakness along which they will tear easily from inside, or with exit holes that only allow a one-way passage out; such features facilitate the escape of the adult insect after it emerges from the pupal skin.

Puparium

Some pupae remain inside the exoskeleton of the final larval instar and this last larval "shell" is called a puparium (plural, puparia). Flies of the group Muscomorpha have puparia, as do members of the order Strepsiptera, and the Hemipteran family Aleyrodidae.

Drosophila Embryogenesis

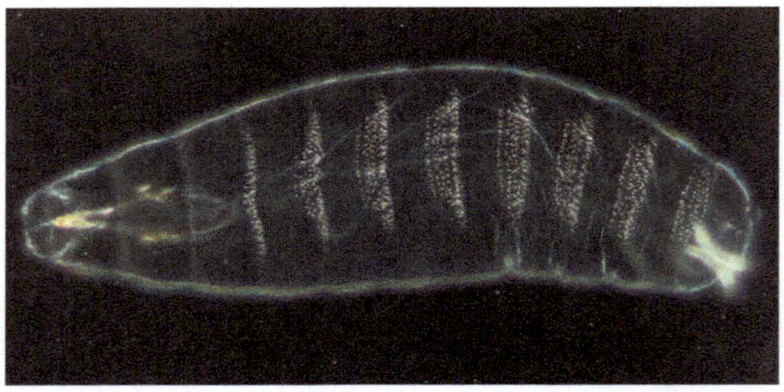

Ventral view of repeating denticle bands on the cuticle of a 22-hour-old embryo. The head is on the left.

Drosophila embryogenesis, the process by which *Drosophila* (fruit fly) embryos form, is a favorite model system for genetics and developmental biology. The study of its embryogenesis unlocked the century-long puzzle of how development was controlled, creating the field of evolutionary developmental biology. The small size, short generation time, and large brood size make it ideal for genetic studies. Transparent embryos facilitate developmental studies. *Drosophila melanogaster* was introduced into the field of genetic experiments by Thomas Hunt Morgan in 1909.

Life Cycle

Drosophila display a holometabolous method of development, meaning that they have three distinct stages of their post-embryonic life cycle, each with a radically different body plan: larva, pupa and finally, adult. The machinery necessary for the function and smooth transition between these three phases develops during embryogenesis. During embryogenesis, the larval stage fly will develop and hatch at a stage of its life known as the first larval instar. Cells that will produce adult structures are put aside in imaginal discs. During the pupal stage, the larval body breaks down as the imaginal disks grow and produce the adult body. This process is called *complete metamorphosis*. About 24 hours after fertilization, an egg hatches into a larva, which undergoes three molts taking about 5.5 to 6 days, after which it is called a pupa. The pupa metamorphoses into an adult fly, which takes about 3.5 to 4.5 days. The entire growth process from egg to adult fly takes an estimated 10 to 12 days to complete at 25 °C.

The mother fly produces oocytes that already have anterior-posterior and dorsal-ventral axes defined by maternal activities.

Embryogenesis in *Drosophila* is unique among model organisms in that cleavage occurs in a multinucleate syncytium (strictly a coenocyte). Early on, 256 nuclei migrate to the perimeter of the egg, creating the syncytial blastoderm. The germ line segregates from the somatic cells through the formation of pole cells at the posterior end of the embryo. After thirteen mitotic divisions and about 4 hours after fertilization, an estimated 6,000 nuclei accumulate in the unseparated cytoplasm of the oocyte before they migrate to the surface and are encompassed by plasma membranes to form cells surrounding the yolk sac producing a cellular blastoderm.

Like other triploblastic metazoa, gastrulation leads to the formation of three germ layers: the endoderm, mesoderm, and ectoderm. The mesoderm invaginates from the ventral furrow (VF), as does the ectoderm that will give rise to the midgut. The pole cells are internalized by a different route.

Germ band elongation involves many rearrangements of cells, and the appearance of distinct differences in the cells of the three germ bands and various regions of the embryo. The posterior region (including the hindgut) expands and extends towards the anterior pole along the dorsal side of the embryo. At this time, segments of the embryo become visible, creating a striped arrangement along the anterior-posterior axis. The earliest signs of segmentation appear during this phase with the formation of *parasegmental furrows*. This is also when the tracheal pits form, the first signs of structures for breathing.

Germ band retraction returns the hindgut to the dorsal side of the posterior pole and coincides with overt segmentation. The remaining stages involve the internalization of the nervous system (ectoderm) and the formation of internal organs (mainly mesoderm).

Anterior-posterior Axis Patterning in Drosophila

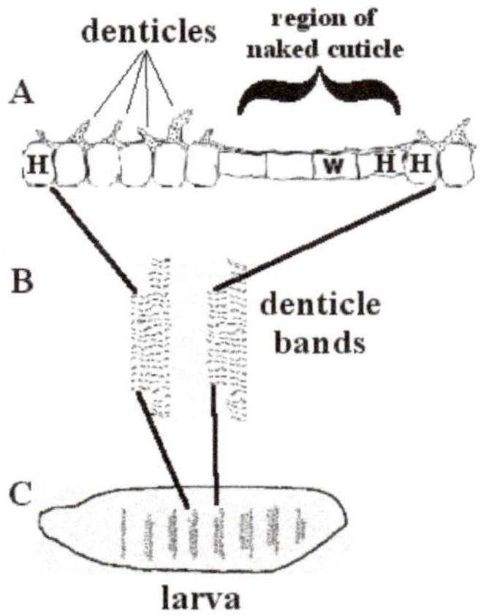

The abdominal cuticular segments of the *Drosophila* embryo consist of repeating denticle bands separated by naked cuticle.

One of the best understood examples of pattern formation is the patterning along the future head to tail (antero-posterior) axis of the fruit fly *Drosophila melanogaster*. There are three fundamental types of genes that give way to the developmental structure of the fly: maternal effect genes, segmentation genes, and homeotic genes. The development of *Drosophila* is particularly well studied, and it is representative of a major class of animals, the insects or insecta. Other multicellular organisms sometimes use similar mechanisms for axis formation, although the relative importance of signal transfer between the earliest cells of many developing organisms is greater than in the example described here.

Maternal Effect Genes

The building-blocks of anterior-posterior axis patterning in *Drosophila* are laid out during egg formation (oogenesis), well before the egg is fertilized and deposited. The maternal effect genes are responsible for the polarity of the egg and of the embryo. The developing egg (oocyte) is polarized by differentially localized mRNA molecules.

The genes that code for these mRNAs, called maternal effect genes, encode for proteins that get translated upon fertilization to establish concentration gradients that span the egg. Bicoid and Hunchback are the maternal effect genes that are most important for patterning of anterior parts (head and thorax) of the Drosophila embryo. Nanos and Caudal are maternal effect genes that are important in the formation of more posterior abdominal segments of the Drosophila embryo.

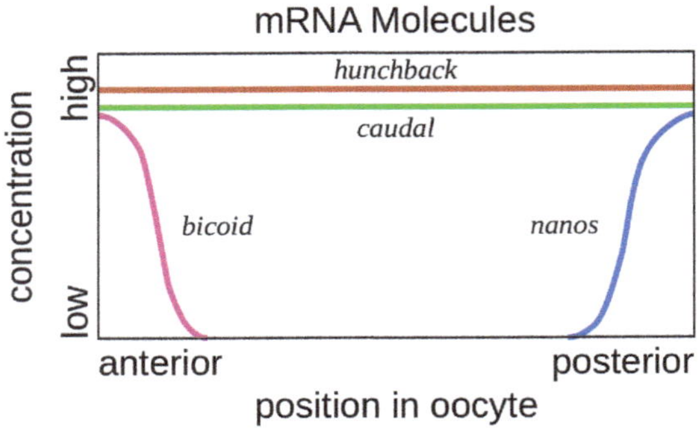

Figure 1. mRNA distributions

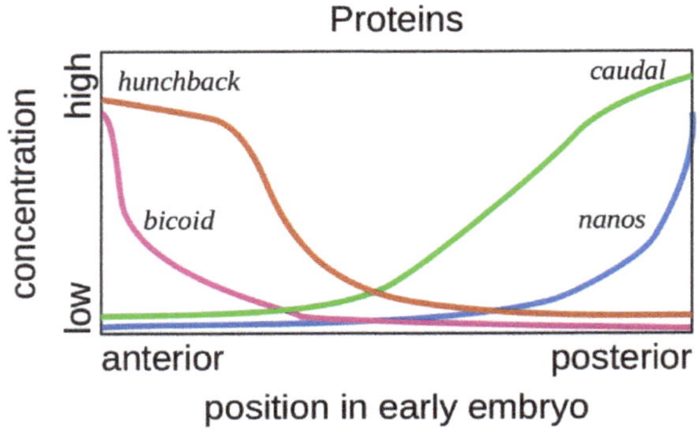

Figure 2. Protein distributions

In embryos from *bicoid* mutant mothers, the head and thoracic structures are converted to the abdomen making the embryo with posterior structures on both ends, a lethal phenotype.

Cytoskeletal elements such as microtubules are polarized within the oocyte and can be used to allow the localization of mRNA molecules to specific parts of the cell. Maternally synthesized *bicoid* mRNAs attach to microtubules and are concentrated at the anterior ends of forming *Drosophila* eggs. In unfertilized eggs, transcripts are still strictly localized at the tip, but immediately after fertilization, a small mRNA gradient is formed in the anterior 20% of the eggs. Another report documents a mRNA gradient up to 40%. *nanos* mRNA also attaches to a *Drosophila* egg's cytoskeleton but is concentrated at the posterior end of the egg. *hunchback* and *caudal* mRNAs lack special location control systems and are fairly evenly spread throughout the entire interior of the egg cells.

It has been shown that the dsRNA-binding protein STAUFEN (STAU1) is responsible for guiding bicoid, nanos and other proteins, which play a role in forming the anterior-posterior axis, to the correct regions of the embryo to build gradients. When the mRNAs from the maternal effect genes are translated into proteins, a Bicoid protein gradient forms at the anterior end of the egg. Nanos protein forms a gradient at the posterior end. The Bicoid protein blocks translation of *caudal* mRNA so Caudal protein is of lower concentration at the anterior part of the embryo and at higher concentration at the posterior part of the embryo. This is of opposite direction of the Bicoid protein. The caudal protein then activates later to turn genes on to form the posterior structures during the segmentation phase. Nanos protein creates a posterior-to-anterior slope and is a morphogen that helps in abdomen formation. Nanos protein, in complex with Pumilio protein, binds to the *hunchback* mRNA and blocks its translation in the posterior end of *Drosophila* embryos.

The Bicoid, Hunchback, and Caudal proteins are transcription factors. The Bicoid protein is a morphogen as well. The Nanos protein is a translational repressor protein. Bicoid has a DNA-binding homeodomain that binds both DNA and the *nanos* mRNA. Bicoid binds a specific RNA sequence in the 3' untranslated region, called the Bicoid 3'-UTR regulatory element, of *caudal* mRNA and blocks translation.

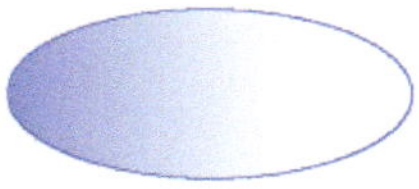

Fig 3. bicoid mRNA + protein gradient

Hunchback protein levels in the early embryo are significantly augmented by new *hunchback* gene transcription and translation of the resulting zygotically produced mRNA. During early *Drosophila* embryogenesis, there are nuclear divisions without cell division. The many nuclei that are produced distribute themselves around the periphery of the cell cytoplasm. Gene expression in these nuclei is regulated by the Bicoid, Hunchback, and Caudal proteins. For example, Bicoid acts as a transcriptional activator of *hunchback* gene transcription. In order for development to continue, Hunchback is needed in an area that is declining in amount from anterior to posterior. This is created by the Nanos protein whose existence is at a declining slope from posterior to anterior ends.

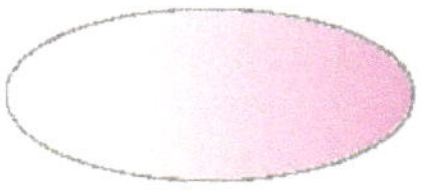

Fig 4. Nanos protein gradient

Gap Genes

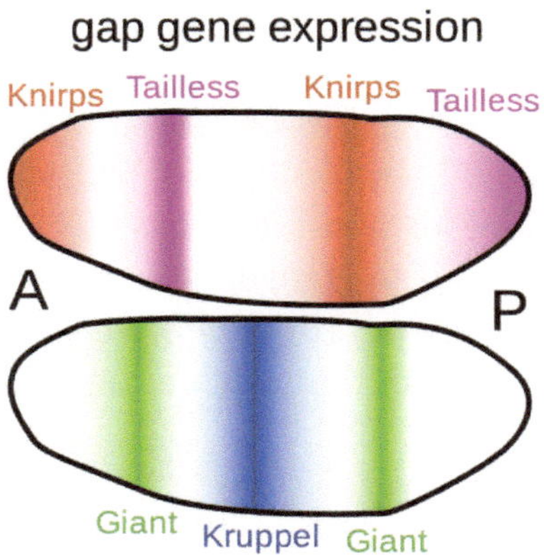

Figure 5. Gap genes

The other important function of the gradients of Bicoid, Hunchback, and Caudal proteins is in the transcriptional regulation of other zygotically expressed proteins. Many of these are the protein products derived from members of the "gap" family of developmental control genes. *giant, huckebein, hunchback, knirps, Krüppel* and *tailless* are all gap genes. Their expression patterns in the early embryo are determined by the maternal effect gene products and shown in the diagrams on the right side of this page. The gap genes are part of a larger family called the segmentation genes. These genes establish the segmented body plan of the embryo along the anterior-posterior axis. The segmentation genes specify 14 *parasegments* that are closely related to the final anatomical segments. The gap genes are the first layer of a hierarchical cascade of the segmentation control genes.

Additional Segmentation Genes

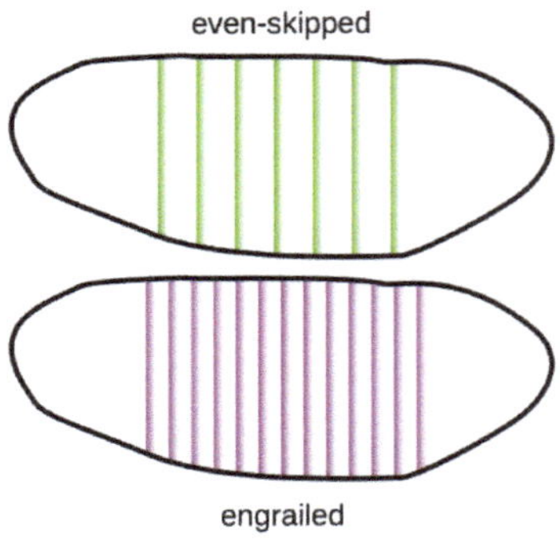

Figure 6. Pair rule

Two additional classes of segmentation genes are expressed after the gap gene products. The pair-rule genes are expressed in striped patterns of seven bands perpendicular to the anterior-posterior axis (see Figure 6, *even-skipped*). These patterns of expression are established within the syncytial blastoderm. After these initial patterning events, cell membranes form around the nuclei of the syncytial blastoderm converting it to a cellular blastoderm.

The expression patterns of the final class of segmentation genes, the segment polarity genes, are then fine-tuned by interactions between the cells of adjacent parasegments. The Engrailed protein is a transcription factor (yellow in Figure 7) that is expressed in one row of cells at the edge of each parasegment. This expression pattern is initiated by the pair-rule genes (like *even-skipped*) that code for transcription factors that regulate the *engrailed* gene's transcription in the syncytial blastoderm.

Cells that make Engrailed can make the cell-to-cell signaling protein Hedgehog (green in Figure 7). The motion of Hedgehog is limited by its lipid modification, and so Hedgehog activates a thin stripe of cells anterior to the Engrailed-expressing cells. Only cells to one side of the Engrailed-expressing cells are competent to respond to Hedgehog because they express the receptor protein Patched (blue in Figure 7). Cells with activated Patched receptor make the Wingless protein (red in Figure 7). Wingless is a secreted protein that acts on the adjacent rows of cells by activating its cell surface receptor, Frizzled.

Wingless acts on Engrailed-expressing cells to stabilize Engrailed expression after the cellular blastoderm forms. The Naked cuticle protein is induced by Wingless to limit the number of rows of cells that express Engrailed. The short-range, reciprocal signaling by Hedgehog and Wingless, held in check by the Patched and Naked proteins, stabilizes the boundary between each segment. The Wingless protein is called "wingless" because of the phenotype of some *wingless* mutants. Wingless and Hedgehog also function in multiple tissues later in embryogenesis and also during metamorphosis.

The transcription factors that are coded for by segmentation genes regulate yet another family of developmental control genes, the homeotic selector genes. These genes exist in two ordered groups on *Drosophila* chromosome 3. The order of the genes on the chromosome reflects the order that they are expressed along the anterior-posterior axis of the developing embryo. The Antennapedia group of homeotic selector genes includes *labial*, *antennapedia*, *sex combs reduced*, *deformed*, and *proboscipedia*. Labial and Deformed proteins are expressed in head segments where they activate the genes that define head features. Sex-combs-reduced and Antennapedia specify the properties of thoracic segments. The bithorax group of homeotic selector genes control the specializations of the third thoracic segment and the abdominal segments. Mutations in some homeotic genes can often be lethal and the cycle of life will end at embryogenesis.

In 1995, the Nobel Prize for Physiology or Medicine was awarded for studies concerning the genetic control of early embryonic development to Christiane Nüsslein-Volhard, Edward B. Lewis and Eric Wieschaus. Their research on genetic screening for embryo patterning mutants revealed the role played in early embryologic development by *homeobox genes* like *bicoid*. An example of a homeotic mutation is the so-called Antennapedia mutation. In *Drosophila*, antennae and legs are created by the same basic "program", they only differ in a single transcription factor. If this transcription factor is damaged, the fly grows legs on its head instead of antennae. See images of this "antennapedia" mutant and others, at FlyBase. Another example is in the bithorax complex. If nonlethal mutations occur in this complex, it can cause the fly to have two sets of wings, instead of one pair of wings and one pair of halteres, which aid in balance in flight.

Dorsal-ventral Axis

Formation of the dorsal-ventral axis is dependent on the ventral nuclear concentration of a maternally synthesized transcription factor called Dorsal. The determination of the dorsal side of the embryo occurs during oogenesis when the oocyte nucleus moves along microtubules from the posterior to the anterior-dorsal margin of the oocyte. The nucleus expresses a protein called Gurken which is secreted locally and thus only activates follicle cells in the dorsal region by interacting with the Torpedo receptor. This inhibits the production of Pipe protein and thus follicular cells expressing Pipe are on the ventral side. Pipe activates an extracellular protease cascade in the perivitelline space between the follicle cells and the egg which results in the cleavage of the Toll-ligand Spätzle and activation of the Toll signaling cascade on the ventral side. Dorsal protein is present throughout embryonic cytoplasm but bound to Cactus which prevents it from translocating to the nucleus. Toll signaling results in the degradation of Cactus which allows Dorsal to enter the nuclei on the ventral side of the blastoderm. Overall, a difference in the localization of the oocyte nucleus becomes a difference in the signaling state of the surrounding follicle cells which then signal to the resulting blastoderm nuclei.

Once in the nucleus, Dorsal activates different genes depending upon its nuclear concentration. This process sets up a gradient between the ventral and dorsal side of the blastoderm embryo with the repression or induction of Dorsal target genes being differentially regulated. At the ventral end of the embryo, blastoderm nuclei exposed to high concentrations of dorsal protein induce the transcription of the transcription factors *twist* and *snail* while repressing *zerknüllt* and *decapentaplegic*. This results in the formation of the mesoderm. In the lateral regions of the embryo, low nuclear concentrations of Dorsal lead to the expression of *rhomboid* which identifies future neuroectoderm. More dorsally, active Dpp signaling represses *rhomboid* thus confining it to the lateral blastoderm nuclei. At the dorsal side of the embryo, blastoderm nuclei where this is little or no nuclear dorsal protein express *zerknüllt, tolloid,* and *decapentaplegic* (Dpp). This leads to the specification of non-neural ectoderm and later in the blastula stage to anmioserosa. The ventral activity of the TGF-β family signaling protein Dpp is maintained by the expression of the secreted Dpp-agonist Sog (short gastrulation) in the neuroectoderm. Sog binds to and prevents Dpp from diffusing to the ventral side of the embryo and through the cleavage of Sog by Tolloid also enables a sharpening of the Dpp gradient on the dorsal side. The DV axis of *Drosophila* is due to the interaction of two gradients – a ventral concentration of nuclear Dorsal and a dorsal concentration of Dpp activity.

Insect Reproductive System

Most insects reproduce oviparously, i.e. by laying eggs. The eggs are produced by the female in a pair of ovaries. Sperm, produced by the male in one testis or more commonly two, is transmitted to the female during mating by means of external genitalia. The sperm is stored within the female in one or more spermathecae. At the time of fertilization, the eggs travel along oviducts to be fertilized by the sperm and are then expelled from the body ("laid"), in most cases via an ovipositor.

Internal

Female

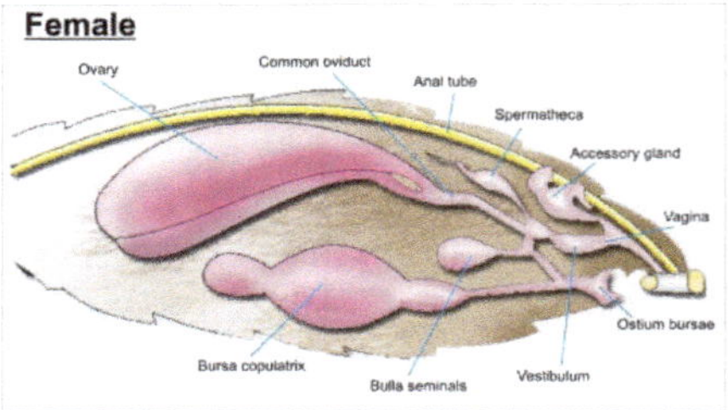

The female genitalia of Lepidoptera

Female insects are able to make eggs, receive and store sperm, manipulate sperm from different males, and lay eggs. Their reproductive systems are made up of a pair of ovaries, accessory glands, one or more spermathecae, and ducts connecting these parts. The ovaries make eggs and accessory glands produce the substances to help package and lay the eggs. Spermathecae store sperm for varying periods of time and, along with portions of the oviducts, can control sperm use. The ducts and spermathecae are lined with a cuticle.

The ovaries are made up of a number of egg tubes, called ovarioles, which vary in size and number by species. The number of eggs that the insect is able to make varies according to the number of ovarioles, with the rate at which eggs develop being also influenced by ovariole design. In meroistic ovaries, the eggs-to-be divide repeatedly and most of the daughter cells become helper cells for a single oocyte in the cluster. In panoistic ovaries, each egg-to-be produced by stem germ cells develops into an oocyte; there are no helper cells from the germ line. Production of eggs by panoistic ovaries tends to be slower than that by meroistic ovaries.

Accessory glands or glandular parts of the oviducts produce a variety of substances for sperm maintenance, transport, and fertilization, as well as for protection of eggs. They can produce glue and protective substances for coating eggs or tough coverings for a batch of eggs called oothecae. Spermathecae are tubes or sacs in which sperm can be stored between the time of mating and the time an egg is fertilized. Paternity testing of insects has revealed that some, and probably many, female insects use the spermatheca and various ducts to control or bias sperm used in favor of some males over others.

Male

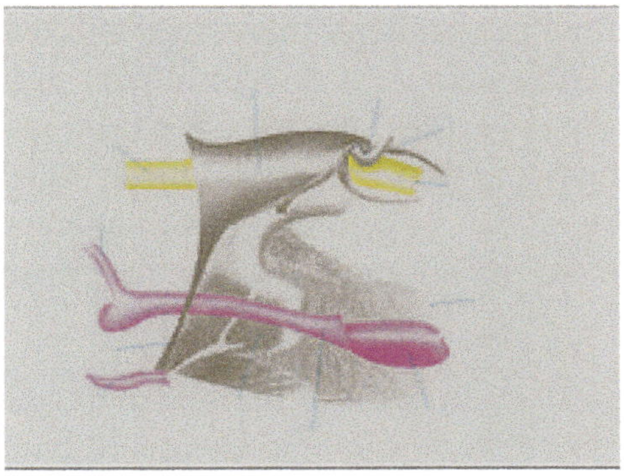

Male genitalia of Lepidoptera

The main component of the male reproductive system is the testis, suspended in the body cavity by tracheae and the fat body. The more primitive apterygote insects have a single testis, and in some lepidopterans the two maturing testes are secondarily fused into one structure during the later stages of larval development, although the ducts leading from them remain separate. However, most male insects have a pair of testes, inside of which are sperm tubes or follicles that are enclosed within a membranous sac. The follicles connect to the vas deferens by the vas efferens, and the two tubular vasa deferentia connect to a median ejaculatory duct that leads to the outside. A portion of the vas deferens is often enlarged to form the seminal vesicle, which stores the sperm before they are discharged into the female. The seminal vesicles have glandular linings that secrete nutrients for nourishment and maintenance of the sperm. The ejaculatory duct is derived from an invagination of the epidermal cells during development and, as a result, has a cuticular lining. The terminal portion of the ejaculatory duct may be sclerotized to form the intromittent organ, the aedeagus. The remainder of the male reproductive system is derived from embryonic mesoderm, except for the germ cells, or spermatogonia, which descend from the primordial pole cells very early during embryogenesis.

External

The ground plan of the abdomen of an adult insect typically consists of 11–12 segments and is less strongly sclerotized than the head or thorax. Each segment of the abdomen is represented by a sclerotized tergum, sternum, and perhaps a pleurite. Terga are separated from each other and from the adjacent sterna or pleura by a membrane. Spiracles are located in the pleural area. Variation of this ground plan includes the fusion of terga or terga and sterna to form continuous dorsal or ventral shields or a conical tube. Some insects bear a sclerite in the pleural area called a laterotergite. Ventral sclerites are sometimes called laterosternites. During the embryonic stage of many insects and the postembryonic stage of primitive insects, 11 abdominal segments are present. In modern insects there is a tendency toward reduction in the number of the abdominal segments, but the primitive number of 11 is maintained during embryogenesis. Variation in abdominal segment number is considerable. If the Apterygota are considered to be indicative of the ground plan for pterygotes, confusion reigns: adult Protura have 12 segments, Collembola have 6. The orthopteran family Acrididae has 11 segments, and a fossil specimen of Zoraptera has a 10-segmented abdomen.

Generally, the first seven abdominal segments of adults (the pregenital segments) are similar in structure and lack appendages. However, apterygotes (bristletails and silverfish) and many immature aquatic insects have abdominal appendages. Apterygotes possess a pair of styles; rudimentary appendages that are serially homologous with the distal part of the thoracic legs. And, mesally, one or two pairs of protrusible (or exsertile) vesicles on at least some abdominal segments. These vesicles are derived from the coxal and trochanteral endites (inner annulated lobes) of the ancestral abdominal appendages. Aquatic larvae and nymphs may have gills laterally on some to most abdominal segments. Of the rest of the abdominal segments consist of the reproductive and anal parts.

The anal-genital part of the abdomen, known as the terminalia, consists generally of segments 8 or 9 to the abdominal apex. Segments 8 and 9 bear the genitalia; segment 10 is visible as a complete segment in many "lower" insects but always lacks appendages; and the small segment 11 is represented by a dorsal epiproct and pair of ventral paraprocts derived from the sternum. A pair of appendages, the cerci, articulates laterally on segment 11; typically these are annulated and filamentous but have been modified (e.g. the forceps of earwigs) or reduced in different insect orders. An annulated caudal filament, the median appendix dorsalis, arises from the tip of the epiproct in apterygotes, most mayflies (Ephemeroptera), and a few fossil insects. A similar structure in nymphal stoneflies (Plecoptera) is of uncertain homology. These terminal abdominal segments have excretory and sensory functions in all insects, but in adults there is an additional reproductive function.

Genitalia

The organs concerned specifically with mating and the deposition of eggs are known collectively as the external genitalia, although they may be largely internal. The components of the external genitalia of insects are very diverse in form and often have considerable taxonomic value, particularly among species that appear structurally similar in other respects. The male external genitalia have been used widely to aid in distinguishing species, whereas the female external genitalia may be simpler and less varied.

The terminalia of adult female insects include internal structures for receiving the male copulatory organ and his spermatozoa, and external structures used for oviposition (egg-laying). Most female insects have an egg-laying tube, or ovipositor; it is absent in termites, parasitic lice, many Plecoptera, and most Ephemeroptera. Ovipositors take two forms:

1. true, or appendicular, formed from appendages of abdominal segments 8 and 9;
1. substitutional, composed of extensible posterior abdominal segments.

Antimicrobial Peptides

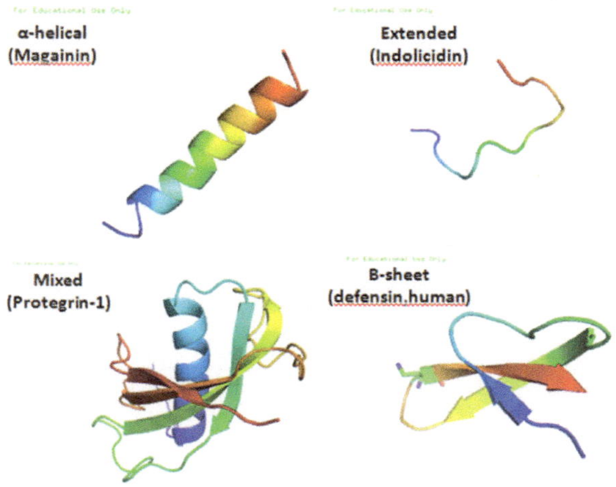

Various structures of antimicrobial peptides

Antimicrobial peptides (AMPs), also called host defense peptides (HDPs) are part of the innate immune response found among all classes of life. Fundamental differences exist between prokaryotic and eukaryotic cells that may represent targets for antimicrobial peptides. These peptides are potent, broad spectrum antibiotics which demonstrate potential as novel therapeutic agents. Antimicrobial peptides have been demonstrated to kill Gram negative and Gram positive bacteria, enveloped viruses, fungi and even transformed or cancerous cells. Unlike the majority of conventional antibiotics it appears as though antimicrobial peptides may also have the ability to enhance immunity by functioning as immunomodulators.

Marine fish sources have high levels of antimicrobial compounds with in vivo testing confirming the efficacy of fish peptides used in food/feed ingredients.

Structure

Antimicrobial peptides are a unique and diverse group of molecules, which are divided into subgroups on the basis of their amino acid composition and structure. Antimicrobial peptides are generally between 12 and 50 amino acids. These peptides include two or more positively charged residues provided by arginine, lysine or, in acidic environments, histidine, and a large proportion (generally >50%) of hydrophobic residues. The secondary structures of these molecules follow 4 themes, including i) α-helical, ii) β-stranded due to the presence of 2 or more disulfide bonds, iii) β-hairpin or loop due to the presence of a single disulfide bond and/or cyclization of the peptide chain, and iv) extended. Many of these peptides are unstructured in free solution, and fold into their final configuration upon partitioning into biological membranes. It contains hydrophilic amino acid residues aligned along one side and hydrophobic amino acid residues aligned along the opposite side of a helical molecule. This amphipathicity of the antimicrobial peptides allows them to partition into the membrane lipid bilayer. The ability to associate with membranes is a definitive feature of antimicrobial peptides although membrane permeabilization is not necessary. These peptides have a variety of antimicrobial activities ranging from membrane permeabilization to action on a range of cytoplasmic targets.

Type	characteristic	AMPs
Anionic peptides	rich in glutamic and aspartic acids	Maximin H5 from amphibians, Dermcidin from humans
Linear cationic α-helical peptides	lack in cysteine	Cecropins, andropin, moricin, ceratotoxin and melittin from insects, Magainin, dermaseptin, bombinin, brevinin-1,esculentins and buforin II from amphibians, CAP18 from rabbits, LL37 from humans
Cationic peptide enriched for specific amino acid	rich in proline, arginine, phenylalanine, glycine, tryptophan	abaecin, apidaecins from honeybees, prophenin from pigs, indolicidin from cattle.
Anionic and cationic peptides that contain cysteine and form disulfide bonds	contain 1~3 disulfide bond	1 bond:brevinins, 2 bonds:protegrin from pig, tachyplesins from horseshoe crabs, 3 bonds:defensins from humans, more than 3:drosomycin in fruit flies

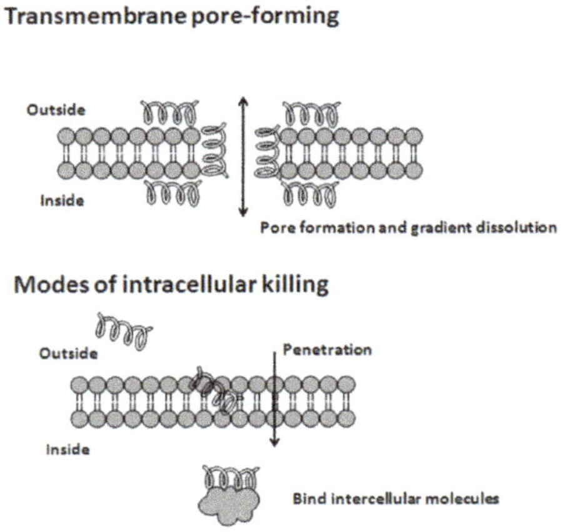

The modes of action by Antimicrobial peptides

Activities

The modes of action by which antimicrobial peptides kill microbes are varied, and may differ for different bacterial species. Some antimicrobial peptides kill both bacteria and fungi, e.g., psoriasin kills *E. coli* and several filamentous fungi. The cytoplasmic membrane is a frequent target, but peptides may also interfere with DNA and protein synthesis, protein folding, and cell wall synthesis. The initial contact between the peptide and the target organism is electrostatic, as most bacterial surfaces are anionic, or hydrophobic, such as in the antimicrobial peptide Piscidin. Their amino acid composition, amphipathicity, cationic charge and size allow them to attach to and insert into membrane bilayers to form pores by 'barrel-stave', 'carpet' or 'toroidal-pore' mechanisms. Alternately, they may penetrate into the cell to bind intracellular molecules which are crucial to cell living. Intracellular binding models includes inhibition of cell wall synthesis, alteration of the cytoplasmic membrane, activation of autolysin, inhibition of DNA, RNA, and protein synthesis, and inhibition of certain enzymes. However, in many cases, the exact mechanism of killing is not known. One emerging technique for the study of such mechanisms is dual polarisation interferometry. In contrast to many conventional antibiotics these peptides appear to be bactericidal instead of bacteriostatic. In general the antimicrobial activity of these peptides is determined by measuring the minimal inhibitory concentration (MIC), which is the lowest concentration of drug that inhibits bacterial growth.

Immunomodulation

In addition to killing bacteria directly they have been demonstrated to have a number of immunomodulatory functions that may be involved in the clearance of infection, including the ability to alter host gene expression, act as chemokines and/or induce chemokine production, inhibiting lipopolysaccharide induced pro-inflammatory cytokine production, promoting wound healing, and modulating the responses of dendritic cells and cells of the adaptive immune response. Animal models indicate that host defence peptides are crucial for both prevention and clearance of infection. It appears as though many peptides initially isolated as and termed "antimicrobial peptides" have been shown to have more significant alternative functions in vivo (e.g. hepcidin).

Mode of Action

Several methods have been used to determine the mechanisms of antimicrobial peptide activity. In particular, solid-state NMR studies have provided an atomic-level resolution explanation of membrane disruption by antimicrobial peptides.

Methods	Applications
Microscopy	to visualize the effects of antimicrobial peptides on microbial cells
Atomic emission spectroscopy	to detect loss of intracellular potassium (an indication that bacterial membrane integrity has been compromised)
Fluorescent dyes	to measure ability of antimicrobial peptides to permeabilize membrane vesicles
Ion channel formation	to assess the formation and stability of an antimicrobial-peptide-induced pore
Circular dichroism and orientated circular dichroism	to measure the orientation and secondary structure of an antimicrobial peptide bound to a lipid bilayer
Dual Polarization Interferometry	to measure the different mechanisms of antimicrobial peptides
Solid-state NMR spectroscopy	to measure the secondary structure, orientation and penetration of antimicrobial peptides into lipid bilayers in the biologically relevant LIQUID-CRYSTALLINE STATE
Neutron and X-ray diffraction	to measure the diffraction patterns of peptide-induced pores within membranes in oriented multilayers or liquids

Therapeutic Potential

These peptides are excellent candidates for development as novel therapeutic agents and complements to conventional antibiotic therapy because they generally have a broad range of activity, are bactericidal as opposed to bacteriostatic and require a short contact time to induce killing. A number of naturally occurring peptides and their derivatives have been developed as novel anti-infective therapies for conditions as diverse as oral mucositis, lung infections associated with cystic fibrosis (CF), cancer, and skin and wound infections. Pexiganan has been shown to be useful to treat infection related diabetic foot ulcer.

A major limitation to the therapeutic potential is the possibility of bacteria developing resistance to the peptides, and particularly if that produces a resistance to the body's own immune system use of those peptides. That is, providing a lot of the peptides as a therapeutic agent makes it easier for resistance to evolve; unlike antibiotic resistance, however, resistance to antimicrobial peptides mimicking those produced by humans can make the bacteria more resistant to the body's own immune system rather than just the antibiotic.

Antimicrobial peptides have been successively incorporated into topical therapeutics. A major challenge associated with systemic delivery of an antimicrobial peptides is their susceptibility to proteolytic degradation. That is, the peptides are quickly broken down when introduced in the bloodstream.

Anticancer

Some of the cecropins (e.g. cecropin A, and cecropin B) have anticancer properties and are called anticancer peptides (ACPs). Hybrid ACPs based on Cecropin A have been studied for anticancer properties.

Selectivity

In the competition of bacterial cells and host cells with the antimicrobial peptides, antimicrobial peptides will preferentially interact with the bacterial cell to the mammalian cells, which enables them to kill microorganisms without being significantly toxic to mammalian cells. Selectivity is a very important feature of the antimicrobial peptides and it can guarantee their function as antibiotics in host defense systems.

Factors

There are some factors that are closely related to the selectivity property of antimicrobial peptides, among which the cationic property contributes most. Since the surface of the bacterial membranes is more negatively charged than mammalian cells, antimicrobial peptides will show different affinities towards the bacterial membranes and mammalian cell membranes.

In addition, there are also other factors that will affect the selectivity. It's well known that cholesterol is normally widely distributed in the mammalian cell membranes as a membrane stabilizing agents but absent in bacterial cell membranes; and the presence of these cholesterols will also generally reduce the activities of the antimicrobial peptides, due either to stabilization of the lipid bilayer or to interactions between cholesterol and the peptide. So the cholesterol in mammalian cells will protect the cells from attack by the antimicrobial peptides.

Besides, the transmembrane potential is well-known to affect peptide-lipid interactions. There's an inside-negative transmembrane potential existing from the outer leaflet to the inner leaflet of the cell membranes and this inside-negative transmembrane potential will facilitate membrane permeabilization probably by facilitating the insertion of positively charged peptides into membranes. By comparison, the transmembrane potential of bacterial cells is more negative than that of normal mammalian cells, so bacterial membrane will be prone to be attacked by the positively charged antimicrobial peptides.

Similarly, it is also believed that increasing ionic strength, which in general reduces the activity of most antimicrobial peptides, contributes partially to the selectivity of the antimicrobial peptides by weakening the electrostatic interactions required for the initial interaction.

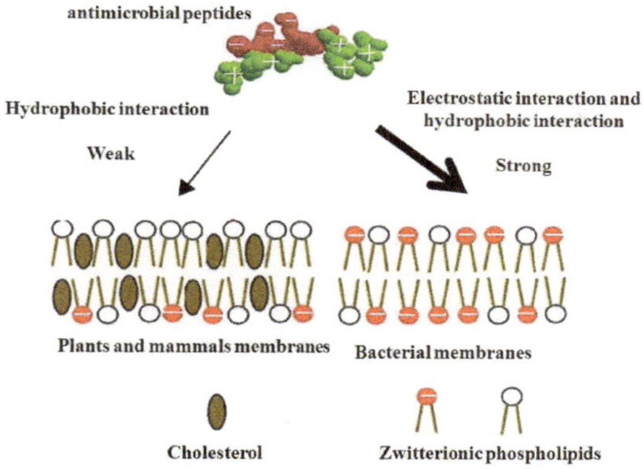

Molecular Basis of Cell Selectivity of Antimicrobial Peptides

Mechanism

The cell membranes of bacteria are rich in acidic phospholipids, such as phosphatidylglycerol and cardiolipin. These phospholipid headgroups are heavily negatively charged. Therefore, the outmost leaflets of the bilayer which is exposed to the outside of the bacterial membranes are more attractive to the attack of the positively charged antimicrobial peptides. So the interaction between the positive charges of antimicrobial peptides and the negatively charged bacterial membranes is mainly the electrostatic interactions, which is the major driving force for cellular association. In addition, since antimicrobial peptides form structures with a positively charged face as well as a hydrophobic face, there are also some hydrophobic interactions between the hydrophobic regions of the antimicrobial peptides and the zwitterionic phospholipids (electrically neutral) surface of the bacterial membranes, which act only as a minor effect in this case.

In contrast, the outer part of the membranes of plants and mammals is mainly composed of lipids without any net charges since most of the lipids with negatively charged headgroups are principally sequestered into the inner leaflet of the plasma membranes. Thus in the case of mammalian cells, the outer surfaces of the membranes are usually made of zwitterionic phosphatidylcholine and sphingomyelin, even though a small portion of the membrane's outer surfaces contain some negatively charged gangliosides. Therefore, the hydrophobic interaction between the hydrophobic face of amphipathic antimicrobial peptides and the zwitterionic phospholipids on the cell surface of mammalian cell membranes plays a major role in the formation of peptide-cell binding. However, the hydrophobic interaction is relatively weak when compared to the electrostatic interaction, thus, the antimicrobial peptides will preferentially interact with bacterial membranes.

Dual polarisation interferometry has been used *in vitro* to study and quantify the association to headgroup, insertion into the bilayer, pore formation and eventual disruption of the membrane.

Control

A lot of effort has been put into controlling cell selectivity. For example, Katsumi tried to modify and optimize the physicochemical parameters of the peptides to control the selectivities, including net charge, helicity, hydrophobicity per residue (H), hydrophobic moment (μ) and the angle subtended by the positively charged polar helix face (Φ). Other mechanisms like the introduction of D-amino acids and fluorinated amino acids in the hydrophobic phase are believed to break the secondary structure and thus reduce hydrophobic interaction with mammalian cells. Wan L Z, el found that Pro→Nlys substitution in Pro-containing β-turn antimicrobial peptides was a promising strategy for the design of new small bacterial cell-selective antimicrobial peptides with intracellular mechanisms of action. Nadezhda V el suggested that direct attachment of magainin to the substrate surface decreased nonspecific cell binding and led to improved detection limit for bacterial cells such as Salmonella and E. coli.

Bacterial Resistance

Bacteria use various resistance strategies to avoid antimicrobial peptide killing. Some microorganisms alter net surface charges. *Staphylococcus aureus* transports D-alanine from the cytoplasm to the surface teichoic acid which reduces the net negative charge by introducing basic amino groups. *S. aureus* also modifies its anionic membranes via MprF with L-lysine, increasing the positive net charge. The interaction of antimicrobial peptides with membrane targets can be limited by capsule polysaccharide of *Klebsiella pneumoniae*. Alterations occur in Lipid A. *Salmonella* species reduce the fluidity of their outer membrane by increasing hydrophobic interactions between an increased number of Lipid A acyl tails by adding myristate to Lipid A with 2-hydroxymyristate and forming hepta-acylated Lipid A by adding palmitate. The increased hydrophobic moment is thought to retard or abolish antimicrobial peptide insertion and pore formation. The residues undergo alteration in membrane proteins. In some Gram-negative bacteria, alteration in the production of outer membrane proteins correlates with resistance to killing by antimicrobial peptides. Nontypeable Hemophilus influenzae transports AMPs into the interior of the cell, where they are degraded. And H. influenzae remodels its membranes to make it appear as if the bacterium has already been successfully attacked by AMPs, protecting it from being attacked by more AMPs. ATP-binding cassette transporters import antimicrobial peptides and the resistance-nodulation cell-division efflux pump exports antimicrobial peptides. Both transporters have been associated with antimicrobial peptide resistance. Bacteria produce proteolytic enzymes,which may degrade antimicrobial peptides leading to their resistance. Outer membrane vesicles produced by Gram-negative bacteria bind the antimicrobial peptides and sequester them away from the cells, thereby protecting the cells. The outer membrane vesicles are also known to contain various proteases, peptidases and other lytic enzymes, which may have a role in degrading the extracellular peptide and nucleic acid molecules, which if allowed to reach to the bacterial cells may be dangerous for the cells. Cyclic-di-GMP signaling had also been involved in the regulation of antimicrobial peptide resistance in *Pseudomonas aeruginosa*

While these examples show that resistance can evolve naturally, there is increasing concern that using pharmaceutical copies of antimicrobial peptides can make resistance happen more often and faster. In some cases, resistance to these peptides used as a pharmaceutical to treat medical problems can lead to resistance, not only to the medical application of the peptides, but to the physiological function of those peptides. Further research is needed to determine if this will lead to greater harm than benefit from the use of certain antimicrobial peptides.

Examples

Antimicrobal peptides are produced by all known species, including peptides from bacteria, from fungi, from hydra, insects (mastoparan, poneratoxin, cecropin, moricin, melittin and others), frogs (magainin, dermaseptin and others), and mammals (for example, cathelicidins, defensins and protegrins). Research has increased in recent years to develop artificially-engineered mimics of antimicrobial peptides such as SNAPPs, in part due to the prohibitive cost of producing naturally-derived AMPs.

Bioinformatics

Several bioinformatic databases exist to catalogue antimicrobial peptides such as CAMP, CAMP release 2 (Collection of sequences and structures of antimicrobial peptides) the Antimicrobial Peptide Database, LAMP, BioPD and ADAM (A Database of Anti-Microbial peptides). The Antimicrobial peptide databases may be divided into two categories on the basis of the source of peptides it contains, as specific databases and general databases. These databases have various tools for antimicrobial peptides analysis and prediction. For example, CAMP contains AMP prediction, feature calculator, BLAST search, clustalW, VAST, PRATT, Helical wheel etc. In addition, ADAM allows users tosearch or browse through AMP sequence-structure relationships.

References

- McGavin, George C (2001). Essential Entomology: An Order-by-Order Introduction. Oxford: Oxford University Press. ISBN 9780198500025.
- Triplehorn, Charles A; Johnson, Norman F (2005). Borror and DeLong's introduction to the study of insects. (7th ed.). Australia: Thomson, Brooks/Cole. ISBN 9780030968358.
- Gullan, P.J.; P.S. Cranston (2005). The Insects: An Outline of Entomology (3 ed.). Oxford: Blackwell Publishing. pp. 61–65. ISBN 1-4051-1113-5.
- Borror, D. J.; DeLong, Dwight M.; Triplehorn, Charles A. (2004). Introduction to the Study of Insects (Sixth ed.). New York: Holt, Rinehart & Winston. ISBN 0-03-096835-6.
- Preston–Mafham, Rod; Preston–Mafham, Ken (1993). The encyclopedia of land invertebrate behaviour (Illustrated ed.). MIT Press. p. 113. ISBN 978-0-262-16137-4. Retrieved 16 November 2010.
- Boggs, Carol L.; Watt, Ward B.; Ehrlich, Paul R. (2003). Butterflies: ecology and evolution taking flight (Illustrated ed.). University of Chicago Press. p. 739. ISBN 978-0-226-06318-8. Retrieved 16 November 2010.
- Scoble, Malcolm J. (1992). The Lepidoptera: Form, Function and Diversity. Oxford: Oxford University Press. ISBN 0-19-854031-0.
- Gullan, P. J.; Cranston, P. S. (2000). The Insects: An Outline of Entomology (2nd ed.). London: Blackwell. ISBN 0-632-05343-7.
- Resh, Vincent H.; Ring T. Carde (July 1, 2009). Encyclopedia of Insects (2 ed.). U. S. A.: Academic Press. ISBN 0-12-374144-0.
- Gullan, P.J.; P.S. Cranston (2005). The Insects: An Outline of Entomology (3 ed.). Oxford: Blackwell Publishing. pp. 22–48. ISBN 1-4051-1113-5.

Anatomy: An Overview

5

Anatomy studies the physical structure of organisms and their composition. Comparative anatomy studies the similarities and differences that are found in different species. The topics elucidated in the section are homology, convergent evolution, connective tissue, muscle tissue, nervous tissue etc. This section on anatomy offers an insightful focus, keeping in mind the subject matter.

Anatomy

Anatomy is the branch of biology concerned with the study of the structure of organisms and their parts. Anatomy is inherently tied to embryology, comparative anatomy, evolutionary biology, and phylogeny, as these are the processes by which anatomy is generated over immediate (embryology) and long (evolution) timescales. Human anatomy is one of the basic essential sciences of medicine.

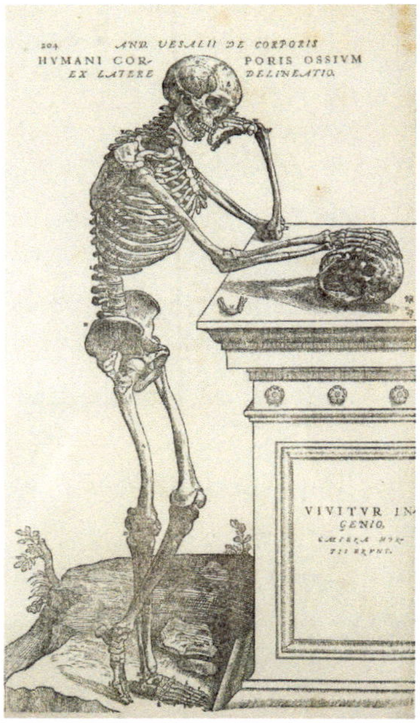

One of the large, detailed illustrations in Andreas Vesalius's De humani corporis fabrica 16th century, marking the rebirth of anatomy

The discipline of anatomy is divided into macroscopic and microscopic anatomy. Macroscopic anatomy, or gross anatomy, is the examination of an animal's body parts using unaided eyesight. Gross anatomy also includes the branch of superficial anatomy. Microscopic anatomy involves the use of optical instruments in the study of the tissues of various structures, known as histology, and also in the study of cells.

The history of anatomy is characterized by a progressive understanding of the functions of the organs and structures of the human body. Methods have also improved dramatically, advancing from the examination of animals by dissection of carcasses and cadavers (corpses) to 20th century medical imaging techniques including X-ray, ultrasound, and magnetic resonance imaging.

Anatomy and physiology, which study (respectively) the structure and function of organisms and their parts, make a natural pair of related disciplines, and they are often studied together.

Definition

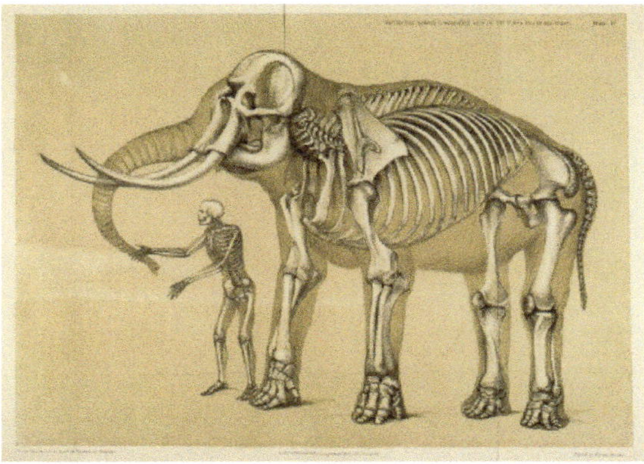

Human compared to elephant frame. Benjamin Waterhouse Hawkins, 1860

Anatomy is the scientific study of the structure of organisms including their systems, organs and tissues. It includes the appearance and position of the various parts, the materials from which they are composed, their locations and their relationships with other parts. Anatomy is quite distinct from physiology and biochemistry, which deal respectively with the functions of those parts and the chemical processes involved. For example, an anatomist is concerned with the shape, size, position, structure, blood supply and innervation of an organ such as the liver; while a physiologist is interested in the production of bile, the role of the liver in nutrition and the regulation of bodily functions.

The discipline of anatomy can be subdivided into a number of branches including gross or macroscopic anatomy and microscopic anatomy. Gross anatomy is the study of structures large enough to be seen with the naked eye, and also includes superficial anatomy or surface anatomy, the study by sight of the external body features. Microscopic anatomy is the study of structures on a microscopic scale, including histology (the study of tissues), and embryology (the study of an organism in its immature condition).

Anatomy can be studied using both invasive and non-invasive methods with the goal of obtaining information about the structure and organization of organs and systems. Methods used include dissection, in which a body is opened and its organs studied, and endoscopy, in which a video camera-equipped instrument is inserted through a small incision in the body wall and used to explore the internal organs and other structures. Angiography using X-rays or magnetic resonance angiography are methods to visualize blood vessels.

The term "anatomy" is commonly taken to refer to human anatomy. However, substantially the same structures and tissues are found throughout the rest of the animal kingdom and the term also includes the anatomy of other animals. The term *zootomy* is also sometimes used to specifically refer to animals. The structure and tissues of plants are of a dissimilar nature and they are studied in plant anatomy.

Animal Tissues

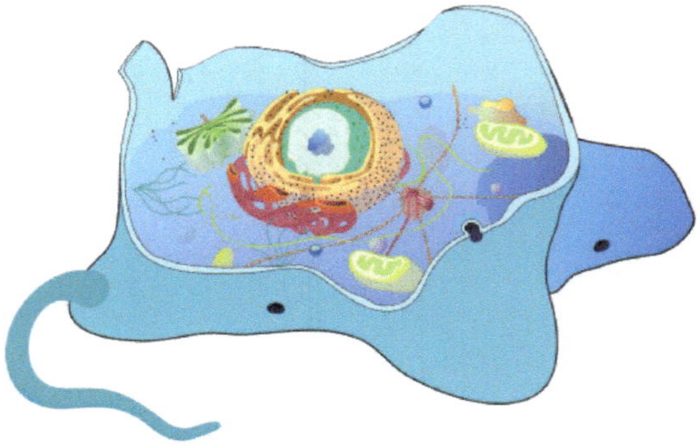

Stylized cutaway diagram of an animal cell (with flagella)

The kingdom Animalia or metazoa, contains multicellular organisms that are heterotrophic and motile (although some have secondarily adopted a sessile lifestyle). Most animals have bodies differentiated into separate tissues and these animals are also known as eumetazoans. They have an internal digestive chamber, with one or two openings; the gametes are produced in multicellular sex organs, and the zygotes include a blastula stage in their embryonic development. Metazoans do not include the sponges, which have undifferentiated cells.

Unlike plant cells, animal cells have neither a cell wall nor chloroplasts. Vacuoles, when present, are more in number and much smaller than those in the plant cell. The body tissues are composed of numerous types of cell, including those found in muscles, nerves and skin. Each typically has a cell membrane formed of phospholipids, cytoplasm and a nucleus. All of the different cells of an animal are derived from the embryonic germ layers. Those simpler invertebrates which are formed from two germ layers of ectoderm and endoderm are called diploblastic and the more developed animals whose structures and organs are formed from three germ layers are called triploblastic. All of a triploblastic animal's tissues and organs are derived from the three germ layers of the embryo, the ectoderm, mesoderm and endoderm.

Animal tissues can be grouped into four basic types: connective, epithelial, muscle and nervous tissue.

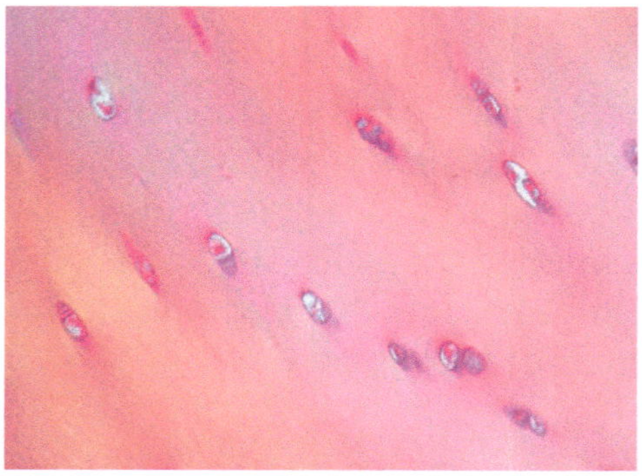

Hyaline cartilage at high magnification (H&E stain)

Connective Tissue

Connective tissues are fibrous and made up of cells scattered among inorganic material called the extracellular matrix. Connective tissue gives shape to organs and holds them in place. The main types are loose connective tissue, adipose tissue, fibrous connective tissue, cartilage and bone. The extracellular matrix contains proteins, the chief and most abundant of which is collagen. Collagen plays a major part in organizing and maintaining tissues. The matrix can be modified to form a skeleton to support or protect the body. An exoskeleton is a thickened, rigid cuticle which is stiffened by mineralization, as in crustaceans or by the cross-linking of its proteins as in insects. An endoskeleton is internal and present in all developed animals, as well as in many of those less developed.

Epithelium

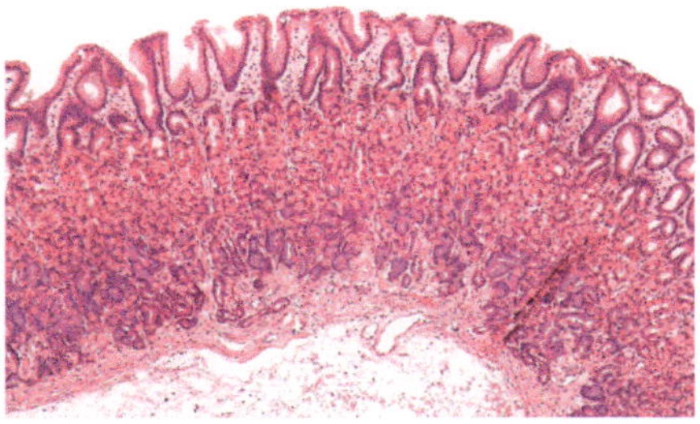

Gastric mucosa at low magnification (H&E stain)

Epithelial tissue is composed of closely packed cells, bound to each other by cell adhesion molecules, with little intercellular space. Epithelial cells can be squamous (flat), cuboidal or columnar and rest on a basal lamina, the upper layer of the basement membrane, the lower layer is the reticular lamina lying next to the connective tissue in the extracellular matrix secreted by the epithelial cells. There are many different types of epithelium, modified to suit a particular function. In the respiratory tract there is a type of ciliated epithelial lining; in the small intestine there are microvilli on the epithelial lining and in the large intestine there are intestinal villi. Skin consists of an outer layer of keratinized stratified squamous epithelium that covers the exterior of the vertebrate body. Keratinocytes make up to 95% of the cells in the skin. The epithelial cells on the external surface of the body typically secrete an extracellular matrix in the form of a cuticle. In simple animals this may just be a coat of glycoproteins. In more advanced animals, many glands are formed of epithelial cells.

Muscle Tissue

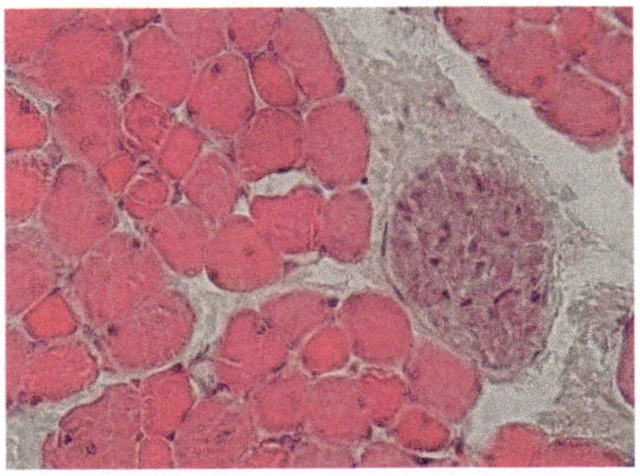

Cross section through skeletal muscle and a small nerve at high magnification (H&E stain)

Muscle cells (myocytes) form the active contractile tissue of the body. Muscle tissue functions to produce force and cause motion, either locomotion or movement within internal organs. Muscle is formed of contractile filaments and is separated into three main types; smooth muscle, skeletal muscle and cardiac muscle. Smooth muscle has no striations when examined microscopically. It contracts slowly but maintains contractibility over a wide range of stretch lengths. It is found in such organs as sea anemone tentacles and the body wall of sea cucumbers. Skeletal muscle contracts rapidly but has a limited range of extension. It is found in the movement of appendages and jaws. Obliquely striated muscle is intermediate between the other two. The filaments are staggered and this is the type of muscle found in earthworms that can extend slowly or make rapid contractions. In higher animals striated muscles occur in bundles attached to bone to provide movement and are often arranged in antagonistic sets. Smooth muscle is found in the walls of the uterus, bladder, intestines, stomach, oesophagus, respiratory airways, and blood vessels. Cardiac muscle is found only in the heart, allowing it to contract and pump blood round the body.

Nervous Tissue

Nervous tissue is composed of many nerve cells known as neurons which transmit information. In some slow-moving radially symmetrical marine animals such as ctenophores and cnidarians (including sea anemones and jellyfish), the nerves form a nerve net, but in most animals they are organized longitudinally into bundles. In simple animals, receptor neurons in the body wall cause a local reaction to a stimulus. In more complex animals, specialized receptor cells such as chemoreceptors and photoreceptors are found in groups and send messages along neural networks to other parts of the organism. Neurons can be connected together in ganglia. In higher animals, specialized receptors are the basis of sense organs and there is a central nervous system (brain and spinal cord) and a peripheral nervous system. The latter consists of sensory nerves that transmit information from sense organs and motor nerves that influence target organs. The peripheral nervous system is divided into the somatic nervous system which conveys sensation and controls voluntary muscle, and the autonomic nervous system which involuntarily controls smooth muscle, certain glands and internal organs, including the stomach.

Vertebrate Anatomy

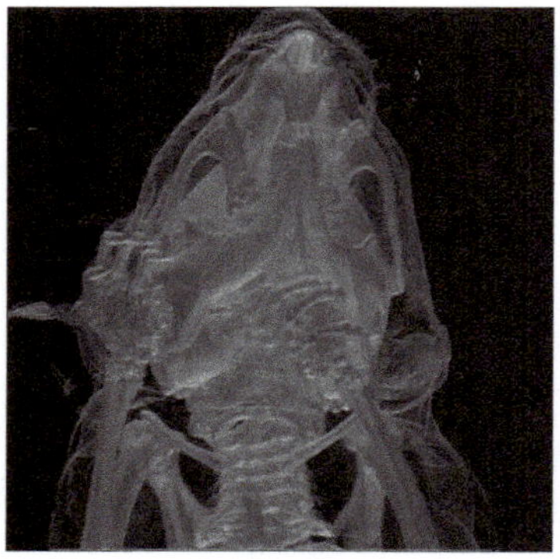

Mouse skull

All vertebrates have a similar basic body plan and at some point in their lives, (mostly in the embryonic stage), share the major chordate characteristics; a stiffening rod, the notochord; a dorsal hollow tube of nervous material, the neural tube; pharyngeal arches; and a tail posterior to the anus. The spinal cord is protected by the vertebral column and is above the notochord and the gastrointestinal tract is below it. Nervous tissue is derived from the ectoderm, connective tissues are derived from mesoderm, and gut is derived from the endoderm. At the posterior end is a tail which continues the spinal cord and vertebrae but not the gut. The mouth is found at the anterior end of the animal, and the anus at the base of the tail. The defining characteristic of a vertebrate is the vertebral column, formed in the development of the segmented series of vertebrae. In most vertebrates the notochord becomes the nucleus pulposus of the intervertebral discs. However, a few vertebrates, such as the sturgeon and the coelacanth retain the notochord into adulthood. Jawed vertebrates are typified by paired appendages, fins or legs, which may be secondarily lost. The limbs of vertebrates are considered to be homologous because the same underlying skeletal structure was inherited from their last common ancestor. This is one of the arguments put forward by Charles Darwin to support his theory of evolution.

Fish Anatomy

Cutaway diagram showing various organs of a fish

The body of a fish is divided into a head, trunk and tail, although the divisions between the three are not always externally visible. The skeleton, which forms the support structure inside the fish, is either made of cartilage, in cartilaginous fish, or bone in bony fish. The main skeletal element is the vertebral column, composed of articulating vertebrae which are lightweight yet strong. The ribs attach to the spine and there are no limbs or limb girdles. The main external features of the fish, the fins, are composed of either bony or soft spines called rays, which with the exception of the caudal fins, have no direct connection with the spine. They are supported by the muscles which compose the main part of the trunk. The heart has two chambers and pumps the blood through the respiratory surfaces of the gills and on round the body in a single circulatory loop. The eyes are adapted for seeing underwater and have only local vision. There is an inner ear but no external or middle ear. Low frequency vibrations are detected by the lateral line system of sense organs that run along the length of the sides of fish, and these respond to nearby movements and to changes in water pressure.

Sharks and rays are basal fish with numerous primitive anatomical features similar to those of ancient fish, including skeletons composed of cartilage. Their bodies tend to be dorso-ventrally flattened, they usually have five pairs of gill slits and a large mouth set on the underside of the head. The dermis is covered with separate dermal placoid scales. They have a cloaca into which the urinary and genital passages open, but not a swim bladder. Cartilaginous fish produce a small number of large, yolky eggs. Some species are ovoviviparous and the young develop internally but others are oviparous and the larvae develop externally in egg cases.

The bony fish lineage shows more derived anatomical traits, often with major evolutionary changes from the features of ancient fish. They have a bony skeleton, are generally laterally flattened, have five pairs of gills protected by an operculum, and a mouth at or near the tip of the snout. The dermis is covered with overlapping scales. Bony fish have a swim bladder which helps them maintain a constant depth in the water column, but not a cloaca. They mostly spawn a large number of small eggs with little yolk which they broadcast into the water column.

Amphibian Anatomy

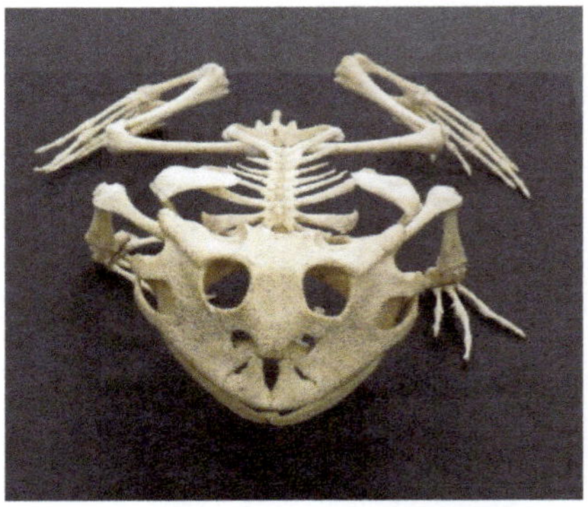

Skeleton of Surinam horned frog (Ceratophrys cornuta)

Amphibians are a class of animals comprising frogs, salamanders and caecilians. They are tetrapods, but the caecilians and a few species of salamander have either no limbs or their limbs are much reduced in size. Their main bones are hollow and lightweight and are fully ossified and the vertebrae interlock with each other and have articular processes. Their ribs are usually short and may be fused to the vertebrae. Their skulls are mostly broad and short, and are often incompletely ossified. Their skin contains little keratin and lacks scales, but contains many mucous glands and in some species, poison glands. The hearts of amphibians have three chambers, two atria and one ventricle. They have a urinary bladder and nitrogenous waste products are excreted primarily as urea. Amphibians breathe by means of buccal pumping, a pump action in which air is first drawn into the buccopharyngeal region through the nostrils. These are then closed and the air is forced into the lungs by contraction of the throat. They supplement this with gas exchange through the skin which needs to be kept moist.

Plastic model of a frog

In frogs the pelvic girdle is robust and the hind legs are much longer and stronger than the fore-limbs. The feet have four or five digits and the toes are often webbed for swimming or have suction pads for climbing. Frogs have large eyes and no tail. Salamanders resemble lizards in appearance; their short legs project sideways, the belly is close to or in contact with the ground and they have a long tail. Caecilians superficially resemble earthworms and are limbless. They burrow by means of zones of muscle contractions which move along the body and they swim by undulating their body from side to side.

Reptile Anatomy

Skeleton of a diamondback rattlesnake

Reptiles are a class of animals comprising turtles, tuataras, lizards, snakes and crocodiles. They are tetrapods, but the snakes and a few species of lizard either have no limbs or their limbs are much reduced in size. Their bones are better ossified and their skeletons stronger than those of amphibians. The teeth are conical and mostly uniform in size. The surface cells of the epidermis are modified into horny scales which create a waterproof layer. Reptiles are unable to use their skin for respiration as do amphibians and have a more efficient respiratory system drawing air into their lungs by expanding their chest walls. The heart resembles that of the amphibian but there is a septum which more completely separates the oxygenated and deoxygenated bloodstreams. The reproductive system is designed for internal fertilization, with a copulatory organ present in most species. The eggs are surrounded by amniotic membranes which prevents them from drying out and are laid on land, or develop internally in some species. The bladder is small as nitrogenous waste is excreted as uric acid.

Turtles are notable for their protective shells. They have an inflexible trunk encased in a horny carapace above and a plastron below. These are formed from bony plates embedded in the dermis which are overlain by horny ones and are partially fused with the ribs and spine. The neck is long and flexible and the head and the legs can be drawn back inside the shell. Turtles are vegetarians and the typical reptile teeth have been replaced by sharp, horny plates. In aquatic species, the front legs are modified into flippers.

Tuataras superficially resemble lizards but the lineages diverged in the Triassic period. There is one living species, *Sphenodon punctatus*. The skull has two openings (fenestrae) on either side and the jaw is rigidly attached to the skull. There is one row of teeth in the lower jaw and this fits between the two rows in the upper jaw when the animal chews. The teeth are merely projections of bony material from the jaw and eventually wear down. The brain and heart are more primitive than those of other reptiles, and the lungs have a single chamber and lack bronchi. The tuatara has a well-developed parietal eye on its forehead.

Lizards have skulls with only one fenestra on each side, the lower bar of bone below the second fenestra having been lost. This results in the jaws being less rigidly attached which allows the mouth to open wider. Lizards are mostly quadrupeds, with the trunk held off the ground by short, sideways-facing legs, but a few species have no limbs and resemble snakes. Lizards have moveable eyelids, eardrums are present and some species have a central parietal eye.

Snakes are closely related to lizards, having branched off from a common ancestral lineage during the Cretaceous period, and they share many of the same features. The skeleton consists of a skull, a hyoid bone, spine and ribs though a few species retain a vestige of the pelvis and rear limbs in the form of pelvic spurs. The bar under the second fenestra has also been lost and the jaws have extreme flexibility allowing the snake to swallow its prey whole. Snakes lack moveable eyelids, the eyes being covered by transparent "spectacle" scales. They do not have eardrums but can detect ground vibrations through the bones of their skull. Their forked tongues are used as organs of taste and smell and some species have sensory pits on their heads enabling them to locate warm-blooded prey.

Crocodilians are large, low-slung aquatic reptiles with long snouts and large numbers of teeth. The head and trunk are dorso-ventrally flattened and the tail is laterally compressed. It undulates from side to side to force the animal through the water when swimming. The tough keratinized scales provide body armour and some are fused to the skull. The nostrils, eyes and ears are elevated above the top of the flat head enabling them to remain above the surface of the water when the animal is floating. Valves seal the nostrils and ears when it is submerged. Unlike other reptiles, crocodilians have hearts with four chambers allowing complete separation of oxygenated and deoxygenated blood.

Bird Anatomy

Part of a wing. Albrecht Dürer, c. 1500–1512

Birds are tetrapods but though their hind limbs are used for walking or hopping, their front limbs are wings covered with feathers and adapted for flight. Birds are endothermic, have a high metabolic rate, a light skeletal system and powerful muscles. The long bones are thin, hollow and very light. Air sac extensions from the lungs occupy the centre of some bones. The sternum is wide and usually has a keel and the caudal vertebrae are fused. There are no teeth and the narrow jaws are adapted into a horn-covered beak. The eyes are relatively large, particularly in nocturnal species such as owls. They face forwards in predators and sideways in ducks.

The feathers are outgrowths of the epidermis and are found in localized bands from where they fan out over the skin. Large flight feathers are found on the wings and tail, contour feathers cover the bird's surface and fine down occurs on young birds and under the contour feathers of water birds. The only cutaneous gland is the single uropygial gland near the base of the tail. This produces an oily secretion that waterproofs the feathers when the bird preens. There are scales on the legs, feet and claws on the tips of the toes.

Mammal Anatomy

Mammals are a diverse class of animals, mostly terrestrial but some are aquatic and others have evolved flapping or gliding flight. They mostly have four limbs but some aquatic mammals have no limbs or limbs modified into fins and the forelimbs of bats are modified into wings. The legs of most mammals are situated below the trunk, which is held well clear of the ground. The bones of mammals are well ossified and their teeth, which are usually differentiated, are coated in a layer of prismatic enamel. The teeth are shed once (milk teeth) during the animal's lifetime or not at all, as is the case in cetaceans. Mammals have three bones in the middle ear and a cochlea in the inner ear. They are clothed in hair and their skin contains glands which secrete sweat. Some of these glands are specialized as mammary glands, producing milk to feed the young. Mammals breathe with lungs and have a muscular diaphragm separating the thorax from the abdomen which helps them draw air into the lungs. The mammalian heart has four chambers and oxygenated and deoxygenated blood are kept entirely separate. Nitrogenous waste is excreted primarily as urea.

Mammals are amniotes, and most are viviparous, giving birth to live young. The exception to this are the egg-laying monotremes, the platypus and the echidnas of Australia. Most other mammals have a placenta through which the developing foetus obtains nourishment, but in marsupials, the foetal stage is very short and the immature young is born and finds its way to its mother's pouch where it latches on to a nipple and completes its development.

Human Anatomy

Modern anatomic technique showing sagittal sections of the head as seen by a MRI scan.

In the human, the development of skilled hand movements and increased brain size is likely to have evolved simultaneously.

Humans have the overall body plan of a mammal. Humans have a head, neck, trunk (which includes the thorax and abdomen), two arms and hands, and two legs and feet.

Generally, students of certain biological sciences, paramedics, prosthetists and orthotists, physiotherapists, occupational therapists, nurses, and medical students learn gross anatomy and microscopic anatomy from anatomical models, skeletons, textbooks, diagrams, photographs, lectures and tutorials, and in addition, medical students generally also learn gross anatomy through practical experience of dissection and inspection of cadavers. The study of microscopic anatomy (or histology) can be aided by practical experience examining histological preparations (or slides) under a microscope.

Human anatomy, physiology and biochemistry are complementary basic medical sciences, which are generally taught to medical students in their first year at medical school. Human anatomy can be taught regionally or systemically; that is, respectively, studying anatomy by bodily regions such as the head and chest, or studying by specific systems, such as the nervous or respiratory systems. The major anatomy textbook, Gray's Anatomy, has been reorganized from a systems format to a regional format, in line with modern teaching methods. A thorough working knowledge of anatomy is required by physicians, especially surgeons and doctors working in some diagnostic specialties, such as histopathology and radiology.

Academic anatomists are usually employed by universities, medical schools or teaching hospitals. They are often involved in teaching anatomy, and research into certain systems, organs, tissues or cells.

Invertebrate Anatomy

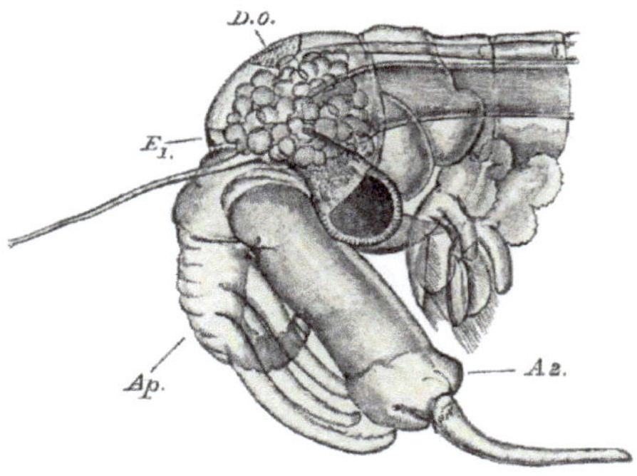

Head of a male *Daphnia*, a planktonic crustacean

Invertebrates constitute a vast array of living organisms ranging from the simplest unicellular eukaryotes such as *Paramecium* to such complex multicellular animals as the octopus, lobster and dragonfly. They constitute about 95% of the animal species. By definition, none of these creatures has a backbone. The cells of single-cell protozoans have the same basic structure as those of multicellular animals but some parts are specialized into the equivalent of tissues and organs. Locomotion is often provided by cilia or flagella or may proceed via the advance of pseudopodia, food may be gathered by phagocytosis, energy needs may be supplied by photosynthesis and the cell may be supported by an endoskeleton or an exoskeleton. Some protozoans can form multicellular colonies.

Metazoans are multicellular organism, different groups of cells of which have separate functions. The most basic types of metazoan tissues are epithelium and connective tissue, both of which are present in nearly all invertebrates. The outer surface of the epidermis is normally formed of epithelial cells and secretes an extracellular matrix which provides support to the organism. An endoskeleton derived from the mesoderm is present in echinoderms, sponges and some cephalopods.

Exoskeletons are derived from the epidermis and is composed of chitin in arthropods (insects, spiders, ticks, shrimps, crabs, lobsters). Calcium carbonate constitutes the shells of molluscs, brachiopods and some tube-building polychaete worms and silica forms the exoskeleton of the microscopic diatoms and radiolaria. Other invertebrates may have no rigid structures but the epidermis may secrete a variety of surface coatings such as the pinacoderm of sponges, the gelatinous cuticle of cnidarians (polyps, sea anemones, jellyfish) and the collagenous cuticle of annelids. The outer epithelial layer may include cells of several types including sensory cells, gland cells and stinging cells. There may also be protrusions such as microvilli, cilia, bristles, spines and tubercles.

Marcello Malpighi, the father of microscopical anatomy, discovered that plants had tubules similar to those he saw in insects like the silk worm. He observed that when a ring-like portion of bark was removed on a trunk a swelling occurred in the tissues above the ring, and he unmistakably interpreted this as growth stimulated by food coming down from the leaves, and being captured above the ring.

Arthropod Anatomy

Arthropods comprise the largest phylum in the animal kingdom with over a million known invertebrate species.

Insects possess segmented bodies supported by a hard-jointed outer covering, the exoskeleton, made mostly of chitin. The segments of the body are organized into three distinct parts, a head, a thorax and an abdomen. The head typically bears a pair of sensory antennae, a pair of compound eyes, one to three simple eyes (ocelli) and three sets of modified appendages that form the mouthparts. The thorax has three pairs of segmented legs, one pair each for the three segments that compose the thorax and one or two pairs of wings. The abdomen is composed of eleven segments, some of which may be fused and houses the digestive, respiratory, excretory and reproductive systems. There is considerable variation between species and many adaptations to the body parts, especially wings, legs, antennae and mouthparts.

Spiders a class of arachnids have four pairs of legs; a body of two segments—a cephalothorax and an abdomen. Spiders have no wings and no antennae. They have mouthparts called chelicerae which are often connected to venom glands as most spiders are venomous. They have a second pair of appendages called pedipalps attached to the cephalothorax. These have similar segmentation to the legs and function as taste and smell organs. At the end of each male pedipalp is a spoon-shaped cymbium that acts to support the copulatory organ.

Other Branches of Anatomy

- Superficial or surface anatomy is important as the study of anatomical landmarks that can be readily seen from the exterior contours of the body. It enables physicians or veterinary surgeons to gauge the position and anatomy of the associated deeper structures. Superficial is a directional term that indicates that structures are located relatively close to the surface of the body.
- Comparative anatomy relates to the comparison of anatomical structures (both gross and microscopic) in different animals.
- Artistic anatomy relates to anatomic studies for artistic reasons.

History

Ancient

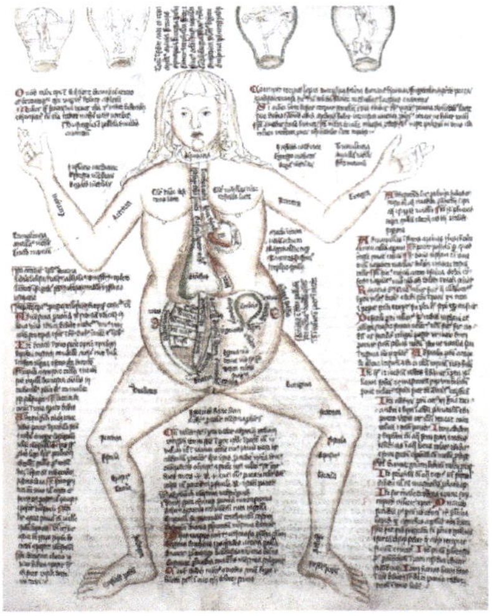

Image of early rendition of anatomy findings

Ancient Greek anatomy and physiology underwent great changes and advances throughout the early medieval world. Over time, this medical practice expanded by a continually developing understanding of the functions of organs and structures in the body. Phenomenal anatomical observations of the human body were made, which have contributed towards the understanding of the brain, eye, liver, reproductive organs and the nervous system.

The city of Alexandria was the stepping-stone for Greek anatomy and physiology. Alexandria not only housed the biggest library for medical records and books of the liberal arts in the world during the time of the Greeks, but was also home to many medical practitioners and philosophers. Great patronage of the arts and sciences from the Ptolemy rulers helped raise Alexandria up, further rivalling the cultural and scientific achievements of other Greek states.

Some of the most striking advances in early anatomy and physiology took place in Hellenistic Alexandria. Two of the most famous Greek anatomists and physiologists of the third century were Herophilus and Erasistratus. These two physicians helped pioneer human dissection for medical research. They also conducted vivisections on the cadavers of condemned criminals, which was considered taboo until the Renaissance – Herophilus was recognized as the first person to perform systematic dissections. Herophilus became known for his anatomical works making impressing contributions to many branches of anatomy and many other aspects of medicine. Some of the works included classifying the system of the pulse, the discovery that human arteries had thicker walls then veins, and that the atria were parts of the heart. Herophilus's knowledge of the human body has provided vital input towards understanding the brain, eye, liver, reproductive organs and nervous system, and characterizing the course of disease. Erasistratus accurately described the structure of the brain, including the cavities and membranes, and made a distinction between its cerebrum and cerebellum During his study in Alexandria, Erasistratus was particularly concerned with studies of the circulatory and nervous systems. He was able to distinguish the sensory and the motor nerves in the human body and believed that air entered the lungs and heart, which was then carried throughout the body. His distinction between the arteries and veins – the arteries carrying the air through the body, while the veins carried the blood from the heart was a great anatomical discovery. Erasistratus was also responsible for naming and describing the function of the epiglottis and the valves of the heart, including the tricuspid. During the third century, Greek physicians were able to differentiate nerves from blood vessels and tendons and to realize that the nerves convey neural impulses. It was Herophilus who made the point that damage to motor nerves induced paralysis. Herophilus named the meninges and ventricles in the brain, appreciated the division between cerebellum and cerebrum and recognized that the brain was the "seat of intellect" and not a "cooling chamber" as propounded by Aristotle Herophilus is also credited with describing the optic, oculomotor, motor division of the trigeminal, facial, vestibulocochlear and hypoglossal nerves

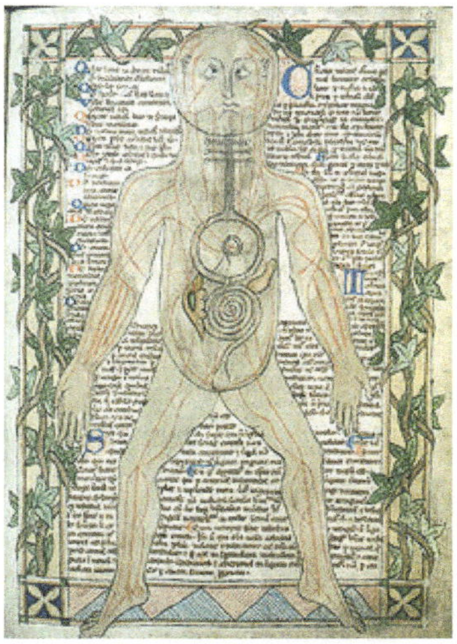

13th century anatomical illustration

Great feats were made during the third century in both the digestive and reproductive systems. Herophilus was able to discover and describe not only the salivary glands, but the small intestine and liver. He showed that the uterus is a hollow organ and described the ovaries and uterine tubes. He recognized that spermatozoa were produced by the testes and was the first to identify the prostate gland.

In 1600 BCE, the Edwin Smith Papyrus, an Ancient Egyptian medical text, described the heart, its vessels, liver, spleen, kidneys, hypothalamus, uterus and bladder, and showed the blood vessels diverging from the heart. The Ebers Papyrus (c. 1550 BCE) features a "treatise on the heart", with vessels carrying all the body's fluids to or from every member of the body.

The anatomy of the muscles and skeleton is described in the *Hippocratic Corpus*, an Ancient Greek medical work written by unknown authors. Aristotle described vertebrate anatomy based on animal dissection. Praxagoras identified the difference between arteries and veins. Also in the 4th century BCE, Herophilos and Erasistratus produced more accurate anatomical descriptions based on vivisection of criminals in Alexandria during the Ptolemaic dynasty.

In the 2nd century, Galen of Pergamum, an anatomist, clinician, writer and philosopher, wrote the final and highly influential anatomy treatise of ancient times. He compiled existing knowledge and studied anatomy through dissection of animals. He was one of the first experimental physiologists through his vivisection experiments on animals. Galen's drawings, based mostly on dog anatomy, became effectively the only anatomical textbook for the next thousand years. His work was known to Renaissance doctors only through Islamic Golden Age medicine until it was translated from the Greek some time in the 15th century.

Medieval to Early Modern

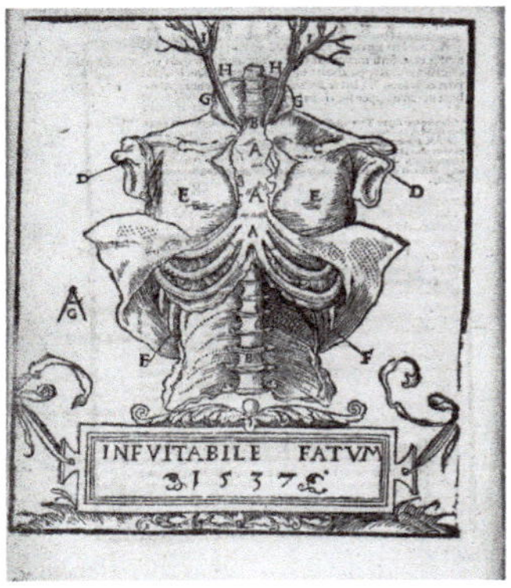

Mondino de Luzzi, *Anathomia*, 1541

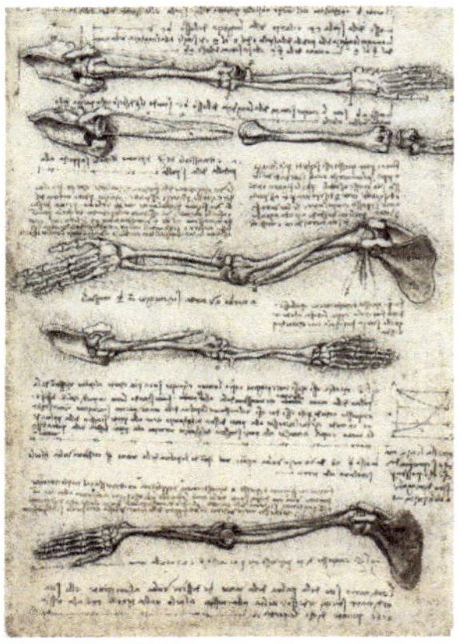

Anatomical study of the arm, by Leonardo da Vinci, (about 1510)

Anatomical chart in Vesalius's *Epitome*, 1543

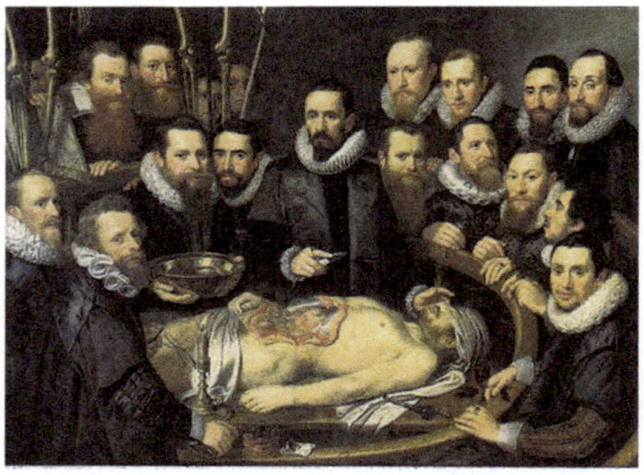

Michiel Jansz van Mierevelt – Anatomy lesson of Dr. Willem van der Meer, 1617

Anatomy developed little from classical times until the sixteenth century; as the historian Marie Boas writes, "Progress in anatomy before the sixteenth century is as mysteriously slow as its development after 1500 is startlingly rapid". Between 1275 and 1326, the anatomists Mondino de Luzzi,

Alessandro Achillini and Antonio Benivieni at Bologna carried out the first systematic human dissections since ancient times. Mondino's *Anatomy* of 1316 was the first textbook in the medieval rediscovery of human anatomy. It describes the body in the order followed in Mondino's dissections, starting with the abdomen, then the thorax, then the head and limbs. It was the standard anatomy textbook for the next century.

Leonardo da Vinci (1452–1519) was trained in anatomy by Andrea del Verrocchio. He made use of his anatomical knowledge in his artwork, making many sketches of skeletal structures, muscles and organs of humans and other vertebrates that he dissected.

Andreas Vesalius (1514–1564) (Latinized from Andries van Wezel), professor of anatomy at the University of Padua, is considered the founder of modern human anatomy. Originally from Brabant, Vesalius published the influential book *De humani corporis fabrica* ("the structure of the human body"), a large format book in seven volumes, in 1543. The accurate and intricately detailed illustrations, often in allegorical poses against Italianate landscapes, are thought to have been made by the artist Jan van Calcar, a pupil of Titian.

In England, anatomy was the subject of the first public lectures given in any science; these were given by the Company of Barbers and Surgeons in the 16th century, joined in 1583 by the Lumleian lectures in surgery at the Royal College of Physicians.

Late Modern

In the United States, medical schools began to be set up towards the end of the 18th century. Classes in anatomy needed a continual stream of cadavers for dissection and these were difficult to obtain. Philadelphia, Baltimore and New York were all renowned for body snatching activity as criminals raided graveyards at night, removing newly buried corpses from their coffins. A similar problem existed in Britain where demand for bodies became so great that grave-raiding and even anatomy murder were practised to obtain cadavers. Some graveyards were in consequence protected with watchtowers. The practice was halted in Britain by the Anatomy Act of 1832, while in the United States, similar legislation was enacted after the physician William S. Forbes of Jefferson Medical College was found guilty in 1882 of "complicity with resurrectionists in the despoliation of graves in Lebanon Cemetery".

The teaching of anatomy in Britain was transformed by Sir John Struthers, Regius Professor of Anatomy at the University of Aberdeen from 1863 to 1889. He was responsible for setting up the system of three years of "pre-clinical" academic teaching in the sciences underlying medicine, including especially anatomy. This system lasted until the reform of medical training in 1993 and 2003. As well as teaching, he collected many vertebrate skeletons for his museum of comparative anatomy, published over 70 research papers, and became famous for his public dissection of the Tay Whale. From 1822 the Royal College of Surgeons regulated the teaching of anatomy in medical schools. Medical museums provided examples in comparative anatomy, and were often used in teaching. Ignaz Semmelweis investigated puerperal fever and he discovered how it was caused. He noticed that the frequently fatal fever occurred more often in mothers examined by medical students than by midwives. The students went from the dissecting room to the hospital ward and examined women in childbirth. Semmelweis showed that when the trainees washed their hands in chlorinated lime before each clinical examination, the incidence of puerperal fever among the mothers could be reduced dramatically.

An electron microscope from 1973

Before the era of modern medical procedures, the main means for studying the internal structure of the body were palpation and dissection. It was the advent of microscopy that opened up an understanding of the building blocks that constituted living tissues. Technical advances in the development of achromatic lenses increased the resolving power of the microscope and around 1839, Matthias Jakob Schleiden and Theodor Schwann identified that cells were the fundamental unit of organization of all living things. Study of small structures involved passing light through them and the microtome was invented to provide sufficiently thin slices of tissue to examine. Staining techniques using artificial dyes were established to help distinguish between different types of tissue. The fields of cytology and histology developed from here in the late 19th century. The invention of the electron microscope brought a great advance in resolution power and allowed research into the ultrastructure of cells and the organelles and other structures within them. About the same time, in the 1950s, the use of X-ray diffraction for studying the crystal structures of proteins, nucleic acids and other biological molecules gave rise to a new field of molecular anatomy.

Short wavelength electromagnetic radiation such as X-rays can be passed through the body and used in medical radiography to view interior structures that have different degrees of opaqueness. Nowadays, modern techniques such as magnetic resonance imaging, computed tomography, fluoroscopy and ultrasound imaging have enabled researchers and practitioners to examine organs, living or dead, in unprecedented detail. They are used for diagnostic and therapeutic purposes and provide information on the internal structures and organs of the body to a degree far beyond the imagination of earlier generations.

Comparative Anatomy

Comparative anatomy is the study of similarities and differences in the anatomy of different species. It is closely related to evolutionary biology and phylogeny (the evolution of species).

Comparative anatomy has long served as evidence for evolution; it indicates that various organisms share a common ancestor. Also, it assists scientists in classifying organisms based on similar characteristics of their anatomical structures. A common example of comparative anatomy is the similar bone structures in forelimbs of cats, whales, bats, and humans. All of these appendages consist of the same basic parts; yet, they serve completely different functions. The skeletal parts which form a structure used for swimming, such as a fin, would not be ideal to form a wing, which is better-suited for flight. One explanation for the forelimbs' similar composition is descent with modification. Through random mutations and natural selection, each organism's anatomical structures gradually adapted to suit their respective habitats. The rules for development of *special* characteristics which differ significantly from general homology were listed by Karl Ernst von Baer (the Baer laws).

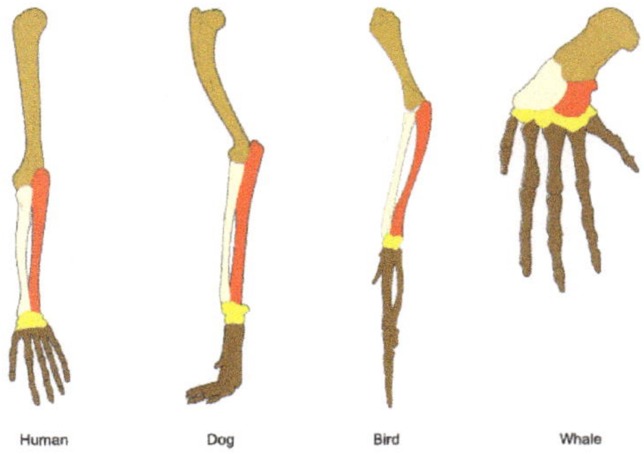

Comparative anatomy studies similarities and differences in organisms. The image shows homologous bones in the upper limb of various vertebrates.

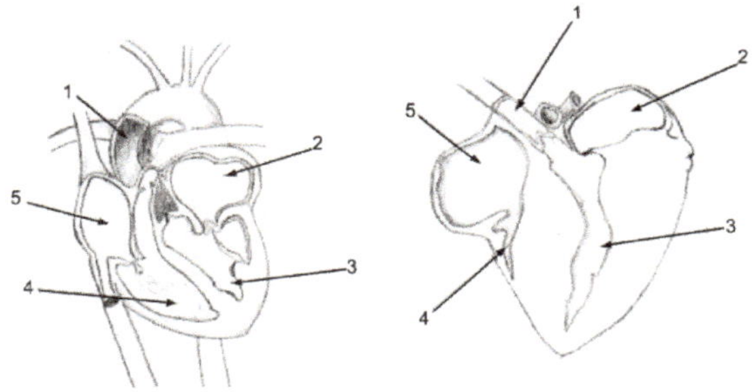

The human heart (left) and chicken heart (right) share many similar characteristics. Avian hearts pump faster than mammalian hearts. Due to the faster heart rate, the muscles surrounding the ventricles of the chicken heart are thicker. Both hearts are labeled with the following parts: 1. Ascending Aorta 2. Left Atrium 3. Left Ventricle 4. Right Ventricle 5. Right Atrium.

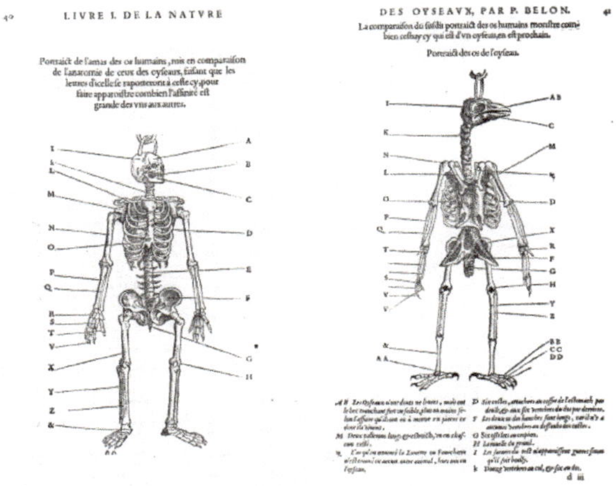

A comparison of the skeleton of humans and birds in Belon's *L'Histoire de la nature des oyseaux (Natural History of Birds)*, 1555.

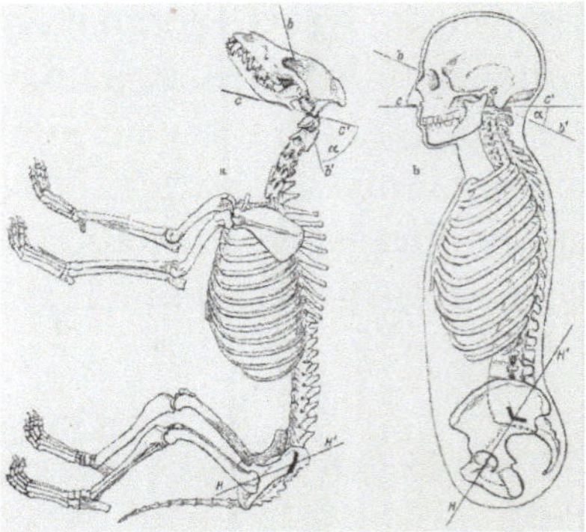

A 20th century illustration comparing the bone-structures of a human and a dog.

Two major concepts of comparative anatomy are:

1. Homologous structures - structures (body parts/anatomy) which are similar in different species because the species have common descent. They may or may not perform the same function. An example is the forelimb structure shared by cats and whales.
2. Analogous structures - structures similar in different organisms because they evolved in a *similar environment*, rather than were inherited from a recent common ancestor. They usually serve the same or similar purposes. An example is the streamlined torpedo body shape of porpoises and sharks. So even though they evolved from different ancestors, porpoises and sharks developed analogous structures as a result of their evolution in the same aquatic environment.

History

The first specifically anatomical investigation separate from a surgical or medical procedure is associated by early commentators with Alcmaeon of Croton. Pierre Belon, a French naturalist born in 1517, conducted research and held discussions on dolphin embryos as well as the comparisons between the skeletons of birds to the skeletons of humans. His research later led to what is referred to as modern comparative anatomy.

A posthumous painting of Andreas Vesalius

Around the same time, Andreas Vesalius was also making some strides of his own. A young anatomist of Flemish descent made famous by a penchant for amazing charts, he was systematically investigating and correcting the anatomical knowledge of the Greek physician Galen. He noticed that many of Galen's observations were not even based on actual humans. Instead, they were based on animals such as oxen.Up until that point, Galen and his teachings had been the authority on human anatomy. The irony is that Galen himself had emphasized the fact that you should make your own observations instead of using those of another. But this advice was lost during the numerous translations of his work. As Vesalius began to uncover these mistakes, other physicians of the time began to trust their own observations more than Galen. An interesting observation made by some of these physicians was the presence of homologous structures in a wide variety of animals which included humans. These observations were later used by Darwin as he formed his theory of Natural Selection.

A drawing by Edward Tyson

Edward Tyson is regarded as the founder of comparative anatomy. He is credited with determining that marine mammals are, in fact, mammals. Also, he concluded that chimpanzees are more similar to humans than to monkeys because of their arms. Marco Aurelio Severino also compared various animals, including birds, in his *Zootomia democritaea*, one of the first works of comparative anatomy. In the 18th and 19th century, great anatomists like George Cuvier, Richard Owen and Thomas Henry Huxley revolutionized our understanding of the basic build and systematics of vertebrates, laying the foundation for Charles Darwin's work on evolution. An example of a 20th-century comparative anatomist is Victor Negus, who worked on the structure and evolution of the larynx. Until the advent of genetic techniques like DNA sequencing, comparative anatomy together with embryology were the primary tools for understanding phylogeny, as exemplified by the work of Alfred Romer.

In contemporary acedemia, comparative anatomy is still taught and used, particularly in the field of paleontology.

Homology (Biology)

In the context of biology, homology is the existence of shared ancestry between a pair of structures, or genes, in different taxa. A common example of homologous structures in evolutionary biology are the wings of bats and the arms of primates. Evolutionary theory explains the existence of homologous structures adapted to different purposes as the result of descent with modification from a common ancestor.

In the context of sexual differentiation—the process of development of the differences between males and females from an undifferentiated fertilized egg — the male and female organs are homologous if they develop from the same embryonic tissue. A typical example is the ovaries of female humans and the testicles of male humans.

In the context of morphological differentiation, organs that developed in the same embryological manner from similar origins, such as from matching primordia in successive segments of the same organism, may be said to be homologous. Examples include the legs of a centipede, the maxillary palp and labial palp of an insect, and the spinous processes of successive vertebrae in a vertebral column. In contrast, a sesamoid bone such as the patella is not homologous to a neighbouring skeletal bone such as the femur.

Sequence homology between protein or DNA sequences is similarly defined in terms of shared ancestry. Two segments of DNA can have shared ancestry because of either a speciation event (orthologs) or a duplication event (paralogs). Homology among proteins or DNA is inferred from their sequence similarity. Significant similarity is strong evidence that two sequences are related by divergent evolution from a common ancestor. Alignments of multiple sequences are used to discover the homologous regions.

Definition

Homology is the relationship between biological structures or sequences that are derived from a common ancestor. Homologous traits of organisms are therefore explained by descent from a common ancestor. The opposite of homologous organs are analogous organs which do similar jobs in two taxa that were not present in the last common ancestor but rather evolved separately. An example of an analogous trait would be the wings of bats and birds, which evolved independently in each lineage separately after diverging from ancestors with forelimbs not used as wings (terrestrial mammals and theropod dinosaurs, respectively).

A structure can be homologous at one level, but only analogous at another. For example, the bird and bat wings are analogous as wings, but homologous as forelimbs because the organ served as a forearm (not a wing) in the last common ancestor of tetrapods. Homology can also be described at the level of the gene. In genetics homology can refer to both the gene (DNA) and the corresponding protein product. It has been suggested that some behaviors might be homologous, based on either shared behavior across related taxa or common origins of the behavior in an individual's development, though this remains controversial.

Anatomical Homology

Evolutionary ancestry means that structures evolved from some structure in a common ancestor; for example, the wings of bats and the arms of primates are homologous in this sense. Developmental ancestry means that structures arose from the same tissue in embryonal development; the ovaries of female humans and the testicles of male humans are homologous in this sense.

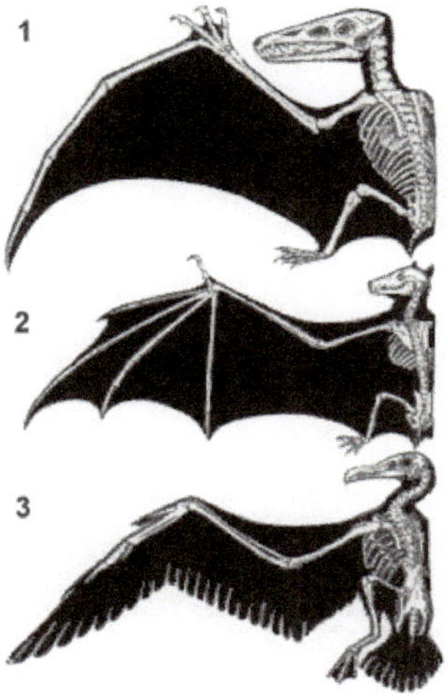

The wings of pterosaurs (1), bats (2) and birds (3) are analogous as wings, but homologous as forelimbs.

In evolutionary developmental biology (evo-devo), homology can also be partial. New structures can evolve through the combination of multiple developmental pathways, or parts of them. As a result, hybrid or mosaic structures can evolve that exhibit partial homologies. For example, certain compound leaves of flowering plants are partially homologous both to leaves and shoots because their development has evolved from a combination of leaf and shoot development.

Determining Homology

Systematists identify two forms of homology: primary homology is that initially conjectured by a researcher based on similar structure or anatomical connections, who states a hypothesis that two characters share an ancestry; secondary homology is implied by parsimony analysis, where a character that only occurs once on a tree is taken to be homologous. As implied in this definition, many cladists consider homology to be synonymous with synapomorphy, a shared derived character or trait state that distinguishes a clade from other organisms.

Homologous Structures in other Phyla

Introductory discussions of homology commonly limit themselves to the limbs of tetrapod vertebrates, occasionally touching on other structures, such as modified teeth as in whales and elephants. Homologies provide the fundamental basis for all biological classification, although some may be highly counter-intuitive. The embryonic body segments (somites) of different arthropod taxa are an example:

Somite	Trilobite	Spider (Chelicerata)	Centipede (Uniramia)	Insect (Uniramia)	Shrimp (Crustacea)
1	antennae	chelicerae (jaws and fangs)	antennae	antennae	1st antennae
2	1st legs	pedipalps	-	-	2nd antennae
3	2nd legs	1st legs	mandibles	mandibles	mandibles (jaws)
4	3rd legs	2nd legs	1st maxillae	1st maxillae	1st maxillae
5	4th legs	3rd legs	2nd maxillae	2nd maxillae	2nd maxillae
6	5th legs	4th legs	collum (no legs)	1st legs	1st legs

7	6th legs	-	1st legs	2nd legs	2nd legs
8	7th legs	-	2nd legs	3rd legs	3rd legs
9	8th legs	-	3rd legs	-	4th legs
10	9th legs	-	4th legs	-	5th legs

In many plants, defensive or storage structures are made by modifications of the development of primary leaves, stems, and roots.

Primary organs	Defensive structures	Storage structures
Leaves	Spines	Swollen leaves (e.g. succulents)
Stems	Thorns	Tubers (e.g. potato), rhizomes (e.g. ginger), fleshy stems (e.g. cacti)
Roots	-	Root tubers (e.g. sweet potato), taproot (e.g. carrot)

Sequence Homology

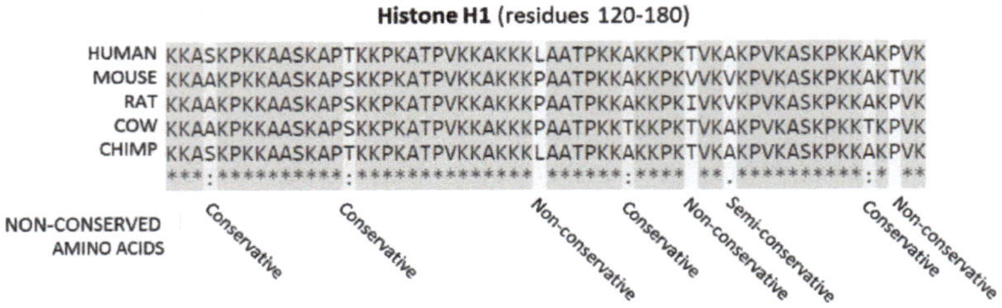

A sequence alignment, produced by ClustalO, of mammalian histone proteins. Sequences are the amino acids for residues 120-180 of the proteins. Residues that are conserved across all sequences are highlighted in grey. Below the protein sequences is a key denoting conserved sequence (*), conservative mutations (:), semi-conservative mutations (.), and non-conservative mutations ( ).

As with anatomical structures, sequence homology between protein or DNA sequences is defined in terms of shared ancestry. Two segments of DNA can have shared ancestry because of either a speciation event (orthologs) or a duplication event (paralogs). Homology among proteins or DNA is typically inferred from their sequence similarity. Significant similarity is strong evidence that two sequences are related by divergent evolution of a common ancestor. Alignments of multiple sequences are used to indicate which regions of each sequence are homologous. Homologous proteins make up protein families and superfamilies, encoded by gene families.

Homologous sequences are orthologous if they are descended from the same ancestral sequence separated by a speciation event: when a species diverges into two separate species, the copies of a single gene in the two resulting species are said to be orthologous. Orthologs, or orthologous genes, are genes in different species that originated by vertical descent from a single gene of the last common ancestor. The term "ortholog" was coined in 1970 by the molecular evolutionist Walter Fitch.

Homologous sequences are paralogous if they were created by a duplication event within the genome. For gene duplication events, if a gene in an organism is duplicated to occupy two different positions in the same genome, then the two copies are paralogous. Paralogous genes often belong to the same species. They can shape the structure of whole genomes and thus explain genome evolution to a large extent. Examples include the Homeobox (Hox) genes in animals. These genes not only underwent gene duplications within chromosomes but also whole genome duplications. As a result Hox genes in most vertebrates are clustered across multiple chromosomes with the HoxA-D clusters being the best studied.

Homology between Sexes and Forms

The term *homology* is sometimes applied to reproductive structures that share a common embryonic origin, but become spectacularly different between the two sexes in the adult. Those listed below are some of the more commonly cited examples.

Male structure	Female structure	Notes
prostate	skene's gland	-
penis	clitoris	-
-	uterus	homologous to eggshell-depositing organs in reptiles and birds
bulbourethral gland	Bartholin's gland	-

Among insects, the stinger used by infertile female worker honey bees is a modified ovipositor and is thus homologous with ovipositors in other Hymenoptera.

Difference Between Homology and Analogy

Homology is different from analogy, which describes the relation between characters that are apparently similar yet phylogenetically independent. The wings of a maple seed and the wings of an albatross are analogous, and perform somewhat similar functions, but not homologous (they both allow the organism to travel on the wind, but they didn't both develop from the same structure). Analogy is commonly also referred to as homoplasy (convergent evolution), which is further distinguished into parallelism, reversal, and convergence.

Convergent Evolution

Example: Two succulent plant genera, *Euphorbia* and *Astrophytum*, are only distantly related, but the species within each have independently converged on a similar body form.

Convergent evolution is the independent evolution of similar features in species of different lineages. Convergent evolution creates analogous structures that have similar form or function but were not present in the last common ancestor of those groups. The cladistic term for the same phenomenon is homoplasy, from Greek for *same form*. The recurrent evolution of flight is a classic example of convergent evolution. Flying insects, birds, and bats have all evolved the capacity of flight independently. They have "converged" on this useful trait.

Functionally similar features arising through convergent evolution are termed *analogous*, in contrast to *homologous* structures or traits, which have a common origin but not necessarily a similar function. The British anatomist Richard Owen was the first scientist to recognise the fundamental difference between analogies and homologies. Bat and pterosaur wings constitute an example of analogous structures, while the bat wing is homologous to human and other mammal forearms, sharing an ancestral state despite serving different functions. The opposite of convergent evolution is divergent evolution in which related species evolve different traits. On a molecular level, that can happen from random mutation unrelated to adaptive changes.

Convergent evolution is similar to but different from parallel evolution. Parallel evolution occurs when two independent but similar species evolve in the same direction and thus independently acquire similar characteristics; for instance, gliding frogs have evolved in parallel from multiple types of tree frog.

Many instances of convergent evolution are known in plants, including the repeated development of C4 photosynthesis.

Causes

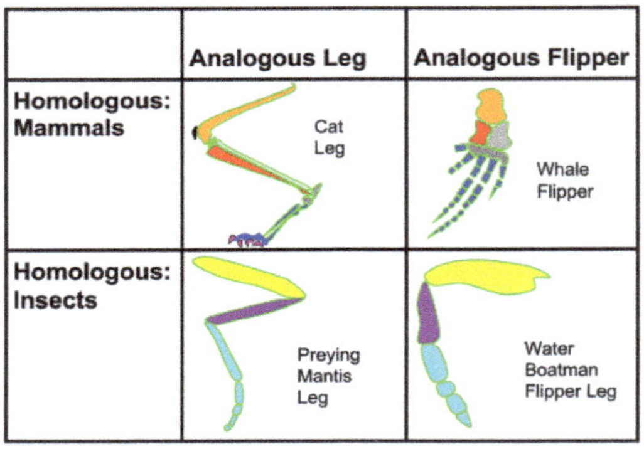

Mammals and insects are part of different homologous and analogous evolutionary groups. In the horizontal direction, the structures are homologous in their morphology, or anatomy, but different in their function due to differences in habitat. In the vertical direction, the structures are analogous in function due to similar lifestyles of organisms but anatomically different since they are part of different groups.

In morphology, analogous traits will often arise where different species live in similar ways and/or similar environment, and so face the same environmental factors. When occupying similar ecological niches (that is, a distinctive way of life) similar problems lead to similar solutions.

In biochemistry, physical and chemical constraints on mechanisms cause some active site arrangements to independently evolve multiple times in separate enzyme superfamilies.

Significance

Convergence has been associated with Darwinian evolution in the popular imagination since at least the 1940s. For example, Elbert A. Rogers argued: "If we lean toward the theories of Darwin might we not assume that man was [just as] apt to have developed in one continent as another?"

In his 1989 book *Wonderful Life*, Stephen Jay Gould argued that if the tape of life were re-wound and played back, life would have taken a very different course. Simon Conway Morris disputes this conclusion, arguing that convergence is a dominant force in evolution, and given that the same environmental and physical constraints are at work, life will inevitably evolve toward an "optimum" body plan, and at some point, evolution is bound to stumble upon intelligence, a trait presently identified with at least primates, corvids, and cetaceans.

Distinctions

Convergent evolution is a topic touched by many different fields of biology, many of which use slightly different nomenclature. This section attempts to clarify some of those terms.

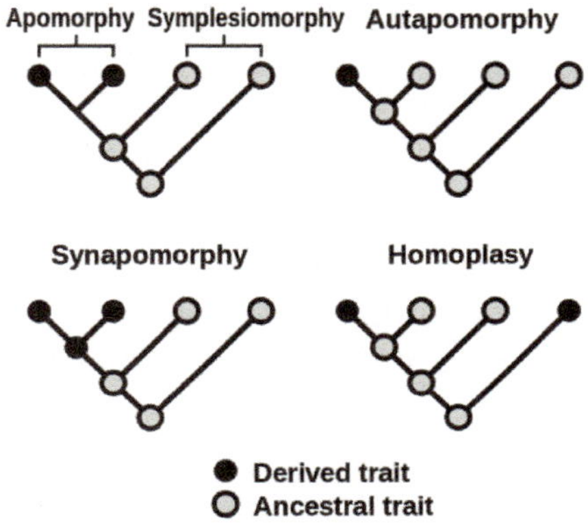

Diagram of cladistic definition of homoplasy, synapomorphy, autapomorphy, apomorphy and plesiomorphy.

Cladistic Definition

In cladistics, a homoplasy or a homoplastic character state is a trait (genetic, morphological etc.) that is shared by two or more taxa because of convergence, parallelism or reversal. Homoplastic character states require extra steps to explain their distribution on a most parsimonious cladogram. Homoplasy is only recognizable when other characters imply an alternative hypothesis of grouping, because in the absence of such evidence, shared features are always interpreted as similarity due to common descent. Homoplasious traits or changes (derived trait values acquired in unrelated organisms in parallel) can be compared with synapomorphy (a derived trait present in all members of a monophyletic clade), autapomorphy (derived trait present in only one member of a clade), or apomorphies, derived traits acquired in all members of a monophyletic clade following divergence where the most recent common ancestor had the *ancestral* trait (the ancestral trait manifesting in paraphyletic species as a plesiomorphy).

Re-evolution vs. Convergent Evolution

In some cases, it is difficult to tell whether a trait has been lost then re-evolved convergently, or whether a gene has simply been 'switched off' and then re-enabled later. Such a re-emerged trait is called an atavism. From a mathematical standpoint, an unused gene (selectively neutral) has a steadily decreasing probability of retaining potential functionality over time. The time scale of this process varies greatly in different phylogenies; in mammals and birds, there is a reasonable probability of remaining in the genome in a potentially functional state for around 6 million years.

Parallel vs. Convergent Evolution

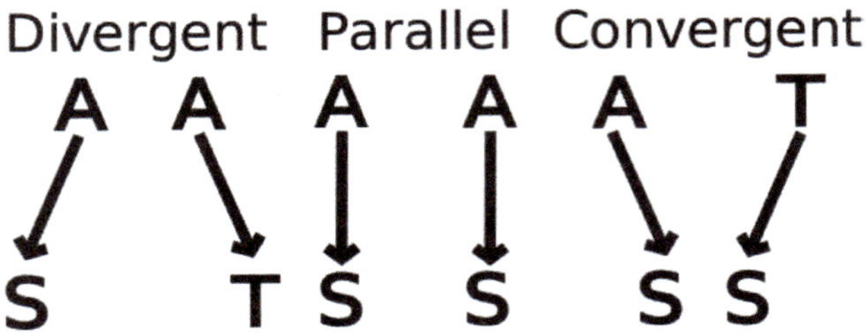

Evolution at an amino acid position. In each case, the left-hand species changes from incorporating alanine (A) at a specific position within a protein in a hypothetical common ancestor deduced from comparison of sequences of several species, and now incorporates serine (S) in its present-day form. The right-hand species may undergo divergent, parallel, or convergent evolution at this amino acid position relative to that of the first species.

For a particular trait, proceeding in each of two lineages from a specified ancestor to a later descendant, parallel and convergent evolutionary trends can be strictly defined and clearly distinguished from one another. However the cutoff point for what is considered convergent and what is considered parallel evolution is assigned somewhat arbitrarily. When two species are similar in a particular character, evolution is defined as parallel if the ancestors were also similar and convergent if they were not. However, this definition is somewhat murky. All organisms share a common ancestor more or less recently, so the question of how far back to look in evolutionary time and how similar the ancestors need to be for one to consider parallel evolution to have taken place is not entirely resolved within evolutionary biology. Some scientists have argued parallel evolution and convergent evolution are more or less indistinguishable from one another. Others have argued that we should not shy away from the gray area and that there are still important distinctions between parallel and convergent evolution.

When the ancestral forms are unspecified or unknown, or the range of traits considered is not clearly specified, the distinction between parallel and convergent evolution becomes more subjective. For instance, the striking example of similar placental and marsupial forms is described by Richard Dawkins in *The Blind Watchmaker* as a case of convergent evolution, because mammals on each continent had a long evolutionary history prior to the extinction of the dinosaurs under which to accumulate relevant differences. Stephen Jay Gould describes many of the same examples as parallel evolution starting from the common ancestor of all marsupials and placentals. Many evolved similarities can be described in concept as parallel evolution from a remote ancestor, with the exception of those where quite different structures are co-opted to a similar function. For example, consider *Mixotricha paradoxa*, a microbe that has assembled a system of rows of apparent cilia and basal bodies closely resembling that of ciliates but that are actually smaller symbiont micro-organisms, or the differently oriented tails of fish and whales. On the converse, any case in which lineages do not evolve together at the same time in the same ecospace might be described as convergent evolution at some point in time.

The definition of a trait is crucial in deciding whether a change is seen as divergent, or as parallel or convergent. In the image above (with a caption that begins "Evolution at an amino acid position"), note that, since serine and threonine possess similar structures with an alcohol side-chain, the example marked *"divergent"* would be termed *"parallel"* if the amino acids were grouped by similarity instead of being considered individually. As another example, if genes in two species independently become restricted to the same region of the animals through regulation by a certain transcription factor, this may be described as a case of parallel evolution — but examination of the actual DNA sequence will probably show only divergent changes in individual base-pair positions, since a new transcription factor binding site can be added in a wide range of places within the gene with similar effect.

A similar situation occurs considering the homology of morphological structures. For example, many insects possess two pairs of flying wings. In beetles, the first pair of wings is hardened into wing covers with little role in flight, while in flies the second pair of wings is condensed into small halteres used for balance. If the two pairs of wings are considered as interchangeable, homologous structures, this may be described as a parallel reduction in the number of wings, but otherwise the two changes are each divergent changes in one pair of wings.

Similar to convergent evolution, evolutionary relay describes how independent species acquire similar characteristics through their evolution in similar ecosystems, but not at the same time (dorsal fins of sharks and ichthyosaurs).

Examples

Enzymes

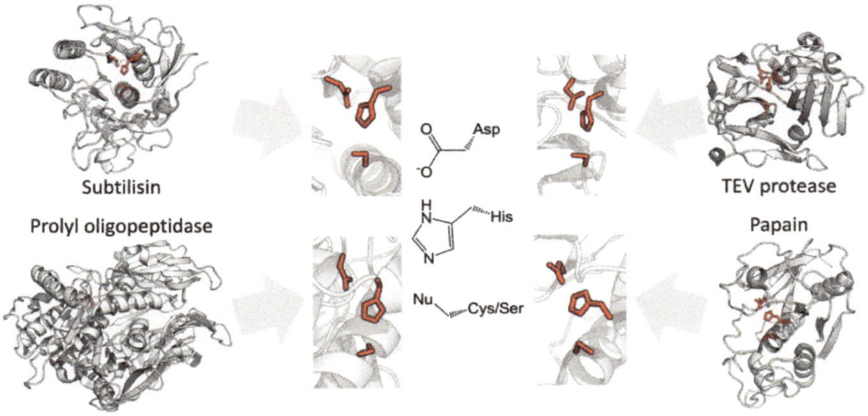

Evolutionary convergence of serine and cysteine protease towards the same catalytic triads organisation of acid-base-nucleophile in different protease superfamilies. Shown are the triads of subtilisin, prolyl oligopeptidase, TEV protease, and papain.

Protease Active Sites

The enzymology of proteases provides some of the clearest examples of convergent evolution. These examples reflect the intrinsic chemical constraints on enzymes, leading evolution to independently converge on equivalent solutions repeatedly.

Serine and cysteine proteases use different amino acid functional groups (alcohol or thiol) as a nucleophile. In order to activate that nucleophile, they orient an acidic and basic residue in a catalytic triad. The chemical and physical constraints on enzyme catalysis have caused identical triad arrangements to have evolved independently over 20 times in different enzyme superfamilies.

Threonine proteases use the amino acid threonine as their catalytic nucleophile. Unlike cysteine and serine, threonine is a secondary alcohol (i.e. has a methyl group). The methyl group of threonine greatly restricts the possible orientations of triad and substrate as the methyl clashes with either the enzyme backbone or histidine base. Consequently, most threonine proteases use an N-terminal threonine in order to avoid such steric clashes. Several evolutionarily independent enzyme superfamilies with different protein folds use the N-terminal residue as a nucleophile. This commonality of active site but difference of protein fold indicates that the active site evolved convergently in those families.

Nucleic Acids

As more sequence data are becoming available, there is growing interest in convergent changes at the level of DNA and amino acids. In 2013 the first genome-wide study of convergence was published. Comparisons of the genomes of echolocating bats and the dolphin identified numerous convergent amino acid substitutions in genes implicated in hearing and vision.

Animals

Bodyplans

A classic comparison is between the marsupial fauna of Australia and the placental mammals of the Old World. The two lineages are clades—that is, they each share a common ancestor that belongs to their own group, and are more closely related to one another than to any other clade—but very similar forms evolved in each isolated population. Many body plans, for instance sabre-toothed cats and flying squirrels, evolved independently in both populations.

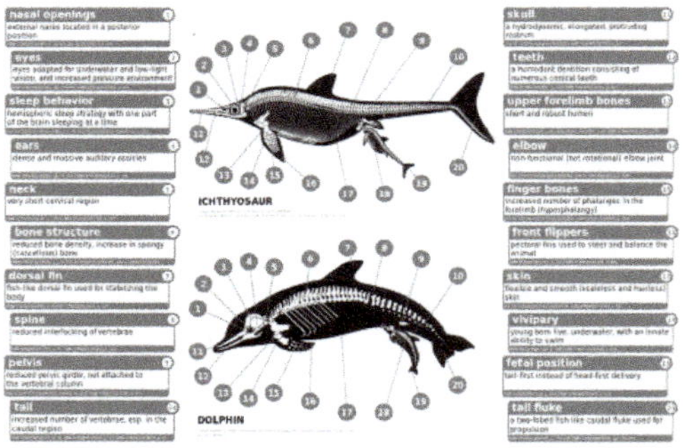

Examples of distinct features shared both by dolphins and derived ichthyopterygians

The same streamlined shape has been converged upon by fish (e.g., herrings), marine mammals (e.g., dolphins) and even the extinct ichthyosaur (Mesozoic era). This bodyplan is an adaptation to being an active predator in a high drag environment. It is also debated whether earless seals and eared seals are a single marine group, or represent two separate episodes of carnivorans turning to a marine environment.

Flight

A classical example of an analogy is the ability to fly in birds and bats. Both groups can move by powered flight, but flight has evolved independently in the two groups. The ability to fly does not make birds and bats close relatives. The ancestors of both bats and birds were terrestrial quadrupeds, and each has independently evolved powered flight via adaptations of their forelimbs. Although both forelimb adaptations are superficially "wing-shaped," they are substantially dissimilar in construction. The bat wing is a membrane stretched across four *extremely* elongated fingers, while the airfoil of the bird wing is made of feathers, which are strongly attached to the forearm (the ulna) and the highly fused bones of the wrist and hand (the carpometacarpus), with only tiny remnants of two fingers remaining, each anchoring a single feather. (Both bats and birds have retained the thumb for specialized functions.) So, while the wings of bats and birds are functionally convergent, they are not anatomically convergent.

Similarly, the extinct pterosaur also shows an independent evolution of vertebrate forelimb to wing. An even more distantly related group with wings is the insects, they not only evolved separately as wings, but from totally different organs, starting from a fundamentally different body-plan.

Opposable Thumbs

Opposable thumbs allowing the grasping of objects are most often associated with primates, like humans, monkeys, apes, and lemurs. Opposable thumbs also evolved in pandas, however, and are completely different in structure, having six fingers including the thumb, which develops from a wrist bone entirely separately from other fingers.

Flying Squirrels and Sugar Gliders

While they look almost the same, flying squirrels and sugar gliders are very different. Flying squirrels are placental mammals and sugar gliders are marsupials, which practically puts them at opposite ends of the mammal lineage. Humans are more closely related to flying squirrels than flying squirrels are to sugar gliders.

Echolocation

As a sensory adaptation, echolocation has evolved several times, in cetaceans (dolphins and whales), as well as bats. Surprisingly, humans actually are capable of echolocation with training.

Eyes

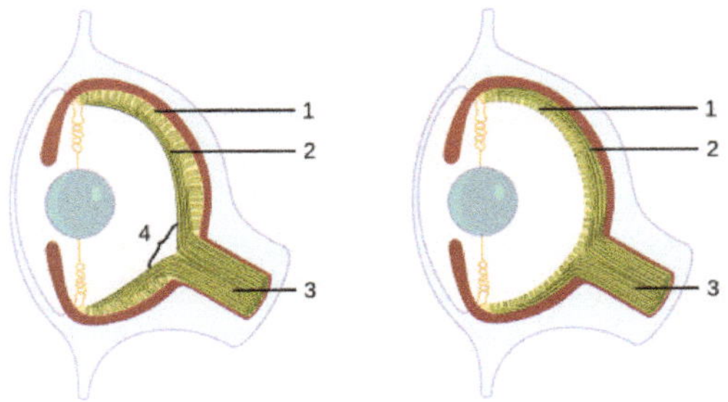

Vertebrates and cephalopods developed the camera eye independently. In the vertebrate version the nerve fibers pass in front of the retina, and there is a blind spot, **4**, where the nerves pass through the retina. In the cephalopod version, the eye is constructed the "right way out", with the nerves attached to the rear of the retina.

One of the best-known examples of convergent evolution is the camera eye of cephalopods (such as squid and octopus), vertebrates (including mammals) and cnidaria (such as jellyfish). Their last common ancestor had at most a simple photoreceptive spot, but a range of processes led to the progressive refinement of camera eyes — with one sharp difference: the cephalopod eye is "wired" in the opposite direction, with blood and nerve vessels entering from the back of the retina, rather than the front as in vertebrates. This means that cephalopods do not have a blind spot.

Insect Mouthparts

Insect mouthparts show many examples of organs which can be used to study convergent evolution in the context of form and function. The mouthparts of different insect groups consist of a set of homologous organs, specialised to the function of dietary intake of that insect group (which can be experimentally quantified). Convergent evolution of many groups of insects led from original biting-chewing mouthparts to different derived function types. They build a proboscis at flower-visiting insects for example, which are able to ingest food very efficiently or biting-sucking mouthparts, showing different function mechanisms at different groups of blood-sucking insects.

Plants

While convergent evolution is often illustrated with animal examples, it has often occurred in plant evolution. For instance, C4 photosynthesis, one of the three major carbon-fixing biochemical processes, has arisen independently up to 40 times. About 7,600 plant species of angiosperms use C_4 carbon fixation, with many monocots including 46% of grasses such as maize and sugar cane, and dicots including a sizeable minority of species in the Chenopodiaceae and the Amaranthaceae.

Good examples of convergence in plants include the evolution of true fruits such as apples. These incorporate one or more ovules and their accessory tissues, but many edible plant-derived tissues commonly regarded as fruits actually arise from very different embryological structures. This implies a convergent process in which genetically unrelated precursors assume a common form under selective pressure, in this case the competition for seed dispersal through consumption by animals.

In myrmecochory, seeds such as those of Chelidonium majus have a hard coating and an attached oil body, an elaiosome, for dispersal by ants.

The emergence of seed dispersal by ants (myrmecochory) has evolved independently more than 100 times and is present in more than 11,000 plant species. It is also considered one of the most dramatic examples of convergent evolution in biology.

Connective Tissue

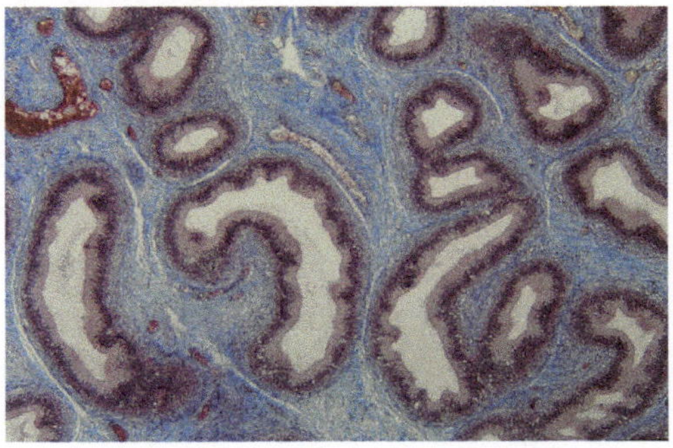

Section of epididymis. Connective tissue (blue) is seen supporting the epithelium (purple)

Connective tissue (CT) is one of the four types of biological tissue that supports, connects or separates different types of tissues and organs in the body. It develops from the mesoderm. The other three types are epithelial, muscle, and nervous tissue. Connective tissue is found in between other tissues everywhere in the body, including the nervous system. In the central nervous system, the three outer membranes (the meninges) that envelop the brain and spinal cord are composed of connective tissue.

All connective tissue consists of three main components: fibers (elastic and collagenous fibers), ground substance and cells. (Not all authorities include blood or lymph as connective tissue.) Blood and lymph lack the fiber component. All are immersed in the body water.

The cells of connective tissue include fibroblasts, adipocytes, macrophages, mast cells and leucocytes.

Structure

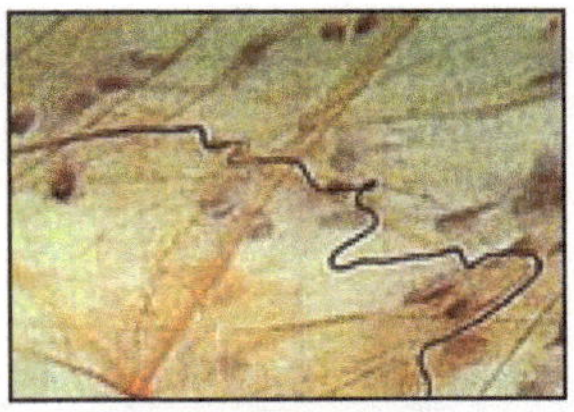

Areolar connective tissue

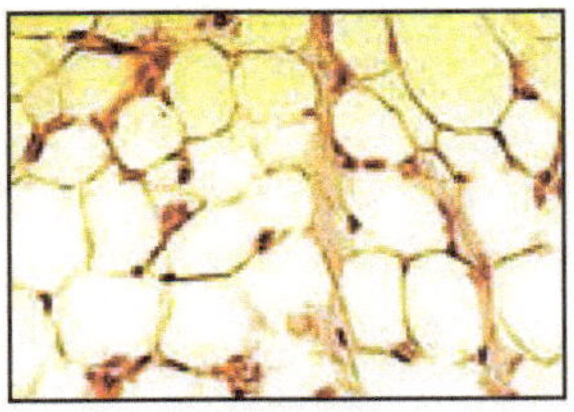

Adipose tissue

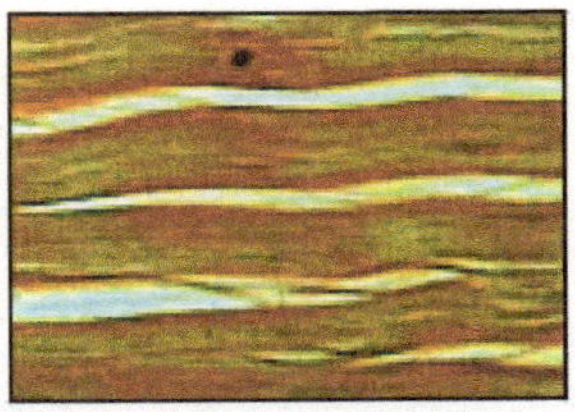

Fibrous connective tissue

Connective tissue can be broadly subdivided into connective tissue proper, and special connective tissue. Connective tissue proper consists of loose connective tissue and dense connective tissue (which is further subdivided into dense regular and dense irregular connective tissues.) Special connective tissue consists of reticular connective tissue, adipose tissue, cartilage, bone, and blood. Other kinds of connective tissues include fibrous, elastic, and lymphoid connective tissues. New vascularised connective tissue that forms in the process of wound healing is termed granulation tissue. Fibroblasts are the cells responsible for the production of some CT.

Type I collagen, is present in many forms of connective tissue, and makes up about 25% of the total protein content of the mammalian body.

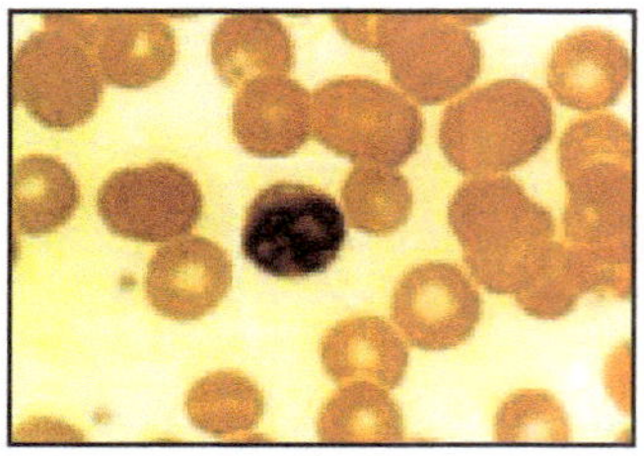
Blood

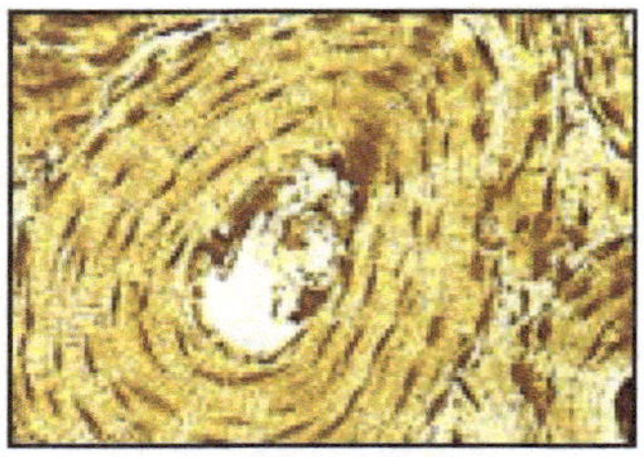
Osseous tissue

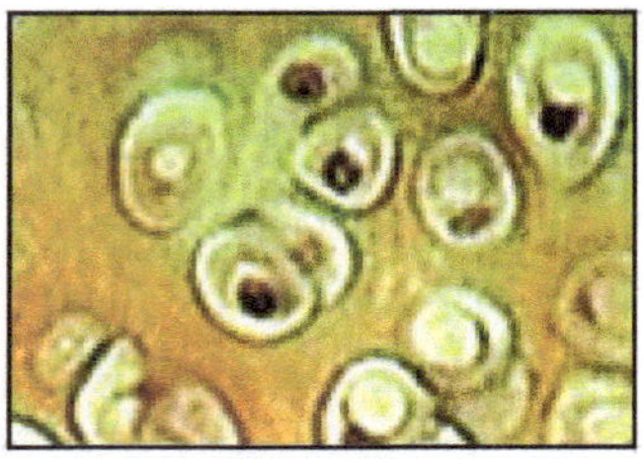
Hyaline cartilage

Characteristics

Characteristics of CT:

- Cells are spread through an extracellular fluid.
- Ground substance - A clear, colorless, and viscous fluid containing glycosaminoglycans and proteoglycans to fix the body water and the collagen fibers in the intercellular spaces. Ground substance slows the spread of pathogens.
- Fibers. Not all types of CT are fibrous. Examples of non-fibrous CT include adipose tissue and blood. Adipose tissue gives "mechanical cushioning" to the body, among other functions. Although there is no dense collagen network in adipose tissue, groups of adipose cells are kept together by collagen fibers and collagen sheets in order to keep fat tissue under compression in place (for example, the sole of the foot). The matrix of blood is plasma.
- Both the ground substance and proteins (fibers) create the matrix for CT.

Types of fibers:

Tissue	Purpose	Components	Location
Collagenous fibers	Bind bones and other tissues to each other	Alpha polypeptide chains	tendon, ligament, skin, cornea, cartilage, bone, blood vessels, gut, and intervertebral disc.
Elastic fibers	Allow organs like arteries and lungs to recoil	Elastic microfibril and elastin	extracellular matrix
Reticular fibers	Form a scaffolding for other cells	Type III collagen	liver, bone marrow, and lymphatic organs

Function

Connective tissue has a wide variety of functions that depend on the types of cells and the different classes of fibers involved. Loose and dense irregular connective tissue, formed mainly by fibroblasts and collagen fibers, have an important role in providing a medium for oxygen and nutrients to diffuse from capillaries to cells, and carbon dioxide and waste substances to diffuse from cells back into circulation. They also allow organs to resist stretching and tearing forces. Dense regular connective tissue, which forms organized structures, is a major functional component of tendons, ligaments and aponeuroses, and is also found in highly specialized organs such as the cornea. Elastic fibers, made from elastin and fibrillin, also provide resistance to stretch forces. They are found in the walls of large blood vessels and in certain ligaments, particularly in the ligamenta flava.

In hematopoietic and lymphatic tissues, reticular fibers made by reticular cells provide the stroma—or structural support—for the parenchyma—or functional part—of the organ.

Mesenchyme is a type of connective tissue found in developing organs of embryos that is capable of differentiation into all types of mature connective tissue. Another type of relatively undifferentiated connective tissue is mucous connective tissue, found inside the umbilical cord.

Various type of specialized tissues and cells are classified under the spectrum of connective tissue, and are as diverse as brown and white adipose tissue, blood, cartilage and bone.Cells of the immune system, such as macrophages, mast cells, plasma cells and eosinophils are found scattered in loose connective tissue, providing the ground for starting inflammatory and immune responses upon the detection of antigens.

Clinical Significance

It is estimated that 1 in 10 people have a connective tissue disorder.Diseases of connective tissue include:

- Connective tissue neoplasms including sarcomas such as hemangiopericytoma and malignant peripheral nerve sheath tumor in nervous tissue.
- Congenital diseases include Marfan syndrome and Ehlers-Danlos Syndrome.
- Myxomatous degeneration – a pathological weakening of connective tissue.
- Mixed connective tissue disease – a disease of the autoimmune system, also undifferentiated connective tissue disease.
- Systemic lupus erythematosus (SLE) – a major autoimmune disease of connective tissue
- Scurvy, caused by a deficiency of vitamin C which is necessary for the synthesis of collagen.

Staining of Connective Tissue

For microscopic viewing, most of the connective tissue staining-techniques, color tissue fibers in contrasting shades. Collagen may be differentially stained by any of the following:

- Van Gieson's stain
- Masson's trichrome stain
- Mallory's trichrome stain
- Aniline blue stain
- Eosin
- Reticulin stain

Epithelium

Epithelium (*epi- + thele + -ium*) is one of the four basic types of animal tissue. The other three types are connective tissue, muscle tissue and nervous tissue. Epithelial tissues line the cavities and surfaces of blood vessels and organs throughout the body.

There are three principal shapes of epithelial cells: squamous, columnar, and cuboidal. These can be arranged in a single layer of cells as simple epithelium, either squamous, columnar, cuboidal, pseudo-stratified columnar or in layers of two or more cells deep as stratified (layered), either squamous, columnar or cuboidal. All glands are made up of epithelial cells. Functions of epithelial cells include secretion, selective absorption, protection, transcellular transport, and sensing.

Epithelial layers contain no blood vessels, so they must receive nourishment via diffusion of substances from the underlying connective tissue, through the basement membrane.

Classification

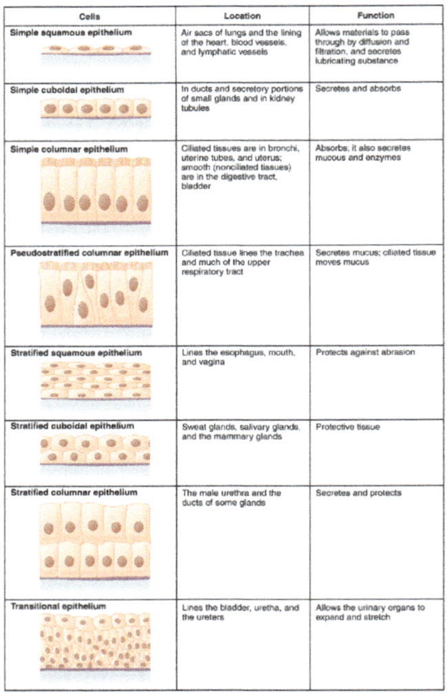

Cells	Location	Function
Simple squamous epithelium	Air sacs of lungs and the lining of the heart, blood vessels, and lymphatic vessels	Allows materials to pass through by diffusion and filtration, and secretes lubricating substance
Simple cuboidal epithelium	In ducts and secretory portions of small glands and in kidney tubules	Secretes and absorbs
Simple columnar epithelium	Ciliated tissues are in bronchi, uterine tubes, and uterus; smooth (nonciliated tissues) are in the digestive tract, bladder	Absorbs; it also secretes mucous and enzymes
Pseudostratified columnar epithelium	Ciliated tissue lines the trachea and much of the upper respiratory tract	Secretes mucus; ciliated tissue moves mucus
Stratified squamous epithelium	Lines the esophagus, mouth, and vagina	Protects against abrasion
Stratified cuboidal epithelium	Sweat glands, salivary glands, and the mammary glands	Protective tissue
Stratified columnar epithelium	The male urethra and the ducts of some glands	Secretes and protects
Transitional epithelium	Lines the bladder, uretha, and the ureters	Allows the urinary organs to expand and stretch

Summary showing different epithelial cells/tissues and their characteristics.

In general, epithelial tissues are classified by the number of their layers and by the shape and function of the cells.

The three principal shapes associated with epithelial cells are—squamous, cuboidal and columnar.

- Squamous epithelium has cells that are wider than their height (flat and scale-like).
- Cuboidal epithelium has cells whose height and width are approximately the same (cube shaped).
- Columnar epithelium has cells taller than they are wide (column-shaped).

By layer, epithelium is classed as either simple epithelium, only one cell thick (unilayered) or stratified epithelium as stratified squamous epithelium, stratified cuboidal epithelium, and stratified columnar epithelium that are two or more cells thick (multi-layered), and both types of layering can be made up of any of the cell shapes. However, when taller simple columnar epithelial cells are viewed in cross section showing several nuclei appearing at different heights, they can be confused with stratified epithelia. This kind of epithelium is therefore described as pseudostratified epithelium.

Transitional epithelium has cells that can change from squamous to cuboidal, depending on the amount of tension on the epithelium.

Simple Epithelium

Simple epithelium is a single layer of cells with every cell in direct contact with the basement membrane that separates it from the underlying connective tissue. In general, it is found where absorption and filtration occur. The thinness of the epithelial barrier facilitates these processes.

In general, simple epithelial tissues are classified by the shape of their cells. The four major classes of simple epithelium are: (1) simple squamous; (2) simple cuboidal; (3) simple columnar; (4) pseudostratified.

(1) simple squamous; which is found lining areas where passive diffusion of gases occur. e.g. skin, walls of capillaries, linings of the pericardial, pleural,and peritoneal cavities, as well as the linings of the alveoli of the lungs.

(2) simple cuboidal: these cells may have secretory, absorptive, or excretory functions. examples include small collecting ducts of kidney,pancreas and salivary gland.

(3) simple columnar; found in areas with extremely high secretive (as in wall of the stomach), or absorptive (as in small intestine) areas. they possess cellular extensions (e.g. microvilli in the small intestine, or cilia found almost exclusively in the female reproductive tract).

(4) pseudostratified epithelium; this is also called respiratory epithelium as it is almost exclusively confined to the larger respiratory airways of the nasal cavity, trachea and bronchi.

Stratified Epithelium

Stratified epithelium differs from simple epithelium in that it is multilayered. It is therefore found where body linings have to withstand mechanical or chemical insult such that layers can be abraded and lost without exposing subepithelial layers. Cells flatten as the layers become more apical, though in their most basal layers the cells can be squamous, cuboidal or columnar.

Stratified epithelia (of columnar, cuboidal or squamous type) can have the following specializations:

Specialization	Description
Keratinized	In this particular case, the most apical layers (exterior) of cells are dead and lose their nucleus and cytoplasm, instead contain a tough, resistant protein called keratin. This specialization makes the epithelium waterproof, so is found in the mammalian skin. The lining of the esophagus is an example of a non-keratinized or "moist" stratified epithelium.

Transitional	Transitional epithelia are found in tissues that stretch and it can appear to be stratified cuboidal when the tissue is not stretched or stratified squamous when the organ is distended and the tissue stretches. It is sometimes called urothelium since it is almost exclusively found in the bladder, ureters and urethra.

Cell Types

The basic cell types are squamous, cuboidal, and columnar classed by their shape.

Type	Description
Squamous	Squamous cells have the appearance of thin, flat plates that can look polygonal when viewed from above. Their name comes from *squāma*, Latin for scale – as on fish or snake skin. The cells fit closely together in tissues; providing a smooth, low-friction surface over which fluids can move easily. The shape of the nucleus usually corresponds to the cell form and helps to identify the type of epithelium. Squamous cells tend to have horizontally flattened, nearly oval shaped nuclei because of the thin flattened form of the cell. Squamous epitheliium is found lining surfaces such as the skin, and alveoli in the lung, enabling simple passive diffusion as also found in the alveolar epithelium in the lungs. Specialized squamous epithelium also forms the lining of cavities such as in blood vessels, as endothelium and in the pericardium, as mesothelium and in other body cavities.
Cuboidal	Cuboidal epithelial cells have a cube-like shape and appear square in cross-section. The cell nucleus is large, spherical and is in the centre of the cell. Cuboidal epithelium is commonly found in secretive tissue such as the exocrine glands, or in absorptive tissue such as the pancreas, the lining of the kidney tubules as well as in the ducts of the glands. The germinal epithelium that covers the female ovary, and the germinal epithelium that lines the walls of the seminferous tubules in the testes are also of the cuboidal type. Cuboidal cells provide protection and may be active in pumping material in or out of the lumen, or passive depending on their location and specialisation. Simple cuboidal epithelium commonly differentiates to form the secretory and duct portions of glands. Stratified cuboidal epithelium protects areas such as the ducts of sweat glands, mammary glands, and salivary glands.
Columnar	Columnar epithelial cells are elongated and column-shaped and have a height of at least four times their width. Their nuclei are elongated and are usually located near the base of the cells. Columnar epithelium forms the lining of the stomach and intestines. The cells here may possess microvilli for maximising the surface area for absorption and these microvilli may form a brush border. Other cells may be ciliated to move mucus in the function of mucociliary clearance. Other ciliated cells are found in the fallopian tubes, the uterus and central canal of the spinal cord. Some columnar cells are specialized for sensory reception such as in the nose, ears and the taste buds. Goblet cells are modified columnar cells and are found between the columnar epithelial cells of the duodenum. They secrete mucus, which acts as a lubricant. Single-layered non-ciliated columnar epithelium tends to indicate an absorptive function. Stratified columnar epithelium is rare but is found in lobar ducts in the salivary glands, the eye, pharynx and sex organs. This consists of a layer of cells resting on at least one other layer of epithelial cells which can be squamous, cuboidal, or columnar.
Pseudostratified	These are simple columnar epithelial cells whose nuclei appear at different heights, giving the misleading (hence "pseudo") impression that the epithelium is stratified when the cells are viewed in cross section. Ciliated pseudostratified epithelial cells have cilia. Cilia are capable of energy dependent pulsatile beating in a certain direction through interaction of cytoskeletal microtubules and connecting structural proteins and enzymes. In the respiratory tract the wafting effect produced causes mucus secreted locally by the goblet cells (to lubricate and to trap pathogens and particles) to flow in that direction (typically out of the body). Ciliated epithelium is found in the airways (nose, bronchi), but is also found in the uterus and Fallopian tubes, where the cilia propel the ovum to the uterus.

Structure

Cells of epithelial tissue are tightly packed and form a continuous sheet. They have almost no intercellular spaces. All epithelia is usually separated from underlying tissues by an extracellular fibrous basement membrane. The lining of the mouth, lung alveoli and kidney tubules all are made of epithelial tissue. The lining of the blood and lymphatic vessels are of a specialised form of epithelium called endothelium.

Location

Epithelium lines both the outside (skin) and the inside cavities and lumina of bodies. The outermost layer of human skin is composed of dead stratified squamous, keratinized epithelial cells.

Tissues that line the inside of the mouth, the esophagus and part of the rectum are composed of nonkeratinized stratified squamous epithelium. Other surfaces that separate body cavities from the outside environment are lined by simple squamous, columnar, or pseudostratified epithelial cells. Other epithelial cells line the insides of the lungs, the gastrointestinal tract, the reproductive and urinary tracts, and make up the exocrine and endocrine glands. The outer surface of the cornea is covered with fast-growing, easily regenerated epithelial cells. A specialised form of epithelium – endothelium forms the inner lining of blood vessels and the heart, and is known as vascular endothelium, and lining lymphatic vessels as lymphatic endothelium. Another type, mesothelium, forms the walls of the pericardium, pleurae, and peritoneum.

In arthropods, the integument, or external "skin", consists of a single layer of epithelial ectoderm from which arises the cuticle, an outer covering of chitin the rigidity of which varies as per its chemical composition.

Basement Membrane

Epithelial tissue rests on a basement membrane, which acts as a scaffolding on which epithelium can grow and regenerate after injuries. Epithelial tissue has a nerve supply, but no blood supply and must be nourished by substances diffusing from the blood vessels in the underlying tissue. The basement membrane acts as a selectively permeable membrane that determines which substances will be able to enter the epithelium.

Cell Junctions

Cell junctions are especially abundant in epithelial tissues. They consist of protein complexes and provide contact between neighbouring cells, between a cell and the extracellular matrix, or they build up the paracellular barrier of epithelia and control the paracellular transport.

Cell junctions are the contact points between plasma membrane and tissue cells. There are mainly 5 different types of cell junctions: tight junctions, adherens junctions, desmosomes, hemidesmosomes, and gap junctions. Tight junctions are a pair of trans-membrane protein fused on outer plasma membrane. Adherens junctions are a plaque (protein layer on the inside plasma membrane) which attaches both cells' microfilaments. Desmosomes attach to the microfilaments of cytoskeleton made up of keratin protein. Hemidesmosomes resemble desmosomes on a section. They are made up of the integrin (a transmembrane protein) instead of cadherin. They attach the epithelial cell to the basement membrane. Gap junctions connect the cytoplasm of two cells and are made up of proteins called connexins (six of which come together to make a connexon).

Development

Epithelial tissues are derived from all of the embryological germ layers:

- from ectoderm (e.g., the epidermis);
- from endoderm (e.g., the lining of the gastrointestinal tract);
- from mesoderm (e.g., the inner linings of body cavities).

However, it is important to note that pathologists do not consider endothelium and mesothelium (both derived from mesoderm) to be true epithelium. This is because such tissues present very different pathology. For that reason, pathologists label cancers in endothelium and mesothelium sarcomas, whereas true epithelial cancers are called carcinomas. Also, the filaments that support these mesoderm-derived tissues are very distinct. Outside of the field of pathology, it is, in general, accepted that the epithelium arises from all three germ layers.

Function

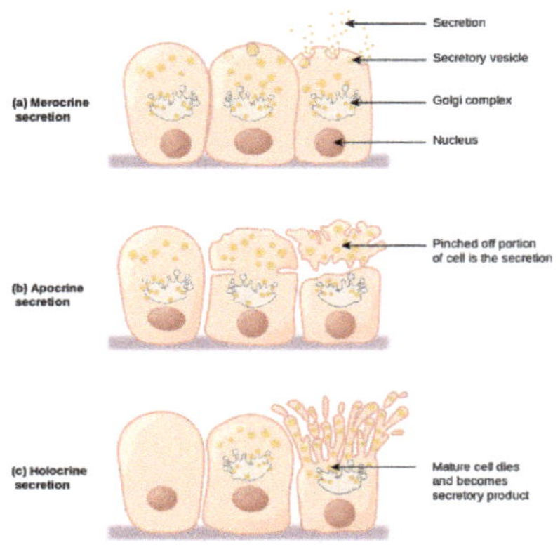

Forms of secretion in glandular tissue

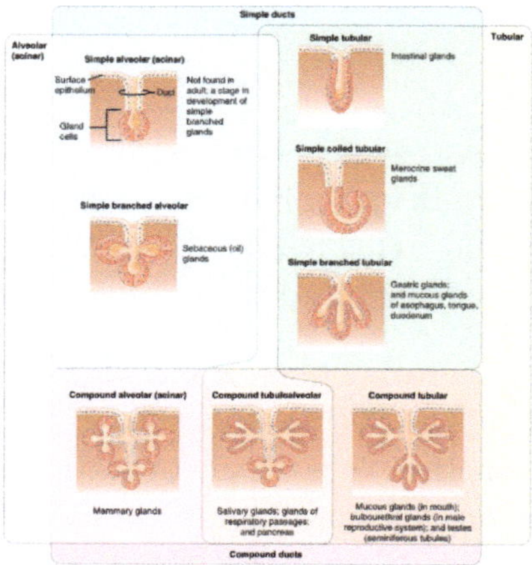

Different characteristics of glands of the body

The primary functions of epithelial tissues are: (1) to protect the tissues that lie beneath it from radiation, desiccation, toxins, invasion by pathogens, and physical trauma; (2) the regulation and exchange of chemicals between the underlying tissues and a body cavity; (3) the secretion of hormones into the blood vascular system, and/or the secretion of sweat, mucus, enzymes, and other products that are delivered by ducts; (4) to provide sensation.

Glandular Tissue

Glandular tissue is the type of epithelium that forms the glands from the infolding of epithelium and subsequent growth in the underlying connective tissue. There are two major classifications of glands: endocrine glands and exocrine glands. Endocrine glands secrete their product into the extracellular space where it is rapidly taken up by the blood vascular system. Exocrine glands secrete their products into a duct that then delivers the product to the lumen of an organ or onto the free surface of the epithelium.

Sensing the Extracellular Environment

"Some epithelial cells are ciliated, especially in respiratory epithelium, and they commonly exist as a sheet of polarised cells forming a tube or tubule with cilia projecting into the lumen." Primary cilia on epithelial cells provide chemosensation, thermoception, and mechanosensation of the extracellular environment by playing "a sensory role mediating specific signalling cues, including soluble factors in the external cell environment, a secretory role in which a soluble protein is released to have an effect downstream of the fluid flow, and mediation of fluid flow if the cilia are motile."

Clinical Significance

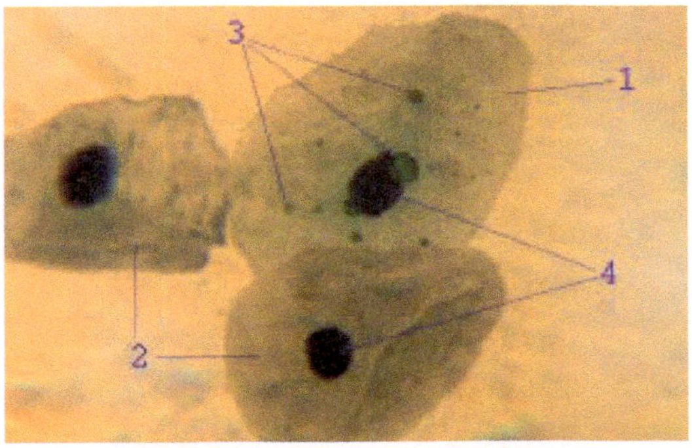

Epithelial cell infected with Chlamydia pneumoniae

The slide shows at (1) an epithelial cell infected by *Chlamydia pneumonia*; their inclusion bodies shown at (3); an uninfected cell shown at (2) and (4) showing the difference between an infected cell nucleus and an uninfected cell nucleus.

Epithelium grown in culture can be identified by examining its morphological characteristics. Epithelial cells tend to cluster together, and have a "characteristic tight pavementlike appearance". But this is not always the case, such as when the cells are derived from a tumor. In these cases, it is often necessary to use certain biochemical markers to make a positive identification. The intermediate filament proteins in the cytokeratin group are almost exclusively found in epithelial cells, and so are often used for this purpose.

Carcinomas develop in epithelial tissues.(Sarcomas develop in connective tissue).

Additional Images

Squamous Epithelium 100x

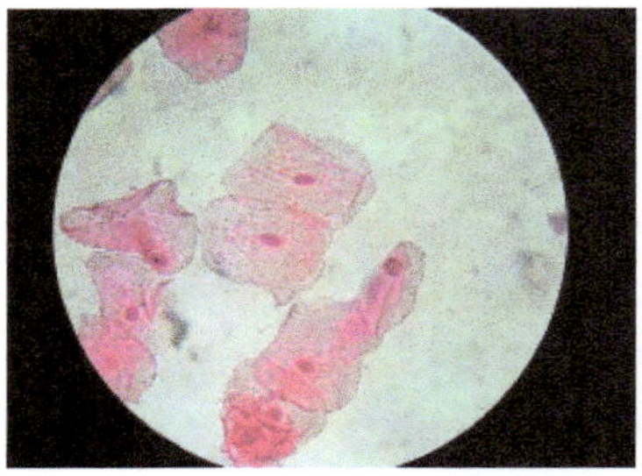

Human cheek cells (Nonkeratinized stratified squamous epithelium) 500x

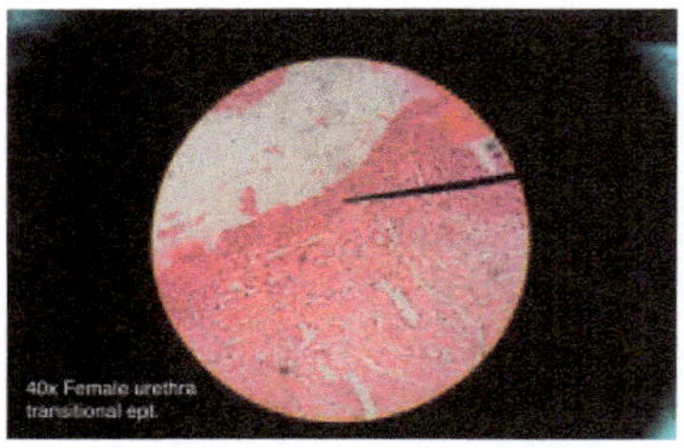

Histology of female urethra showing transitional epithelium

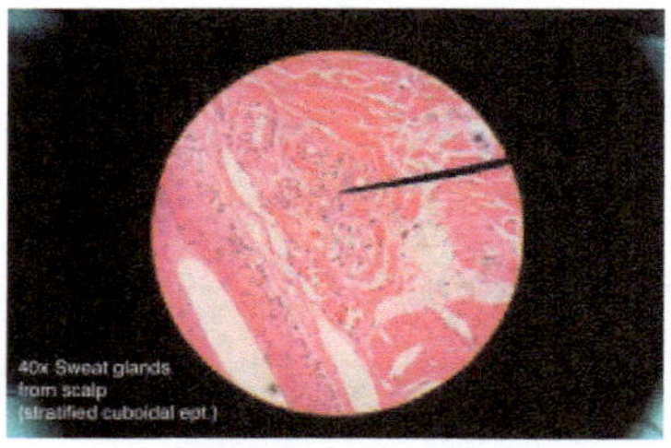

Histology of sweat gland showing stratified cuboidal epithelium

Muscle Tissue

Muscle tissue is a soft tissue that composes muscles in animal bodies, and gives rise to muscles' ability to contract. This is opposed to other components or tissues in muscle such as tendons or perimysium. It is formed during embryonic development through a process known as myogenesis.

Muscle tissue varies with function and location in the body. In mammals the three types are: skeletal or striated muscle; smooth or non-striated muscle; and cardiac muscle, which is sometimes known as semi-striated. Smooth and cardiac muscle contracts involuntarily, without conscious intervention. These muscle types may be activated both through interaction of the central nervous system as well as by receiving innervation from peripheral plexus or endocrine (hormonal) activation. Striated or skeletal muscle only contracts voluntarily, upon influence of the central nervous system. Reflexes are a form of non-conscious activation of skeletal muscles, but nonetheless arise through activation of the central nervous system, albeit not engaging cortical structures until after the contraction has occurred.

The different muscle types vary in their response to neurotransmitters and endocrine substances such as acetyl-choline, noradrenalin, adrenalin, nitric oxide and among others depending on muscle type and the exact location of the muscle.

Sub-categorization of muscle tissue is also possible, depending on among other things the content of myoglobin, mitochondria, myosin ATPase etc.

Structure

Muscle (myocytes) are elongated cells ranging from several millimetres to about 10 centimetres in length and from 10 to 100 micrometres in width. These cells are joined together in tissues that may be either striated or smooth, depending on the presence or absence, respectively, of organized, regularly repeated arrangements of myofibrillar contractile proteins called myofilaments. Striated muscle is further classified as either skeletal or cardiac muscle. Striated muscle is typically subject to conscious control, while smooth muscle is not. Thus, muscle tissue can be described as being one of three different types:

- Skeletal muscle, striated in structure and under voluntary control, is anchored by tendons (or by aponeuroses at a few places) to bone and is used to effect skeletal movement such as locomotion and to maintain posture. (Though postural control is generally maintained as an unconscious reflex—see proprioception—the muscles responsible also react to conscious control like non-postural muscles.) An average adult male is made up of 42% of skeletal muscle and an average adult female is made up of 36% (as a percentage of body mass). It also has striations unlike smooth muscle.
- Smooth muscle, neither striated in structure nor under voluntary control, is found within the walls of organs and structures such as the esophagus, stomach, intestines, bronchi, uterus, urethra, bladder, blood vessels, and the arrector pili in the skin (in which it controls erection of body hair).

In vertebrates, there is a third muscle tissue recognized:

- Cardiac muscle (myocardium), found only in the heart, is a striated muscle similar in structure to skeletal muscle but not subject to voluntary control.

Cardiac and skeletal muscles are "striated" in that they contain sarcomeres and are packed into highly regular arrangements of bundles; smooth muscle has neither. While skeletal muscles are arranged in regular, parallel bundles, cardiac muscle connects at branching, irregular angles (called intercalated discs). Striated muscle contracts and relaxes in short, intense bursts, whereas smooth muscle sustains longer or even near-permanent contractions.

Comparison of Types

	smooth muscle	cardiac muscle	skeletal muscle
Anatomy			
Neuromuscular junction	none	none	present
Fibers	fusiform, short (<0.4 mm)	branching	cylindrical, long (<15 cm)
Mitochondria	few	numerous	many to few (by type)
Nuclei	1	1	>1
Sarcomeres	none	present, max. length 2.6 µm	present, max. length 3.7 µm
Syncytium	none (independent cells)	none (but functional as such)	present
Sarcoplasmic reticulum	little elaborated	moderately elaborated	highly elaborated
ATPase	little	moderate	abundant
Physiology			
Self-regulation	spontaneous action (slow)	yes (rapid)	none (requires nerve stimulus)
Response to stimulus	unresponsive	"all-or-nothing"	"all-or-nothing"
Action potential	yes	yes	yes
Workspace	Force/length curve is variable	the increase in the force/length curve	at the peak of the force/length curve
Response to stimulus	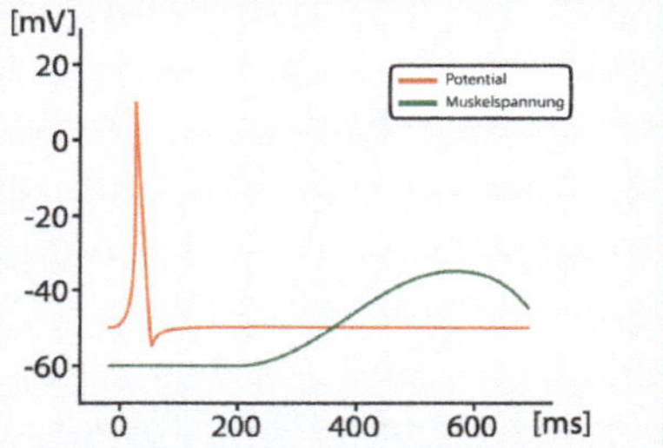	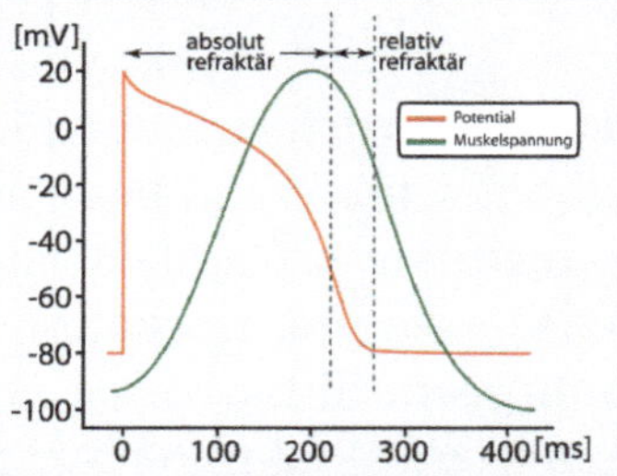	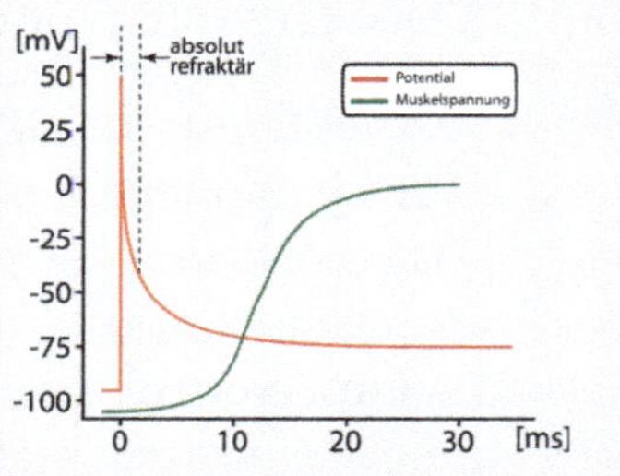

Skeletal Muscle

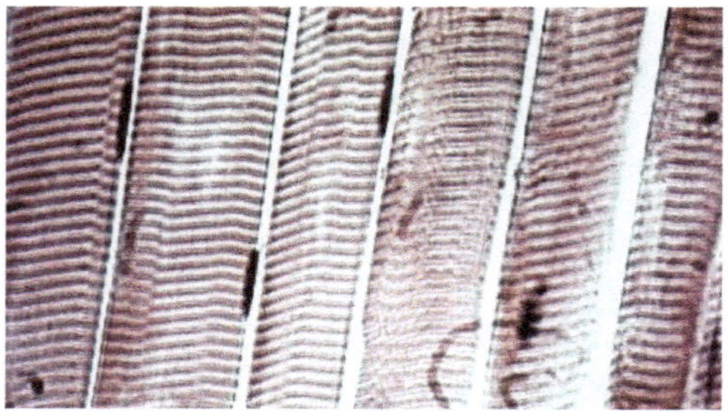

Striated skeletal muscle cells in microscopic view. The myofibers are the straight vertical bands; the horizontal striations (lighter and darker bands) that are visible result from differences in composition and density along the fibrils within the cells. The cigar-like dark patches beside the myofibers are muscle-cell nuclei.

Skeletal muscle is further divided into several subtypes:

- Type I, slow oxidative, slow twitch, or "red" muscle is dense with capillaries and is rich in mitochondria and myoglobin, giving the muscle tissue its characteristic red color. It can carry more oxygen and sustain aerobic activity.

  - Type I muscle fiber are sometimes broken down into Type I and Type Ic categories, as a result of recent research.

- Type II, fast twitch muscle, has three major kinds that are, in order of increasing contractile speed:

  - Type IIa, which, like slow muscle, is aerobic, rich in mitochondria and capillaries and appears red when deoxygenated.

  - Type IIx (also known as type IId), which is less dense in mitochondria and myoglobin. This is the fastest muscle type in humans. It can contract more quickly and with a greater amount of force than oxidative muscle, but can sustain only short, anaerobic bursts of activity before muscle contraction becomes painful (often incorrectly attributed to a build-up of lactic acid). N.B. in some books and articles this muscle in humans was, confusingly, called type IIB.

  - Type IIb, which is anaerobic, glycolytic, "white" muscle that is even less dense in mitochondria and myoglobin. In small animals like rodents this is the major fast muscle type, explaining the pale color of their flesh.

Smooth Muscle

Smooth muscle is an involuntary non-striated muscle. It is divided into two subgroups: the single-unit (unitary) and multiunit smooth muscle. Within single-unit cells, the whole bundle or sheet contracts as a syncytium (i.e. a multinucleate mass of cytoplasm that is not separated into cells). Multiunit smooth muscle tissues innervate individual cells; as such, they allow for fine control and gradual responses, much like motor unit recruitment in skeletal muscle.

Smooth muscle is found within the walls of blood vessels (such smooth muscle specifically being termed vascular smooth muscle) such as in the tunica media layer of large (aorta) and small arteries, arterioles and veins. Smooth muscle is also found in lymphatic vessels, the urinary bladder, uterus (termed uterine smooth muscle), male and female reproductive tracts, gastrointestinal tract, respiratory tract, arrector pili of skin, the ciliary muscle, and iris of the eye. The structure and function is basically the same in smooth muscle cells in different organs, but the inducing stimuli differ substantially, in order to perform individual effects in the body at individual times. In addition, the glomeruli of the kidneys contain smooth muscle-like cells called mesangial cells.

Cardiac Muscle

Cardiac muscle is involuntary, striated muscle that is found in the walls and histological foundation of the heart, specifically the myocardium. Cardiac muscle is one of three major types of muscle, the others being skeletal and smooth muscle. These three types of muscle all form in the process of myogenesis. The cells that constitute cardiac muscle, called cardiomyocytes or myocardiocytes, predominantly contain only one nucleus, although populations with two to four nuclei do exist.The myocardium is the muscle tissue of the heart, and forms a thick middle layer between the outer epicardium layer and the inner endocardium layer.

Coordinated contractions of cardiac muscle cells in the heart propel blood out of the atria and ventricles to the blood vessels of the left/body/systemic and right/lungs/pulmonary circulatory systems. This complex mechanism illustrates systole of the heart.

Cardiac muscle cells, unlike most other tissues in the body, rely on an available blood and electrical supply to deliver oxygen and nutrients and remove waste products such as carbon dioxide. The coronary arteries help fulfill this function.

Function

Skeletal Muscle

1.They carry out movements of the body.

2.They support the body.

3.They maintain the posture of the body.

Smooth Muscle

It is responsible for the contractility of hollow organs, such as blood vessels, the gastrointestinal tract, the bladder.

Cardiac Muscle

Cardiac muscle is the muscle of the heart. It is self-contracting, autonomically regulated and must continue to contract in rhythmic fashion for the whole life of the organism. Hence it has special features.

Nervous Tissue

Nervous tissue or Nerve tissue is the main tissue component of the two parts of the nervous system; the brain and spinal cord of the central nervous system (CNS), and the branching peripheral nerves of the peripheral nervous system (PNS), which regulates and controls bodily functions and activity. It is composed of neurons, or nerve cells, which receive and transmit impulses, and neuroglia, also known as glial cells or more commonly as just *glia*, which assist the propagation of the nerve impulse as well as providing nutrients to the neuron.

Nervous tissue is made up of different types of nerve cells, all of which have an axon, the long stem-like part of the cell that sends action potential signals to the next cell. Bundles of axons make up the nerves.

Functions of the nervous system are sensory input, integration, control of muscles and glands, homeostasis, and mental activity.

Structure

Nervous tissue is composed of neurons, also called nerve cells, and neuroglial cells. Typically, nervous tissue is categorized into four types of tissue. In the central nervous system (CNS), the tissue types found are grey matter and white matter. In the peripheral nervous system (PNS), the tissue types are nerves and ganglia. The tissue is categorized by its neuronal and neuroglial components.

Components

Neurons are cells with specialized features that allow them to receive and facilitate nerve impulses, or action potentials, across their membrane to the next neuron. They possess a large cell body (soma,perikaryon,cyton), with cell projections called dendrites and an axon. Dendrites are thin, branching projections that receive electrochemical signaling (neurotransmitters) to create a change in voltage in the cell. Axons are long projections that carry the action potential away from the cell body toward the next neuron. The bulb-like end of the axon, called the axon terminal, is separated from the dendrite of the following neuron by a small gap called a synaptic cleft. When the action potential travels to the axon terminal, neurotransmitters are released across the synapse and bind to the post-synaptic receptors, continuing the nerve impulse.

Neurons are classified both functionally and structurally.

Functional classification:

- Sensory neurons (afferent): Relay sensory information in the form of an action potential (nerve impulse) from the PNS to the CNS
- Motor neurons (efferent): Relay an action potential out of the CNS to the proper effector (muscles, glands)
- Interneurons: Cells that form connections between neurons and whose processes are limited to a single local area in the brain or spinal cord

Structural classification:

- Multipolar neurons: Have 3 or more processes coming off the soma (cell body). They are the major neuron type in the CNS and include interneurons and motor neurons.
- Bipolar neurons: Sensory neurons that have two processes coming off the soma, one dendrite and one axon
- Pseudounipolar neurons: Sensory neurons that have one process that splits into two branches, forming the axon and dendrite
- Unipolar brush cells: Are excitatory glutamatergic interneurons that have a single short dendrite terminating in a brush-like tuft of dendrioles. These are found in the granular layer of the cerebellum.

Neuroglia encompasses the non-neural cells in nervous tissue that provide various crucial supportive functions for neurons. They are smaller than neurons, and vary in structure according to their function.

Neuroglial cells are classified as follows:

- Microglial cells: Microglia are macrophage cells that make up the primary immune system for the CNS. They are the smallest neuroglial cell.
- Astrocytes: Star-shaped macroglial cells with many processes found in the CNS. They are the most abundant cell type in the brain, and are intrinsic to a healthy CNS.
- Oligodendrocytes: CNS cells with very few processes. They form myelin sheaths on the axons of a neuron, which are lipid-based insulation that increases the speed at which the action potential, can travel down the axon.
- NG2 glia: CNS cells that are distinct from astrocytes, oligodendrocytes, and microglia, and serve as the developmental precursors of oligodendrocytes
- Schwann cells: The PNS equivalent of oligodendrocytes, they help maintain axons and form myelin sheaths in the PNS.
- Satellite glial cell: Line the surface of neuron cell bodies in ganglia (groups of nerve body cells bundled or connected together in the PNS)
- Enteric glia: Found in the enteric nervous system, within the gastrointestinal tract.

Classification of Tissue

In the Central Nervous System:

- Grey matter is composed of cell bodies, dendrites, unmyelinated axons, protoplasmic astrocytes (astrocyte subtype), satellite oligodendrocytes (non-myelinating oligodendrocyte subtype), microglia, and very few myelinated axons.
- White matter is composed of myelinated axons, fibrous astrocytes, myelinating oligodendrocytes, and microglia.

In the Peripheral Nervous System:

- Ganglion tissue is composed of cell bodies, dendrites, and satellite glial cells.
- Nerves are composed of myelinated and unmyelinated axons, Schwann cells surrounded by connective tissue.

The three layers of connective tissue surrounding each nerve are:

- Endoneurium. Each nerve axon, or fiber is surrounded by the endoneurium, which is also called the endoneurial tube, channel or sheath. This is a thin, delicate, protective layer of connective tissue.
- Perineurium. Each nerve fascicle containing one or more axons, is enclosed by the perineurium, a connective tissue having a lamellar arrangement in seven or eight concentric layers. This plays a very important role in the protection and support of the nerve fibers and also serves to prevent the passage of large molecules from the epineurium into a fascicle.
- Epineurium. The epineurium is the outermost layer of dense connective tissue enclosing the (peripheral) nerve.

Function

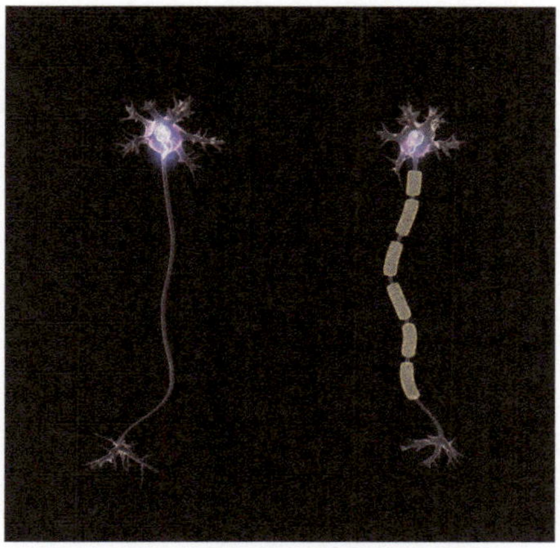

Myelinated axons conduct impulses faster than unmyelinated axons.

The function of nervous tissue is to form the communication network of the nervous system by conducting electric signals across tissue. In the CNS, grey matter, which contains the synapses, is important for information processing. White matter, containing myelinated axons, connects and facilitates nerve impulse between grey matter areas in the CNS. In the PNS, the ganglion tissue, containing the cell bodies and dendrites, contain relay points for nerve tissue impulses. The nerve tissue, containing myelinated axons bundles, carry action potential/nerve impulses.

Clinical Significance

Tumours

Neoplasms (tumours) in nervous tissue include:

- Gliomas (glial cell tumors)

  Gliomatosis cerebri, Oligoastrocytoma, Choroid plexus papilloma, Ependymoma, Astrocytoma (Pilocytic astrocytoma, Glioblastoma multiforme), Dysembryoplastic neuroepithelial tumour, Oligodendroglioma, Medulloblastoma, Primitive neuroectodermal tumor

- Neuroepitheliomatous tumors

  Ganglioneuroma, Neuroblastoma, Atypical teratoid rhabdoid tumor, Retinoblastoma, Esthesioneuroblastoma

- Nerve sheath tumors

  Neurofibroma (Neurofibrosarcoma, Neurofibromatosis), Schwannoma, Neurinoma, Acoustic neuroma, Neuroma

References

- Hyde, Janet Shibley; DeLamater, John D. (June 2010), "Chapter 5" (PDF), Understanding Human Sexuality (11th ed.), New York: McGraw-Hill, p. 103, ISBN 978-0073382821
- Sattler, R. (1994). "Homology, homeosis, and process morphology in plants". In Hall, Brian Keith. Homology:

the hierarchical basis of comparative biology. Academic Press. pp. 423–75. ISBN 0-12-319583-7.

- Page, Roderick D.M.; Holmes, Edward C. (2009). Molecular Evolution: A Phylogenetic Approach. John Wiley & Sons. ISBN 978-1-4443-1336-9.
- Reece, J; Meyers, N; Urry, L; Cain, M; Wasserman, S; Minorsky, P; Jackson, R; Cooke, B. Cambell Biology, 9th Edition. Pearson. p. 472. ISBN 9781442531765.
- Lomolino, M; Riddle, B; Whittaker, R; Brown, J. Biogeography, Fourth Edition. Sinauer Associates, Inc. p. 426. ISBN 978-0-878-93-494-2.
- Thunstad, Erik (2009). Darwins teori, evolusjon gjennom 400 år (in Norwegian). Oslo, Norway: Humanist forlag. p. 404. ISBN 978-82-92622-53-7.
- Conway Morris, Simon (2005). Life's solution: inevitable humans in a lonely universe. Cambridge, UK: Cambridge University Press. doi:10.2277/0521827043. ISBN 0-521-60325-0. OCLC 156902715.
- Ross M, Pawlina W (2011). Histology: A Text and Atlas (6th ed.). Lippincott Williams & Wilkins. pp. 158–197. ISBN 978-0-7817-7200-6.
- Young B, Woodford P, O'Dowd G (2013). Wheater's Functional Histology: A Text and Colour Atlas (6th ed.). Elsevier. p. 65. ISBN 978-0702047473.
- Melfi, Rudy C.; Alley, Keith E., eds. (2000). Permar's oral embryology and microscopic anatomy: a textbook for students in dental hygiene. Lippincott Williams & Wilkins. p. 9. ISBN 978-0-683-30644-6.
- Jenkins, Gail W.; Tortora, Gerard J. (2013). Anatomy and Physiology from Science to Life (3rd ed.). John Wiley & Sons. pp. 110–115. ISBN 978-1-118-12920-3.
- Eroschenko, Victor P. (2008). "Integumentary System". DiFiore's Atlas of Histology with Functional Correlations. Lippincott Williams & Wilkins. pp. 212–234. ISBN 9780781770576.
- McConnell, Thomas H. (2006). The nature of disease: pathology for the health professions. Lippincott Williams & Wilkins. p. 55. ISBN 978-0-7817-5317-3.
- Blerkom, edited by Jonathan Van; Gregory, Linda (2004). Essential IVF : basic research and clinical applications. Boston: Kluwer Academic Publishers. p. 3. ISBN 978-1-4020-7551-3.
- Marieb, Elaine; Hoehn, Katja (2007). Human Anatomy & Physiology (7th ed.). Pearson Benjamin Cummings. p. 317. ISBN 0-8053-5387-9.
- "homogeneous, adj.". OED Online. March 2016. Oxford University Press. http://www.oed.com/view/Entry/88045? (accessed April 09, 2016).

6

Anatomy of Vertebrates and Invertebrates

Vertebrates in fishes include the jawless fish and also the jawed vertebrates. Invertebrates, neuraxis and limbless vertebrates are also explained in the chapter. The chapter discusses the anatomy of vertebrate and invertebrates in a critical manner providing key analysis to the subject matter.

Vertebrate

Vertebrates comprise all species of animals within the subphylum Vertebrata (chordates with backbones). Vertebrates represent the overwhelming majority of the phylum Chordata, with currently about 64,000 species described. Vertebrates include the jawless fish and the jawed vertebrates, which include the cartilaginous fish (sharks and rays) and the bony fish.

A bony fish clade known as the lobe-finned fishes is included with tetrapods, which are further divided into amphibians, reptiles, mammals, and birds. Extant vertebrates range in size from the frog species *Paedophryne amauensis*, at as little as 7.7 mm (0.30 in), to the blue whale, at up to 33 m (108 ft). Vertebrates make up less than five percent of all described animal species; the rest are invertebrates, which lack vertebral columns.

The vertebrates traditionally include the hagfish, which do not have proper vertebrae, though their closest living relatives, the lampreys, do. Hagfish do, however, possess a cranium. For this reason, the vertebrate subphylum is sometimes referred to as "Craniata" when discussing morphology.

Molecular analysis since 1992 has suggested that hagfish are most closely related to lampreys, and so also are vertebrates in a monophyletic sense. Others consider them a sister group of vertebrates in the common taxon of craniata.

Etymology

The word origin of *vertebrate* derives from the Latin word *vertebratus* (Pliny), meaning *joint of the spine*. The Proto-Indo-European language origins are still unclear.

Vertebrate is closely related to the word *vertebra*, which refers to any of the bones or segments of the spinal column.

Anatomy and Morphology

All vertebrates are built along the basic chordate body plan: a stiff rod running through the length of the animal (vertebral column or notochord), with a hollow tube of nervous tissue (the spinal cord) above it and the gastrointestinal tract below.

In all vertebrates, the mouth is found at, or right below, the anterior end of the animal, while the anus opens to the exterior before the end of the body. The remaining part of the body continuing after of the anus forms a tail with vertebrae and spinal cord, but no gut.

Vertebral Column

The defining characteristic of a vertebrate is the vertebral column, in which the notochord (a stiff rod of uniform composition) found in all chordates has been replaced by a segmented series of stiffer elements (vertebrae) separated by mobile joints (intervertebral discs, derived embryonically and evolutionarily from the notochord).

However, a few vertebrates have secondarily lost this anatomy, retaining the notochord into adulthood, such as the sturgeon and *Latimeria*. Jawed vertebrates are typified by paired appendages (fins or legs, which may be secondarily lost), but this is not part of the definition of vertebrates as a whole.

Fossilized skeleton of Diplodocus carnegii, showing an extreme example of the backbone that characterizes the vertebrates.

Gills

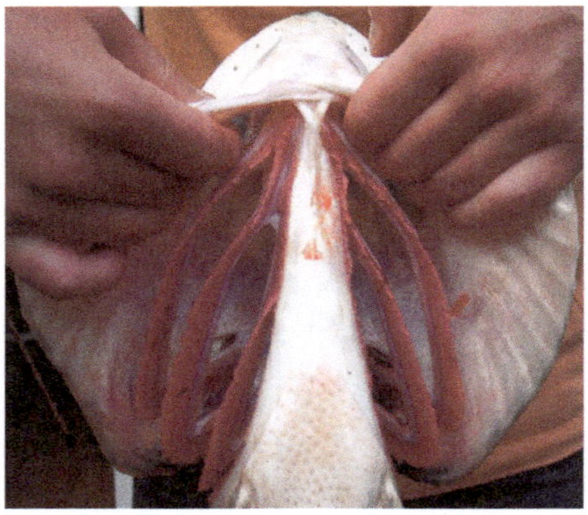

Gill arches bearing gills in a pike

All basal vertebrates breathe with gills. The gills are carried right behind the head, bordering the posterior margins of a series of openings from the pharynx to the exterior. Each gill is supported by a cartilagenous or bony gill arch. The bony fish have three pairs of arches, cartilaginous fish have five to seven pairs, while the primitive jawless fish have seven. The vertebrate ancestor no doubt had more arches, as some of their chordate relatives have more than 50 pairs of gills.

In amphibians and some primitive bony fishes, the larvae bear external gills, branching off from the gill arches. These are reduced in adulthood, their function taken over by the gills proper in fishes and by lungs in most amphibians. Some amphibians retain the external larval gills in adulthood, the complex internal gill system as seen in fish apparently being irrevocably lost very early in the evolution of tetrapods.

While the higher vertebrates lack gills, the gill arches form during fetal development, and lay the basis of essential structures such as jaws, the thyroid gland, the larynx, the *columella* (corresponding to the stapes in mammals) and, in mammals, the malleus and incus.

Central Nervous System

The central nervous system of vertebrates is based on a hollow nerve cord running along the length of the animal. Of particular importance and unique to vertebrates is the presence of neural crest cells. These are progenitors of stem cells, and critical to coordinating the functions of cellular components. Neural crest cells migrate through the body from the nerve cord during development, and initiate the formation of neural ganglia and structures such as the jaws and skull.

The vertebrates are the only chordate group to exhibit cephalisation, the concentration of brain functions in the head. A slight swelling of the anterior end of the nerve cord is found in the lancelet, though it lacks the eyes and other complex sense organs comparable to those of vertebrates. Other chordates do not show any trends towards cephalisation.

A peripheral nervous system branches out from the nerve cord to innervate the various systems. The front end of the nerve tube is expanded by a thickening of the walls and expansion of the central canal of spinal cord into three primary brain vesicles: The prosencephalon (forebrain), mesencephalon (midbrain) and rhombencephalon (hindbrain), further differentiated in the various vertebrate groups. Two laterally placed eyes form around outgrowths from the midbrain, except in hagfish, though this may be a secondary loss. The forebrain is well developed and subdivided in most tetrapods, while the midbrain dominate in many fish and some salamanders. Vesicles of the forebrain are usually paired, giving rise to hemispheres like the cerebral hemispheres in mammals.

The resulting anatomy of the central nervous system, with a single hollow nerve cord topped by a series of (often paired) vesicles, is unique to vertebrates. All invertebrates with well-developed brains, such as insects, spiders and squids, have a ventral rather than dorsal system of ganglions, with a split brain stem running on each side of the mouth or gut.

Evolutionary History

First Vertebrates

The early vertebrate Haikouichthys

Vertebrates originated about 525 million years ago during the Cambrian explosion, which saw the rise in organism diversity. The earliest known vertebrate is believed to be the *Myllokunmingia*. Another early vertebrate is *Haikouichthys ercaicunensis*. Unlike the other fauna that dominated the Cambrian, these groups had the basic vertebrate body plan: a notochord, rudimentary vertebrae, and a well-defined head and tail. All of these early vertebrates lacked jaws in the common sense and relied on filter feeding close to the seabed. A vertebrate group of uncertain phylogeny, small-eel-like conodonts, are known from microfossils of their paired tooth segments from the late Cambrian to the end of the Triassic.

From Fish to Amphibians

Acanthostega, a fish-like early labyrinthodont.

The first jawed vertebrates appeared in the latest Ordovician and became common in the Devonian, often known as the "Age of Fishes". The two groups of bony fishes, the actinopterygii and sarcopterygii, evolved and became common. The Devonian also saw the demise of virtually all jawless fishes, save for lampreys and hagfish, as well as the Placodermi, a group of armoured fish that dominated the entirety of that period since the late Silurian. The Devonian also saw the rise of the first labyrinthodonts, which was a transitional form between fishes and amphibians.

Mesozoic Vertebrates

Amniotes branched from labyrinthodonts in the subsequent Carboniferous period. The Parareptilia and synapsid amniotes were common during the late Paleozoic, while diapsids became dominant during the Mesozoic. In the sea, the bony fishes became dominant. The birds a derived form of dinosaurs evolved in the Jurassic. The demise of the non-avian dinosaurs at the end of the Cretaceous allowed for the expansion of mammals, which had evolved from the therapsids, a group of synapsid amniotes, during the late Triassic Period.

After the Mesozoic

The Cenozoic world has seen great diversification of bony fishes, frogs, birds and mammals.

Over half of all living vertebrate species (about 32,000 species) are fish (non-tetrapod craniates), a diverse set of lineages that inhabit all the world's aquatic ecosystems, from snow minnows (Cypriniformes) in Himalayan lakes at elevations over 4,600 metres (15,100 feet) to flatfishes (order Pleuronectiformes) in the Challenger Deep, the deepest ocean trench at about 11,000 metres (36,000 feet). Fishes of myriad varieties are the main predators in most of the world's water bodies, both freshwater and marine. The rest of the vertebrate species are tetrapods, a single lineage that includes amphibians (with roughly 7,000 species); mammals (with approximately 5,500 species); and reptiles and birds (with about 20,000 species divided evenly between the two classes). Tetrapods comprise the dominant megafauna of most terrestrial environments and also include many partially or fully aquatic groups (e.g., sea snakes, penguins, cetaceans).

Reproductive Systems

Nearly all vertebrates undergo sexual reproduction. They produce haploid gametes by meiosis. The smaller, motile gametes are spermatozoa and the larger, non-motile gametes are ova. These fuse by the process of fertilisation to form diploid zygotes, which develop into new individuals.

Inbreeding

During sexual reproduction, mating with a close relative (inbreeding) often leads to inbreeding depression. Inbreeding depression is considered to be largely due to expression of deleterious recessive mutations. The effects of inbreeding have been studied in many vertebrate species.

In several species of fish, inbreeding was found to decrease reproductive success.

Inbreeding was observed to increase juvenile mortality in 11 small animal species.

A common breeding practice for pet dogs is mating between close relatives (e.g. between half- and full siblings). This practice generally has a negative effect on measures of reproductive success, including decreased litter size and puppy survival.

Incestuous matings in birds result in severe fitness costs due to inbreeding depression (e.g. reduction in hatchability of eggs and reduced progeny survival).

Inbreeding Avoidance

As a result of the negative fitness consequences of inbreeding, vertebrate species have evolved mechanisms to avoid inbreeding. Numerous inbreeding avoidance mechanisms operating prior to mating have been described.

Toads and many other amphibians display breeding site fidelity. Individuals that return to natal ponds to breed will likely encounter siblings as potential mates. Although incest is possible, *Bufo americanus* siblings rarely mate. These toads likely recognize and actively avoid close kins as mates. Advertisement vocalizations by males appear to serve as cues by which females recognize their kin.

Inbreeding avoidance mechanisms can also operate subsequent to copulation. In guppies, a post-copulatory mechanism of inbreeding avoidance occurs based on competition between sperm of rival males for achieving fertilization. In competitions between sperm from an unrelated male and from a full sibling male, a significant bias in paternity towards the unrelated male was observed.

When female sand lizards mate with two or more males, sperm competition within the female's reproductive tract may occur. Active selection of sperm by females appears to occur in a manner that enhances female fitness. On the basis of this selective process, the sperm of males that are more distantly related to the female are preferentially used for fertilization, rather than the sperm of close relatives. This preference may enhance the fitness of progeny by reducing inbreeding depression.

Outcrossing

Mating with unrelated or distantly related members of the same species is generally thought to provide the advantage of masking deleterious recessive mutations in progeny. Vertebrates have evolved numerous diverse mechanisms for avoiding close inbreeding and promoting outcrossing.

Outcrossing as a way of avoiding inbreeding depression, has been especially well studied in birds. For instance, inbreeding depression occurs in the great tit when the offspring are produced as a result of a mating between close relatives. In natural populations of the great tit (*Parus major*), inbreeding is avoided by dispersal of individuals from their birthplace, which reduces the chance of mating with a close relative.

The purple-crowned fairywren females paired with related males may undertake extra-pair matings that can reduce the negative effects of inbreeding. However, there are ecological and demographic constraints on extra pair matings. Nevertheless, 46% of broods produced by incestuously paired females contained extra-pair young.

Southern pied babblers (*Turdoides bicolor*) appear to avoid inbreeding in two ways. The first is through dispersal, and the second is by avoiding familiar group members as mates. Although both males and females disperse locally, they move outside the range where genetically related individuals are likely to be encountered. Within their group, individuals only acquire breeding positions when the opposite-sex breeder is unrelated.

Cooperative breeding in birds typically occurs when offspring, usually males, delay dispersal from their natal group in order to remain with the family to help rear younger kin. Female offspring rarely stay at home, dispersing over distances that allow them to breed independently, or to join unrelated groups.

Parthenogenesis

Parthenogenesis is a natural form of reproduction in which growth and development of embryos occur without fertilization.

Reproduction in squamate reptiles is ordinarily sexual, with males having a ZZ pair of sex determining chromosomes, and females a ZW pair. However, various species, including the Colombian Rainbow boa (*Epicrates maurus*), *Agkistrodon contortrix* (copperhead snake) and *Agkistrodon piscivorus* (cotton mouth snake) can also reproduce by facultative parthenogenesis -that is, they are capable of switching from a sexual mode of reproduction to an asexual mode- resulting in production of WW female progeny. The WW females are likely produced by terminal automixis.

Mole salamanders are an ancient (2.4-3.8 million year-old) unisexual vertebrate lineage. In the polyploid unisexual mole salamander females, a premeiotic endomitotic event doubles the number of chromosomes. As a result, the mature eggs produced subsequent to the two meiotic divisions have the same ploidy as the somatic cells of the female salamander. Synapsis and recombination during meiotic prophase I in these unisexual females is thought to ordinarily occur between identical sister chromosomes and occasionally between homologous chromosomes. Thus little, if any, genetic variation is produced. Recombination between homeologous chromosomes occurs only rarely, if at all. Since production of genetic variation is weak, at best, it is unlikely to provide a benefit sufficient to account for the long-term maintenance of meiosis in these organisms. However, meiosis may have been maintained during evolution by the efficient recombinational repair of DNA damages that meiosis provides, an advantage that could be realized at each generation.

Self-fertilization

The mangrove killifish (*Kryptolebias marmoratus*) produces both eggs and sperm by meiosis and routinely reproduces by self-fertilisation. The capacity for selfing in these fishes has apparently persisted for at least several hundred thousand years. Each individual hermaphrodite normally fertilizes itself when an egg and sperm that it has produced by an internal organ unite inside the fish's body. In nature, this mode of reproduction can yield highly homozygous lines composed of individuals so genetically uniform as to be, in effect, identical to one another. Although inbreeding, especially in the extreme form of self-fertilization, is ordinarily regarded as detrimental because it leads to expression of deleterious recessive alleles, self-fertilization does provide the benefit of *fertilization assurance* (reproductive assurance) at each generation.

Invertebrate

The common fruit fly, Drosophila melanogaster, has been used extensively for research.

Invertebrates are animals that neither possess nor develop a vertebral column (commonly known as a back*bone* or *spine*), derived from the notochord. This includes all animals apart from the subphylum Vertebrata. Familiar examples of invertebrates include insects; crabs, lobsters and their kin; snails, clams, octopuses and their kin; starfish, sea-urchins and their kin; and worms. The majority of animal species are invertebrates; one estimate puts the figure at 97%. Many invertebrate taxa have a greater number and variety of species than the entire subphylum of Vertebrata.

Some of the so-called invertebrates, such as the Chaetognatha, Hemichordata, Tunicata and Cephalochordata are more closely related to the vertebrates than to other invertebrates. This makes the term "invertebrate" paraphyletic and hence almost meaningless for taxonomic purposes.

Etymology

The word "invertebrate" comes from the form of the Latin word *vertebra*, which means a joint in general, and sometimes specifically a joint from the spinal column of a vertebrate. In turn the jointed aspect of *vertebra* derived from the concept of turning, expressed in the root *verto* or *vorto*, to turn. Coupled with the prefix *in-*, meaning "not" or "without".

Taxonomic Significance

The term invertebrates is not always precise among non-biologists since it does not accurately describe a taxon in the same way that Arthropoda, Vertebrata or Manidae do. Each of these terms describes a valid taxon, phylum, subphylum or family. "Invertebrata" is a term of convenience, not a taxon; it has very little circumscriptional significance except within the Chordata. The Vertebrata as a subphylum comprises such a small proportion of the Metazoa that to speak of the kingdom Animalia in terms of "Vertebrata" and "Invertebrata" has limited practicality. In the more formal taxonomy of Animalia other attributes that logically should precede the presence or absence of the vertebral column in constructing a cladogram, for example, the presence of a notochord. That would at least circumscribe the Chordata. However, even the notochord would be a less fundamental criterion than aspects of embryological development and symmetry or perhaps bauplan.

Invertebrates don't have a skeleton of bone, either internal or external. They include hugely varied body plans. Many have fluid-filled, hydrostatic skeletons, like jellyfish or worms. Others have hard exoskeletons, outer shells like those of insects and crustaceans. The most familiar invertebrates include the Protozoa, Porifera, Coelenterata, Platyhelminthes, Nematoda, Annelida, Echinodermata, Mollusca and Arthropoda. Arthropoda include insects, crustaceans and arachnids.

Number of Extant Species

By far the largest number of described invertebrate species are insects. The following table lists the number of described extant species for major invertebrate groups as estimated in the IUCN Red List of Threatened Species, *2014.3*.

Invertebrate group	Image	Estimated number of described species
Insects		1,000,000
Molluscs		85,000
Crustaceans	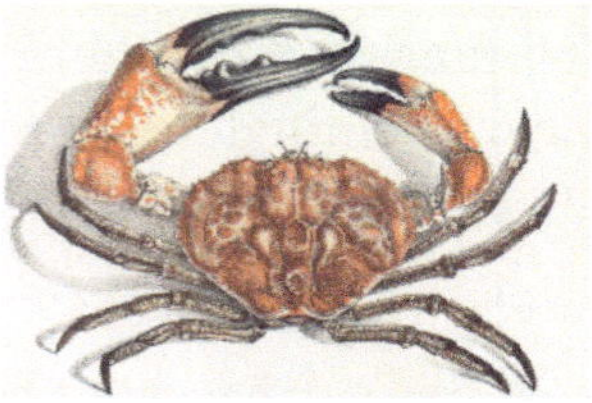	47,000
Corals		2,000

Arachnids		102,248
Velvet worms		165
Horseshoe crabs		4
Others jellyfish, echinoderms, sponges, other worms etc.		68,658
	Totals:	**1,305,075**

The IUCN estimates that 66,178 extant vertebrate species have been described, which means that over 95% of the described animal species in the world are invertebrates.

Characteristics

The trait that is common to all invertebrates is the absence of a vertebral column (backbone): this creates a distinction between invertebrates and vertebrates. The distinction is one of convenience only; it is not based on any clear biologically homologous trait, any more than the common trait of having wings functionally unites insects, bats, and birds, or than not having wings unites tortoises, snails and sponges. Being animals, invertebrates are heterotrophs, and require sustenance in the form of the consumption of other organisms. With a few exceptions, such as the Porifera, invertebrates generally have bodies composed of differentiated tissues. There is also typically a digestive chamber with one or two openings to the exterior.

Morphology and Symmetry

The body plans of most multicellular organisms exhibit some form of symmetry, whether radial, bilateral, or spherical. A minority, however, exhibit no symmetry. One example of asymmetric invertebrates include all gastropod species. This is easily seen in snails and sea snails, which have helical shells. Slugs appear externally symmetrical, but their pneumostome (breathing hole) is located on the right side. Other gastropods develop external asymmetry, such as Glaucus atlanticus that develops asymmetrical cerata as they mature. The origin of gastropod asymmetry is a subject of scientific debate.

Other examples of asymmetry are found in fiddler crabs and hermit crabs. They often have one claw much larger than the other. If a male fiddler loses its large claw, it will grow another on the opposite side after moulting. Sessile animals such as sponges are asymmetrical alongside coral colonies (with the exception of the individual polyps that exhibit radial symmetry); alpheidae claws that lack pincers; and some copepods, polyopisthocotyleans, and monogeneans which parasitize by attachment or residency within the gill chamber of their fish hosts).

Nervous System

Neurons differ in invertebrates from mammalian cells. Invertebrates cells fire in response to similar stimuli as mammals, such as tissue trauma, high temperature, or changes in pH. The first invertebrate in which a neuron cell was identified was the medicinal leech, *Hirudo medicinalis*.

Learning and memory using nociceptors in the sea hare, *Aplysia* has been described. Mollusk neurons are able to detect increasing pressures and tissue trauma.

Neurons have been identified in a wide range of invertebrate species, including annelids, molluscs, nematodes and arthropods.

Respiratory System

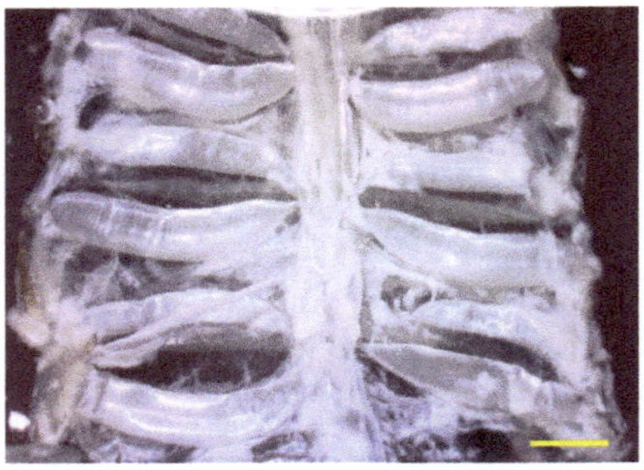

Tracheal system of dissected cockroach. The largest tracheae run across the width of the body of the cockroach and are horizontal in this image. Scale bar, 2 mm.

The tracheal system branches into progressively smaller tubes, here supplying the crop of the cockroach. Scale bar, 2.0 mm.

One type of invertebrate respiratory system is the open respiratory system composed of spiracles, tracheae, and tracheoles that terrestrial arthropods have to transport metabolic gases to and from tissues. The distribution of spiracles can vary greatly among the many orders of insects, but in general each segment of the body can have only one pair of spiracles, each of which connects to an atrium and has a relatively large tracheal tube behind it. The tracheae are invaginations of the cuticular exoskeleton that branch (anastomose) throughout the body with diameters from only a few micrometres up to 0.8 mm. The smallest tubes, tracheoles, penetrate cells and serve as sites of diffusion for water, oxygen, and carbon dioxide. Gas may be conducted through the respiratory system by means of active ventilation or passive diffusion. Unlike vertebrates, insects do not generally carry oxygen in their haemolymph.

A tracheal tube may contain ridge-like circumferential rings of taenidia in various geometries such as loops or helices. In the head, thorax, or abdomen, tracheae may also be connected to air sacs. Many insects, such as grasshoppers and bees, which actively pump the air sacs in their abdomen, are able to control the flow of air through their body. In some aquatic insects, the tracheae exchange gas through the body wall directly, in the form of a gill, or function essentially as normal, via a plastron. Note that despite being internal, the tracheae of arthropods are shed during moulting (ecdysis).

Digestive System

Template:Expanding section

Integumentary System

Reproduction

Like vertebrates, most invertebrates reproduce at least partly through sexual reproduction. They produce specialized reproductive cells that undergo meiosis to produce smaller, motile spermatozoa or larger, non-motile ova. These fuse to form zygotes, which develop into new individuals. Others are capable of asexual reproduction, or sometimes, both methods of reproduction.

Social Interaction

Social behavior is widespread in invertebrates, including cockroaches, termites, aphids, thrips, ants, bees, Passalidae, Acari, spiders, and more. Social interaction is particularly salient in eusocial species but applies to other invertebrates as well.

Insects recognize information transmitted by other insects.

Phyla

The fossil coral Cladocora from the Pliocene of Cyprus

The term invertebrates covers several phyla. One of these are the sponges (Porifera). They were long thought to have diverged from other animals early. They lack the complex organization found in most other phyla. Their cells are differentiated, but in most cases not organized into distinct tissues. Sponges typically feed by drawing in water through pores. Some speculate that sponges are not so primitive, but may instead be secondarily simplified. The Ctenophora and the Cnidaria, which includes sea anemones, corals, and jellyfish, are radially symmetric and have digestive chambers with a single opening, which serves as both the mouth and the anus. Both have distinct tissues, but they are not organized into organs. There are only two main germ layers, the ectoderm and endoderm, with only scattered cells between them. As such, they are sometimes called diploblastic.

The Echinodermata are radially symmetric and exclusively marine, including starfish (Asteroidea), sea urchins, (Echinoidea), brittle stars (Ophiuroidea), sea cucumbers (Holothuroidea) and feather stars (Crinoidea).

The largest animal phylum is also included within invertebrates: the Arthropoda, including insects, spiders, crabs, and their kin. All these organisms have a body divided into repeating segments, typically with paired appendages. In addition, they possess a hardened exoskeleton that is periodically shed during growth. Two smaller phyla, the Onychophora and Tardigrada, are close relatives of the arthropods and share these traits. The Nematoda or roundworms, are perhaps the second largest animal phylum, and are also invertebrates. Roundworms are typically microscopic, and occur in nearly every environment where there is water. A number are important parasites. Smaller phyla related to them are the Kinorhyncha, Priapulida, and Loricifera. These groups have a reduced coelom, called a pseudocoelom. Other invertebrates include the Nemertea or ribbon worms, and the Sipuncula.

Another phylum is Platyhelminthes, the flatworms. These were originally considered primitive, but it now appears they developed from more complex ancestors. Flatworms are acoelomates, lacking a body cavity, as are their closest relatives, the microscopic Gastrotricha. The Rotifera or rotifers, are common in aqueous environments. Invertebrates also include the Acanthocephala or spiny-headed worms, the Gnathostomulida, Micrognathozoa, and the Cycliophora.

Also included are two of the most successful animal phyla, the Mollusca and Annelida. The former, which is the second-largest animal phylum by number of described species, includes animals such as snails, clams, and squids, and the latter comprises the segmented worms, such as earthworms and leeches. These two groups have long been considered close relatives because of the common presence of trochophore larvae, but the annelids were considered closer to the arthropods because they are both segmented. Now, this is generally considered convergent evolution, owing to many morphological and genetic differences between the two phyla.

Among lesser phyla of invertebrates are the Hemichordata, or acorn worms, and the Chaetognatha, or arrow worms. Other phyla include Acoelomorpha, Brachiopoda, Bryozoa, Entoprocta, Phoronida, and Xenoturbellida.

Classification of Invertebrates

Invertebrates can be classified into several main categories, some of which are taxonomically obsolescent or debatable, but still used as terms of convenience.

- *Protozoa* (like the worms, an arbitrary grouping of convenience)
- Sponges (*Porifera*)
- Stinging jellyfish and corals (*Cnidaria*)
- Comb jellies (*Ctenophora*)
- Flatworms (*Platyhelminthes*)
- Round- or threadworms (*Nematoda*)
- segmented worms (*Annelida*)
- Insects, spiders, crabs and their kin (*Arthropoda*)
- Cuttlefish, snails, mussels and their kin (*Mollusca*)
- Starfish, sea-cucumbers and their kin (*Echinodermata*)

History

The earliest animal fossils appear to be those of invertebrates. 665-million-year-old fossils in the Trezona Formation at Trezona Bore, West Central Flinders, South Australia have been interpreted as being early sponges. Some paleontologists suggest that animals appeared much earlier, possibly as early as 1 billion years ago. Trace fossils such as tracks and burrows found in the Tonian era indicate the presence of triploblastic worms, like metazoans, roughly as large (about 5 mm wide) and complex as earthworms.

Around 453 MYA, animals began diversifying, and many of the important groups of invertebrates diverged from one another. Fossils of invertebrates are found in various types of sediment from the Phanerozoic. Fossils of invertebrates are commonly used in stratigraphy.

Classification

Carl Linnaeus divided these animals into only two groups, the Insecta and the now-obsolete Vermes (worms). Jean-Baptiste Lamarck, who was appointed to the position of "Curator of Insecta and Vermes" at the Muséum National d'Histoire Naturelle in 1793, both coined the term "invertebrate" to describe such animals, and divided the original two groups into ten, by splitting Arachnida and Crustacea from the Linnean Insecta, and Mollusca, Annelida, Cirripedia, Radiata, Coelenterata and Infusoria from the Linnean Vermes. They are now classified into over 30 phyla, from simple organisms such as sea sponges and flatworms to complex animals such as arthropods and molluscs.

Significance of the Group

Invertebrates are animals *without* a vertebral column. This has led to the conclusion that *in*vertebrates are a group that deviates from the normal, vertebrates. This has been said to be because researchers in the past, such as Lamarck, viewed vertebrates as a "standard": in Lamarck's theory of evolution, he believed that characteristics acquired through the evolutionary process involved not only survival, but also progression toward a "higher form", to which humans and vertebrates were closer than invertebrates were. Although goal-directed evolution has been abandoned, the distinction of invertebrates and vertebrates persists to this day, even though the grouping has been noted to be "hardly natural or even very sharp." Another reason cited for this continued distinction is that Lamarck created a precedent through his classifications which is now difficult to escape from. It is also possible that some humans believe that, they themselves being vertebrates, the group deserves more attention than invertebrates. In any event, in the 1968 edition of *Invertebrate Zoology*, it is noted that "division of the Animal Kingdom into vertebrates and invertebrates is artificial and reflects human bias in favor of man's own relatives." The book also points out that the group lumps a vast number of species together, so that no one characteristic describes all invertebrates. In addition, some species included are only remotely related to one another, with some more related to vertebrates than other invertebrates .

In Research

For many centuries, invertebrates have been neglected by biologists, in favor of big vertebrates and "useful" or charismatic species. Invertebrate biology was not a major field of study until the work of Linnaeus and Lamarck in the 18th century. During the 20th century, invertebrate zoology became one of the major fields of natural sciences, with prominent discoveries in the fields of medicine, genetics, palaeontology, and ecology. The study of invertebrates has also benefited law enforcement, as arthropods, and especially insects, were discovered to be a source of information for forensic investigators.

Two of the most commonly studied model organisms nowadays are invertebrates: the fruit fly *Drosophila melanogaster* and the nematode *Caenorhabditis elegans*. They have long been the most intensively studied model organisms, and were among the first life-forms to be genetically sequenced. This was facilitated by the severely reduced state of their genomes, but many genes, introns, and linkages have been lost. Analysis of the starlet sea anemone genome has emphasised the importance of sponges, placozoans, and choanoflagellates, also being sequenced, in explaining the arrival of 1500 ancestral genes unique to animals. Invertebrates are also used by scientists in the field of aquatic biomonitoring to evaluate the effects of water pollution and climate change.

Neuraxis

The neuraxis or sometimes neuroaxis is the axis of the central nervous system. It denotes the direction in which the central nervous system lies. During embryological development, the neuraxis is bent by various flexures, contributing to the mature structure of the brain and spinal cord.

Embryonic development can help in understanding the structure of the adult brain because it establishes a framework on which more complex structures can be built. First, the neural tube establishes the anterior–posterior dimension of the nervous system, which is called the neuraxis. The embryonic nervous system in mammals can be said to have a standard arrangement. Humans (and other primates, to some degree) make this complicated by standing up and walking on two legs. The anterior–posterior dimension of the neuraxis overlays the superior–inferior dimension of the body. However, there is a major curve between the brain stem and forebrain, which is called the cephalic flexure. Because of this, the neuraxis starts in an inferior position—the end of the spinal cord—and ends in an anterior position, the front of the cerebrum. This may be confusing, and can be illustrated when looking at a four-legged animal standing up on two legs. Without the flexure in the brain stem, and at the top of the neck, that animal would be looking straight up instead of straight in front.

Additional Images

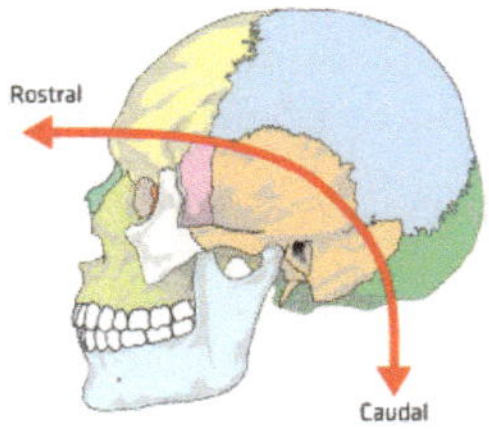

The skull uses different anatomical terminology, due to its embryonic origin of neuraxis.

Limbless Vertebrate

Lizards have evolved limbless forms on a number of occasions. The legless lizard shown above is known as a slowworm (Anguis fragilis).

Many vertebrates have evolved limbless or limb-reduced forms. Reptiles have on a number of occasions evolved into limbless forms - snakes, amphisbaenia, and legless lizards (limb loss in lizards has evolved independently several times, examples include the families Pygopodidae and Dibamidae and species of *Isopachys*, *Anguis*, and *Ophisaurus)*. The same is true of amphibians - caecilians, Sirenidae (a clade of salamanders that are limbless except for atrophied front limbs), Amphiumidae (a clade of salamander with extremely atrophied limbs that appear non-functional) and at least two extinct groups (Aïstopoda and Lysorophia). Legless forms of reptiles and amphibians probably evolved so as to be able to move underground or in water more easily.

There are also a number of fish with elongated bodies that have no or reduced appendage-like fins, for example eels and swamp eels. While hagfish and lamprey also do not have appendage-like fins, they may not have lost them, but simply retained the form that vertebrates had before they evolved limbs.

There are no known limbless species of mammal or bird, although partial limb-loss and reduction has occurred in several groups, including whales and dolphins, sirenians, kiwis, and the extinct moa and elephant birds. Names can be deceiving, as the finless porpoise has two fins, and several species of legless lizards have tiny useless legs.

References

- Liem, Karel F.; Warren Franklin Walker (2001). Functional anatomy of the vertebrates: an evolutionary perspective. Harcourt College Publishers. p. 277. ISBN 978-0-03-022369-3.
- Teng, L. and Labosky P. A. (2006). "Neural crest stem cells" In: Jean-Pierre Saint-Jeannet, Neural Crest Induction and Differentiation, pp. 206-212, Springer Science & Business Media. ISBN 9780387469546.
- Berg, Linda R.; Eldra Pearl Solomon; Diana W. Martin (2004). Biology. Cengage Learning. p. 599. ISBN 978-0-534-49276-2.
- Andersen, Nils Møller; Tom A. Weir (2004). Australian water bugs: their biology and identification (Hemiptera-Heteroptera, Gerromorpha & Nepomorpha). Apollo Books. p. 38. ISBN 978-87-88757-78-1.
- Benton, Michael J. (1 November 2004). Vertebrate Palaeontology (Third ed.). Blackwell Publishing. pp. 33, 455 pp. ISBN 978-0632056378.
- Richards, O. W.; Davies, R.G. (1977). Imms' General Textbook of Entomology: Volume 1: Structure, Physiology and Development Volume 2: Classification and Biology. Berlin: Springer. ISBN 0-412-61390-5.
- Wasserthal, Lutz T. (1998). Chapter 25: The Open Hemolymph System of Holometabola and Its Relation to the Tracheal Space. In "Microscopic Anatomy of Invertebrates". Wiley-Liss, Inc. ISBN 0-471-15955-7.
- Sumich, James L. (2008). Laboratory and Field Investigations in Marine Life. Jones & Bartlett Learning. p. 67. ISBN 978-0-7637-5730-4.
- Langstroth, Lovell; Libby Langstroth; Todd Newberry; Monterey Bay Aquarium (2000). A living bay: the underwater world of Monterey Bay. University of California Press. p. 244. ISBN 978-0-520-22149-9.
- Safra, Jacob E. (2003). The New Encyclopaedia Britannica, Volume 16. Encyclopaedia Britannica. p. 523. ISBN 978-0-85229-961-6.
- Prewitt, Nancy L.; Larry S. Underwood; William Surver (2003). BioInquiry: making connections in biology. John Wiley. p. 289. ISBN 978-0-471-20228-8.

7

Flight in Birds

One of the main modes of locomotion in animals is flight. Flight is used as the primary mode of locomotion by majority of the world's bird species. A keel is a part of the bird's structure; it runs axially along the midline of the sternum and helps in providing an adequate leverage for flight. The aspects explained in the section provide a better understanding of flight in animals.

Bird Flight

Flight is the main mode of locomotion used by most of the world's bird species. Flight assists birds while feeding, breeding, avoiding predators, and migrating.

A flock of domestic pigeons each in a different phase of its flap.

This chapter discusses the mechanics of bird flight, with emphasis on the varied forms of bird's wings. The specifics of hovering, take-off and landing are also examined. Additional adaptations of bird's bodies relating to their flying ability are covered. Finally, theories on the evolution of bird flight are discussed.

Basic Mechanics of Bird Flight

Lift and Drag

The fundamentals of bird flight are similar to those of aircraft, in which the aerodynamic forces sustaining flight are lift and drag. Lift force is produced by the action of air flow on the wing, which is an airfoil. The airfoil is shaped such that the air provides a net upward force on the wing, while the movement of air is directed downward. Additional net lift may come from airflow around the bird's body in some species, especially during intermittent flight while the wings are folded or semi-folded (cf. lifting body).

Aerodynamic drag is the force opposite to the direction of motion, and hence the source of energy loss in flight. The drag force can be separated into two portions, *lift-induced drag*, which is inherent in the wing (primarily arising from wingtip vortices), and *parasitic drag*, including skin friction drag from the friction of air and body surfaces and form drag from the bird's frontal area. The streamlining of bird's body and wings reduces these forces.

Flapping Flight

When a bird flaps, as opposed to gliding, its wings continue to develop lift as before, but the lift is rotated forward to provide thrust, which counteracts drag and increases its speed, which has the effect of also increasing lift to counteract its weight, allowing it to maintain height or to climb. Flapping involves two stages: the down-stroke, which provides the majority of the thrust, and the up-stroke, which can also (depending on the bird's wings) provide some thrust. At each up-stroke the wing is slightly folded inwards to reduce the energetic cost of flapping-wing flight. Birds change the angle of attack continuously within a flap, as well as with speed.

Gliding Flight

Lesser flamingos flying in formation.

When in gliding flight, the upward aerodynamic force on both birds and gliders is equal to the weight. In gliding flight, no propulsion is used; the energy to counteract the energy loss due to aerodynamic drag is either taken from the potential energy of the bird, resulting in a descending flight, or is replaced by rising air currents ("thermals"), referred to as soaring flight.

Wing

A kea in flight.

The bird's forelimbs (the wings) are the key to flight. Each wing has a central vane to hit the wind, composed of three limb bones, the humerus, ulna and radius. The hand, or manus, which ancestrally was composed of five digits, is reduced to three digits (digit II, III and IV or I, II, III depending on the scheme followed), which serves as an anchor for the primaries, one of two groups of flight feathers responsible for the wing's airfoil shape. The other set of flight feathers, behind the carpal joint on the ulna, are called the secondaries. The remaining feathers on the wing are known as coverts, of which there are three sets. The wing sometimes has vestigial claws. In most species these are lost by the time the bird is adult (such as the highly visible ones used for active climbing by hoatzin chicks), but claws are retained into adulthood by the secretarybird, screamers, finfoots, ostriches, several swifts and numerous others, as a local trait, in a few specimens.

Albatrosses have locking mechanisms in the wing joints that reduce the strain on the muscles during soaring flight.

Female birds exposed to predators during ovulation produce chicks that grow their wings faster than chicks produced by predator-free females. Their wings are also longer. Both adaptations may make them better at avoiding avian predators.

Wing Shape and Flight

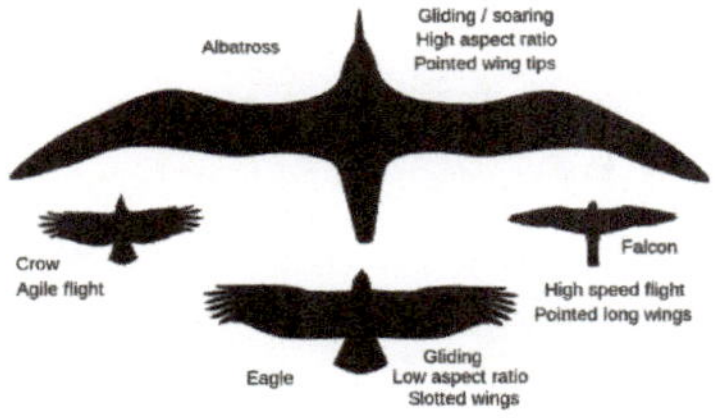

Wing shapes

The shape of the wing is important in determining the flight capabilities of a bird. Different shapes correspond to different trade-offs between advantages such as speed, low energy use, and maneuverability. Two important parameters are the aspect ratio and wing loading. Aspect ratio is the ratio of wingspan to the mean of its chord (or the square of the wingspan divided by wing area). Wing loading is the ratio of weight to wing area.

Most kinds of bird wing can be grouped into four types, with some falling between two of these types. These types of wings are elliptical wings, high speed wings, high aspect ratio wings and soaring wings with slots.

The budgerigar's wings, as seen on this pet female, allow it excellent manoeuvrability.

Elliptical wings

Elliptical wings are short and rounded, having a low aspect ratio, allowing for tight maneuvering in confined spaces such as might be found in dense vegetation. As such they are common in forest raptors (such as *Accipiter* hawks), and many passerines, particularly non-migratory ones (migratory species have longer wings). They are also common in species that use a rapid take off to evade predators, such as pheasants and partridges.

High speed wings

High speed wings are short, pointed wings that when combined with a heavy wing loading and rapid wingbeats provide an energetically expensive high speed. This type of flight is used by the bird with the fastest wing speed, the peregrine falcon, as well as by most of the ducks. The same wing shape is used by the auks for a different purpose; auks use their wings to "fly" underwater.

The peregrine falcon has the highest recorded dive speed of 242 mph (389 km/h). The fastest straight, powered flight is the spine-tailed swift at 105 mph (170 km/h).

A roseate tern uses its low wing loading and high aspect ratio to achieve low speed flight.

High aspect ratio wings

High aspect ratio wings, which usually have low wing loading and are far longer than they are wide, are used for slower flight. This may take the form of almost hovering (as used by kestrels, terns and nightjars) or in soaring and gliding flight, particularly the dynamic soaring used by seabirds, which takes advantage of wind speed variation at different altitudes (wind shear) above ocean waves to provide lift. Low speed flight is also important for birds that plunge-dive for fish.

Soaring wings with deep slots

These wings are favored by larger species of inland birds, such as eagles, vultures, pelicans, and storks. The slots at the end of the wings, between the primaries, reduce the induced drag and wingtip vortices by "capturing" the energy in air flowing from the lower to upper wing surface at the tips, whilst the shorter size of the wings aids in takeoff (high aspect ratio wings require a long taxi to get airborne).

Bounding Flight

Small birds often fly long distances using a technique in which short bursts of flapping are alternated with intervals in which the wings are folded against the body. This is a flight pattern known as "bounding" or "flap-bounding" flight. When the bird's wings are folded, its trajectory is primarily ballistic, with a small amount of body lift. The flight pattern is believed to decrease the energy required by reducing the aerodynamic drag during the ballistic part of the trajectory, and to increase the efficiency of muscle use.

Hovering

The ruby-throated hummingbird can beat its wings 52 times a second.

Several bird species use hovering, one family even specialized in it. True hovering by generating lift through flapping alone rather than by passage through the air demands a lot of energy. This confines the ability to smaller birds. The largest bird able to truly hover is the pied kingfisher, although larger birds can hover for short periods of time. Larger birds that hover for prolonged periods do so by flying into a headwind, allowing them to remain stationary relative to the ground or water. Kestrels, terns and even hawks use this windhovering.

Most birds that hover have high aspect ratio wings that are suited to low speed flying. One major exception to this are the hummingbirds, which are the most accomplished hoverers of all the birds. Hummingbird flight is different from other bird flight in that the wing is extended throughout the whole stroke, the stroke being a symmetrical figure of eight, with the wing producing lift on both the up- and down-stroke. Some hummingbirds can beat their wings 52 times a second, though others do so less frequently.

Take-off and Landing

A male bufflehead runs atop the water while taking off.

Take-off is one of the most energetically demanding aspects of flight, as the bird must generate enough airflow across the wing to create lift. Small birds do this with a simple upward jump. That doesn't work for larger birds, which must take a run up to generate sufficient airflow. Large birds take off by facing into the wind, or, if they can, by perching on a branch or cliff so they can just drop off into the air.

A magpie-goose taking off.

Landing is also a problem for large birds with high wing loads. This problem is dealt with in some species by aiming for a point below the intended landing area (such as a nest on a cliff) then pulling up beforehand. If timed correctly, the airspeed once the target is reached is virtually nil. Landing on water is simpler, and the larger waterfowl species prefer to do so whenever possible, landing into wind and using their feet as skids. To lose height rapidly prior to landing, some large birds such as geese indulge in a rapid alternating series of sideslips or even briefly turning upside down in a maneuver termed as *whiffling*.

Coordinated Formation Flight

A wide variety of birds fly together in a symmetric V-shaped or a J-shaped coordinated formation, also referred to as an "echelon", especially during long distance flight or migration. It is often assumed that birds resort to this pattern of formation flying in order to save energy and improve the aerodynamic efficiency of birds. The birds flying at the tips and at the front would interchange positions in a timely cyclical fashion to spread flight fatigue equally among the flock members.

The wingtips of the leading bird in an echelon create a pair of opposite rotating line vortices. The vortices trailing a bird create an underwash, which creates an induced drag for the bird. At the same time these vortices create an upwash that can aid the flight of following birds. In a 1970 study the authors claimed that each bird in a V formation of 25 members can achieve a reduction of induced drag and as a result increase their range by 71%.

Studies of waldrapp ibis show that birds spatially coordinate the phase of wing flapping and show wingtip path coherence when flying in V positions, thus enabling them to maximally utilise the available energy of upwash over the entire flap cycle. In contrast, birds flying in a stream immediately one behind another do not have wingtip coherence in their flight pattern and their flapping is out of phase, as compared to birds flying in V patterns, so as to avoid the detrimental effects of the downwash due to the leading bird's flight.

Adaptations for Flight

The most obvious adaptation to flight is the wing, but because flight is so energetically demanding birds have evolved several other adaptations to improve efficiency when flying. Birds' bodies are streamlined to help overcome air-resistance. Also, the bird skeleton is hollow to reduce weight, and many unnecessary bones have been lost (such as the bony tail of the early bird *Archaeopteryx*), along with the toothed jaw of early birds, which has been replaced with a lightweight beak. The skeleton's breastbone has also adapted into a large keel, suitable for the attachment of large, powerful flight muscles. The vanes of each feather have hooklets called barbules that zip the vanes of individual feathers together, giving the feathers the strength needed to hold the airfoil (these are often lost in flightless birds). The barbules maintain the shape and function of the feather. Each feather has a major (greater) side and a minor (lesser) side, meaning that the shaft or rachis does not run down the center of the feather. Rather it runs longitudinally of center with the lesser or minor side to the front and the greater or major side to the rear of the feather. This feather anatomy, during flight and flapping of the wings, causes a rotation of the feather in its follicle. The rotation occurs in the up motion of the wing. The greater side points down, letting air slip through the wing. This essentially breaks the integrity of the wing, allowing for a much easier movement in the up direction. The integrity of the wing is reestablished in the down movement, which allows for part of the lift inherent in bird wings. This function is most important in taking off or achieving lift at very low or slow speeds where the bird is reaching up and grabbing air and pulling itself up. At high speeds the air foil function of the wing provides most of the lift needed to stay in flight.

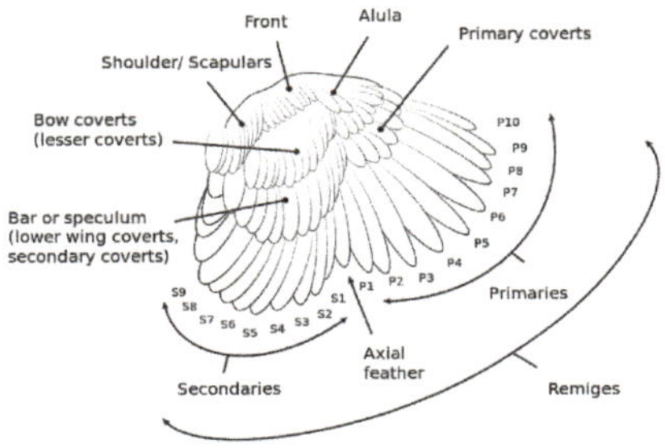

Diagram of the wing of a chicken, top view

The large amounts of energy required for flight have led to the evolution of a unidirectional pulmonary system to provide the large quantities of oxygen required for their high respiratory rates. This high metabolic rate produces large quantities of radicals in the cells that can damage DNA and lead to tumours. Birds, however, do not suffer from an otherwise expected shortened lifespan as their cells have evolved a more efficient antioxidant system than those found in other animals.

Evolution of Bird Flight

Black-legged kittiwakes fly at Cape Hay in the High Arctic.

Most paleontologists agree that birds evolved from small theropod dinosaurs, but the origin of bird flight is one of the oldest and most hotly contested debates in paleontology. The four main hypotheses are:

- *From the trees down*, that birds' ancestors first glided down from trees and then acquired other modifications that enabled true powered flight.
- *From the ground up*, that birds' ancestors were small, fast predatory dinosaurs in which feathers developed for other reasons and then evolved further to provide first lift and then true powered flight.
- *Wing-assisted incline running* (WAIR), a version of "from the ground up" in which birds' wings originated from forelimb modifications that provided *downforce*, enabling the proto-birds to run up extremely steep slopes such as the trunks of trees.
- *Pouncing proavis*, which posits that flight evolved by modification from arboreal ambush tactics.

There has also been debate about whether the earliest known bird, *Archaeopteryx*, could fly. It appears that *Archaeopteryx* had the brain structures and inner-ear balance sensors that birds use to control their flight. *Archaeopteryx* also had a wing feather arrangement like that of modern birds and similarly asymmetrical flight feathers on its wings and tail. But *Archaeopteryx* lacked the shoulder mechanism by which modern birds' wings produce swift, powerful upstrokes; this may mean that it and other early birds were incapable of flapping flight and could only glide. The presence of most fossils in marine sediments in habitats devoid of vegetation has led to the hypothesis that they may have used their wings as aids to run across the water surface in the manner of the basilisk lizards.

From the Trees Down

It is unknown how well *Archaeopteryx* could fly, or if it could even fly at all.

This was the earliest hypothesis, encouraged by the examples of gliding vertebrates such as flying squirrels. It suggests that proto-birds like *Archaeopteryx* used their claws to clamber up trees and glided off from the tops.

Some recent research undermines the "trees down" hypothesis by suggesting that the earliest birds and their immediate ancestors did not climb trees. Modern birds that forage in trees have much more curved toe-claws than those that forage on the ground. The toe-claws of Mesozoic birds and of closely related non-avian theropod dinosaurs are like those of modern ground-foraging birds.

From the Ground Up

Feathers are very common in coelurosaurid dinosaurs (including the early tyrannosauroid *Dilong*). Modern birds are classified as coelurosaurs by nearly all palaeontologists, though not by a few ornithologists. The original functions of feathers may have included thermal insulation and competitive displays. The most common version of the "from the ground up" hypothesis argues that bird's ancestors were small ground-running predators (rather like roadrunners) that used their forelimbs for balance while pursuing prey and that the forelimbs and feathers later evolved in ways that provided gliding and then powered flight. Another "ground upwards" theory argues the evolution of flight was initially driven by competitive displays and fighting: displays required longer feathers and longer, stronger forelimbs; many modern birds use their wings as weapons, and downward blows have a similar action to that of flapping flight. Many of the *Archaeopteryx* fossils come from marine sediments and it has been suggested that wings may have helped the birds run over water in the manner of the *Jesus Christ lizard* (common basilisk).

Most recent attacks on the "from the ground up" hypothesis attempt to refute its assumption that birds are modified coelurosaurid dinosaurs. The strongest attacks are based on embryological analyses, which conclude that birds' wings are formed from digits 2, 3 and 4 (corresponding to the index, middle and ring fingers in humans; the first of a bird's 3 digits forms the alula, which they use to avoid stalling on low-speed flight, for example when landing); but the hands of coelurosaurs are formed by digits 1, 2 and 3 (thumb and first 2 fingers in humans). However these embryological analyses were immediately challenged on the embryological grounds that the "hand" often develops differently in clades that have lost some digits in the course of their evolution, and therefore bird's hands do develop from digits 1, 2 and 3.

Wing-assisted Incline Running

The wing-assisted incline running (WAIR) hypothesis was prompted by observation of young chukar chicks, and proposes that wings developed their aerodynamic functions as a result of the need to run quickly up very steep slopes such as tree trunks, for example to escape from predators. Note that in this scenario birds need *downforce* to give their feet increased grip. But early birds, including *Archaeopteryx*, lacked the shoulder mechanism that modern birds' wings use to produce swift, powerful upstrokes. Since the downforce that WAIR requires is generated by upstrokes, it seems that early birds were incapable of WAIR.

Pouncing Proavis Model

The proavis theory was first proposed by Garner, Taylor, and Thomas in 1999:

We propose that birds evolved from predators that specialized in ambush from elevated sites, using their raptorial hindlimbs in a leaping attack. Drag–based, and later lift-based, mechanisms evolved under selection for improved control of body position and locomotion during the aerial part of the attack. Selection for enhanced lift-based control led to improved lift coefficients, incidentally turning a pounce into a swoop as lift production increased. Selection for greater swooping range would finally lead to the origin of true flight.

The authors believed that this theory had four main virtues:

- It predicts the observed sequence of character acquisition in avian evolution.
- It predicts an *Archaeopteryx*-like animal, with a skeleton more or less identical to terrestrial theropods, with few adaptations to flapping, but very advanced aerodynamic asymmetrical feathers.
- It explains that primitive pouncers (perhaps like *Microraptor*) could coexist with more advanced fliers (like *Confuciusornis* or *Sapeornis*) since they did not compete for flying niches.
- It explains that the evolution of elongated rachis-bearing feathers began with simple forms that produced a benefit by increasing drag. Later, more refined feather shapes could begin to also provide lift.

Uses and Loss of Flight in Modern Birds

Birds use flight to obtain prey on the wing, for foraging, to commute to feeding grounds, and to migrate between the seasons. It is also used by some species to display during the breeding season and to reach safe isolated places for nesting.

Flight is more energetically expensive in larger birds, and many of the largest species fly by soaring and gliding (without flapping their wings) as much as possible. Many physiological adaptations have evolved that make flight more efficient.

Birds that settle on isolated oceanic islands that lack ground-based predators often lose the ability to fly. This illustrates both flight's importance in avoiding predators and its extreme demand for energy.

Keel (Bird Anatomy)

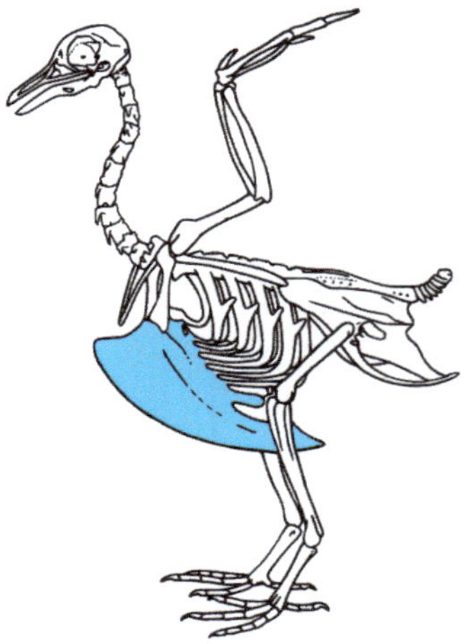

This stylised bird skeleton highlights the keel bone

A keel or carina (plural carinae) in bird anatomy is an extension of the sternum (breastbone) which runs axially along the midline of the sternum and extends outward, perpendicular to the plane of the ribs. The keel provides an anchor to which a bird's wing muscles attach, thereby providing adequate leverage for flight. Keels do not exist on all birds; in particular, some flightless birds lack a keel structure.

Historically, the presence or absence of a pronounced keel structure was used as a broad classification of birds into two orders: Carinatae (from *carina*, "keel"), having a pronounced keel; and ratites (from *ratis*, "raft" — referring to the flatness of the sternum), having a subtle keel structure or lacking one entirely. However, this classification has fallen into disuse as evolutionary studies have shown that many flightless birds have evolved from flighted birds. The current definition of Carinatae now includes all extant birds.

Gliding Flight

Gliding flight is heavier-than-air flight without the use of thrust; the term volplaning also refers to this mode of flight in animals. It is employed by gliding animals and by aircraft such as gliders. This mode of flight involves flying a greater distance horizontally than vertically and therefore can be distinguished from a simple descent like with a round parachute.

Although the human application of gliding flight usually refers to aircraft designed for this purpose, most powered aircraft are capable of gliding without engine power. As with sustained flight, gliding generally requires the application of an airfoil, such as the wings on aircraft or birds, or the gliding membrane on gliding possum. However, gliding can be achieved with a flat (uncambered) wing, as with a simple paper plane, or even with card-throwing. However some aircraft with lifting bodies and animals such as the flying snake can achieve gliding flight without any wings by creating a flattened surface underneath.

Aircraft ("Gliders")

Most winged aircraft can glide to some extent, but there are several types of aircraft designed to glide:

- Glider, also known as a sailplane
- Hang glider
- Paraglider
- Speed glider
- Ram-air parachute
- Rotor kite, if untethered, known as a rotary glider, or gyroglider.
- Military glider
- Paper aeroplane
- Radio-controlled glider
- Rocket glider

The main human application is currently recreational, though during the Second World War military gliders were used for carrying troops and equipment into battle. The types of aircraft that are used for sport and recreation are classified as gliders (sailplanes), hang gliders and paragliders. These two latter types are often foot-launched. The design of all three types enables them to repeatedly climb using rising air and then to glide before finding the next source of lift. When done in gliders (sailplanes), the sport is known as gliding and sometimes as soaring. For foot-launched aircraft, it is known as hang gliding and paragliding. Radio-controlled gliders with fixed wings are also soared by enthusiasts.

In addition to motor gliders, some powered aircraft are designed for routine glides during part of their flight; usually when landing after a period of a powered flight. These include:

- Experimental aircraft such as the North American X-15, which glided back having used their fuel
- Spacecraft such as the Space Shuttles, SpaceShipOne and the Russian Buran

Some aircraft are not primarily designed to glide except in an emergency, for example airliners that have run out of fuel.

Gliding Animals

Birds

A number of animals have separately evolved gliding many times, without any single ancestor. Birds in particular use gliding flight to minimise their use of energy. Large birds are notably adept at gliding, including:

- Albatross
- Condor
- Vulture
- Eagle
- Stork
- Frigatebird

Like recreational aircraft, birds can alternate periods of gliding with periods of soaring in rising air, and so spend a considerable time airborne with a minimal expenditure of energy. The great frigatebird in particular is capable of continuous flights up to several weeks.

Mammals

To assist gliding, some mammals have evolved a structure called the patagium. This is a membranous structure found stretched between a range of body parts. It is most highly developed in bats. For similar reasons to birds, bats can glide efficiently. In bats, the skin forming the surface of the wing is an extension of the skin of the abdomen that runs to the tip of each digit, uniting the forelimb with the body. The patagium of a bat has four distinct parts:

1. Propatagium: the patagium present from the neck to the first digit
2. Dactylopatagium: the portion found within the digits
3. Plagiopatagium: the portion found between the last digit and the hindlimbs
4. Uropatagium: the posterior portion of the body between the two hindlimbs

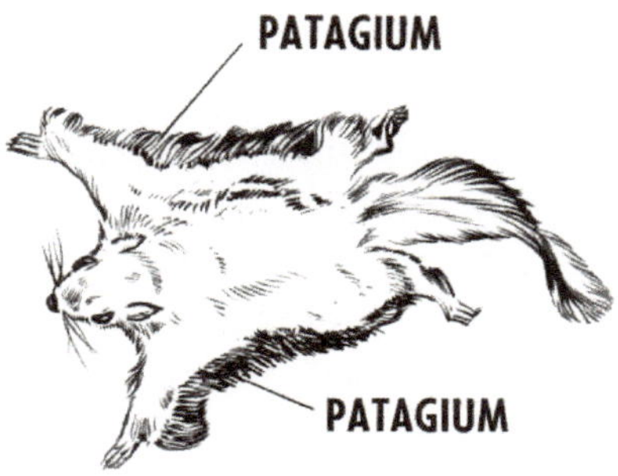

Patagium on a flying squirrel

Other mammals such as gliding possums and flying squirrels also glide using a patagium, but with much poorer efficiency than bats. They cannot gain height. The animal launches itself from a tree, spreading its limbs to expose the gliding membranes, usually to get from tree to tree in rainforests as an efficient means of both locating food and evading predators. This form of arboreal locomotion, is common in tropical regions such as Borneo and Australia, where the trees are tall and widely spaced.

In flying squirrels, the patagium stretches from the fore- to the hind-limbs along the length of each side of the torso. In the sugar glider, the patagia extend between the fifth finger of each hand to the first toe of each foot. This creates an aerofoil enabling them to glide 50 metres or more. This gliding flight is regulated by changing the curvature of the membrane or moving the legs and tail.

Fish, Reptiles, Amphibians and Other Gliding Animals

In addition to mammals and birds, other animals notably flying fish, flying snakes, flying frogs and flying squid also glide.

Flying fish taking off

The flights of flying fish are typically around 50 meters (160 ft), though they can use updrafts at the leading edge of waves to cover distances of up to 400 m (1,300 ft). To glide upward out of the water, a flying fish moves its tail up to 70 times per second. It then spreads its pectoral fins and tilts them slightly upward to provide lift. At the end of a glide, it folds its pectoral fins to re-enter the sea, or drops its tail into the water to push against the water to lift itself for another glide, possibly changing direction. The curved profile of the "wing" is comparable to the aerodynamic shape of a bird wing. The fish is able to increase its time in the air by flying straight into or at an angle to the direction of updrafts created by a combination of air and ocean currents.

Snakes of the genus *Chrysopelea* are also known by the common name "flying snake". Before launching from a branch, the snake makes a J-shape bend. After thrusting its body up and away from the tree, it sucks in its abdomen and flaring out its ribs to turn its body into a "pseudo concave wing", all the while making a continual serpentine motion of lateral undulation parallel to the ground to stabilise its direction in mid-air in order to land safely. Flying snakes are able to glide better than flying squirrels and other gliding animals, despite the lack of limbs, wings, or any other wing-like projections, gliding through the forest and jungle it inhabits with the distance being as great as 100 m. Their destination is mostly predicted by ballistics; however, they can exercise some in-flight attitude control by "slithering" in the air.

Gliding flight has evolved independently among 3,400 species of frogs from both New World (Hylidae) and Old World (Rhacophoridae) families. This parallel evolution is seen as an adaptation to their life in trees, high above the ground. Characteristics of the Old World species include "enlarged hands and feet, full webbing between all fingers and toes, lateral skin flaps on the arms and legs

Forces

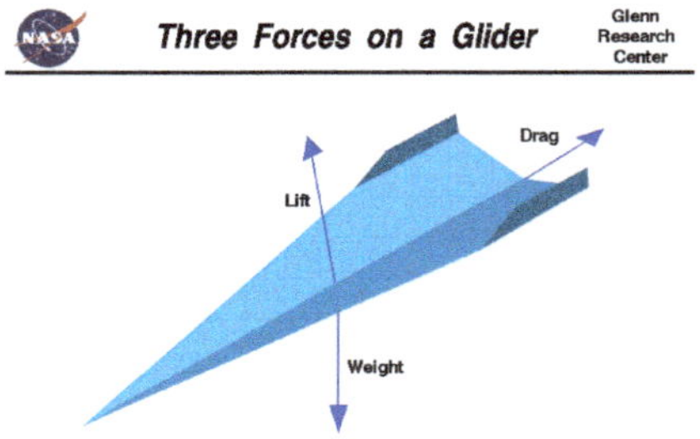

Forces on a gliding animal or aircraft in flight

Three principal forces act on aircraft and animals when gliding:

- weight – gravity acts in the downwards direction
- lift – acts perpendicularly to the vector representing airspeed
- drag – acts parallel to the vector representing the airspeed

As the aircraft or animal descends, the air moving over the wings generates lift. The lift force acts slightly forward of vertical because it is created at right angles to the airflow which comes from slightly below as the glider descends, see angle of attack. This horizontal component of lift is enough to overcome drag and allows the glider to accelerate forward. Even though the weight causes the aircraft to descend, if the air is rising faster than the sink rate, there will be a gain of altitude.

Lift to Drag Ratio

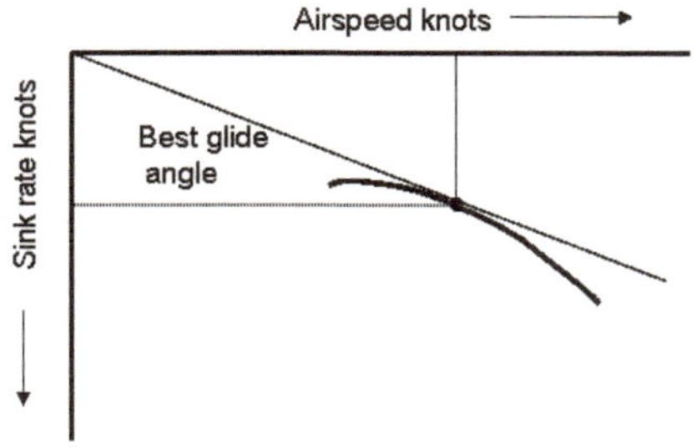

Polar curve showing glide angle for best glide

The lift-to-drag ratio, or *L/D ratio*, is the amount of lift generated by a wing or vehicle, divided by the drag it creates by moving through the air. A higher or more favourable L/D ratio is typically one of the major goals in aircraft design; since a particular aircraft's needed lift is set by its weight, delivering that lift with lower drag leads directly to better fuel economy and climb performance.

The effect of airspeed on the rate of descent can be depicted by a polar curve. These curves show the airspeed where minimum sink can be achieved and the airspeed with the best L/D ratio. The curve is an inverted U-shape. As speeds reduce the amount of lift falls rapidly around the stalling speed. The peak of the 'U' is at minimum drag.

As lift and drag are both proportional to the coefficient of Lift and Drag respectively multiplied by the same factor ($1/2\ \rho_{air}\ v^2 S$), the L/D ratio can be simplified to the Coefficient of lift divided by the coefficient of drag or Cl/Cd, and since both are proportional to the airspeed, the ratio of L/D or Cl/Cd is then typically plotted against angle of attack.

Drag

Induced drag is caused by the generation of lift by the wing. Lift generated by a wing is perpendicular to the relative wind, but since wings typically fly at some small angle of attack, this means that a component of the force is directed to the rear. The rearward component of this force (parallel with the relative wind) is seen as drag. At low speeds an aircraft has to generate lift with a higher angle of attack, thereby leading to greater induced drag. This term dominates the low-speed side of the drag graph, the left side of the U.

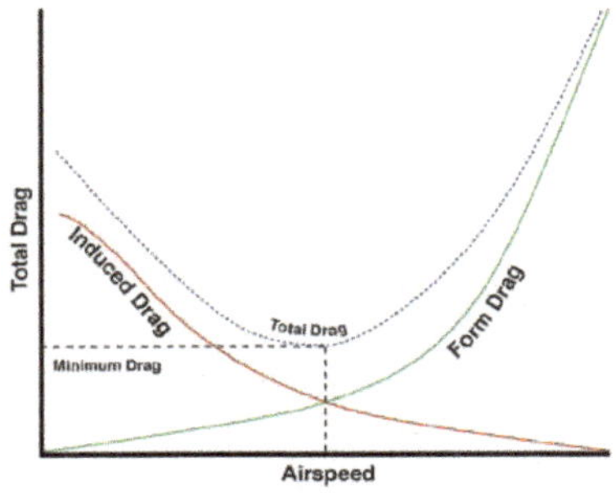

The drag curve

Profile drag is caused by air hitting the wing, and other parts of the aircraft. This form of drag, also known as wind resistance, varies with the square of speed. For this reason profile drag is more pronounced at higher speeds, forming the right side of the drag graph's U shape. Profile drag is lowered primarily by reducing cross section and streamlining.

As lift increases steadily until the critical angle, it is normally the point where the combined drag is at its lowest, that the wing or aircraft is performing at its best L/D.

Designers will typically select a wing design which produces an L/D peak at the chosen cruising speed for a powered fixed-wing aircraft, thereby maximizing economy. Like all things in aeronautical engineering, the lift-to-drag ratio is not the only consideration for wing design. Performance at high angle of attack and a gentle stall are also important.

Minimising drag is of particular interest in the design and operation of high performance glider (sailplane)s, the largest of which can have glide ratios approaching 60 to 1, though many others have a lower performance; 25:1 being considered adequate for training use.

Glide Ratio

When flown at a constant speed in still air a glider moves forwards a certain distance for a certain distance downwards. The ratio of the distance forwards to downwards is called the *glide ratio*. The glide ratio (E) is numerically equal to the lift-to-drag ratio under these conditions; but is not necessarily equal during other manoeuvres, especially if speed is not constant. A glider's glide ratio varies with airspeed, but there is a maximum value which is frequently quoted. Glide ratio usually varies little with vehicle loading; a heavier vehicle glides faster, but nearly maintains its glide ratio.

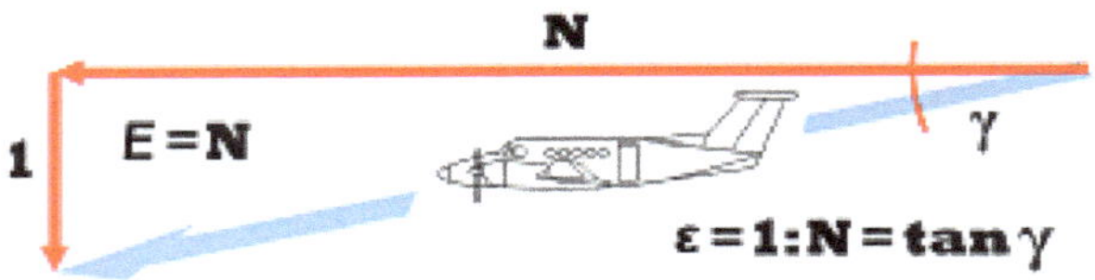

Glide ratio (or "finesse") is the cotangent of the downward angle, the *glide angle* (γ). Alternatively it is also the forward speed divided by sink speed (unpowered aircraft):

Glide number (ε) is the reciprocal of glide ratio but sometime it's confused.

Examples

Flight article	Scenario	L/D ratio/ glide ratio
Modern sailplane	Gliding	40–60 (depending on span)
Great frigatebird	Soaring over the ocean	15-22 at typical speeds
Hang glider		15
Gimli Glider	Boeing 767-200 with fuel exhaustion	~12
Paraglider	High performance model	11
Helicopter	Autorotation	4
Powered parachute	Rectangular/elliptical parachute	3.6/5.6
Space Shuttle	Approach	4.5
Hypersonic Technology Vehicle 2	Equilibrium hypersonic gliding estimate	2.6
Wingsuit	Gliding	2.5

Northern flying squirrel	Gliding	1.98
Space Shuttle	Hypersonic	1
Apollo CM	Reentry	0.368

Importance of the Glide Ratio in Gliding Flight

Although the best glide ratio is important when measuring the performance of a gliding aircraft, its glide ratio at a range of speeds also determines its success.

Pilots sometimes fly at the aircraft's best L/D by precisely controlling airspeed and smoothly operating the controls to reduce drag. However the strength of the likely next lift and the strength of the wind also affects the optimal speed to fly. To achieve higher speed across country, gliders (sailplanes) are often loaded with water ballast to increase the airspeed and so reach the next area of lift sooner. This has little effect on the glide angle but increases rate of sink (and airspeed in proportion) because the heavier aircraft achieves optimal L/D at a higher airspeed.

If the air is rising faster than the rate of sink, the aircraft will climb. At lower speeds an aircraft may have a worse glide ratio but it will also have a lower rate of sink. A low airspeed also improves its ability to turn tightly in centre of the rising air where the rate of ascent is greatest. A sink rate of approximately 1.0 m/s is the most that a practical hang glider or paraglider could have before it would limit the occasions that a climb was possible to only when there was strongly rising air. Gliders (sailplanes) have minimum sink rates of between 0.4 and 0.6 m/s depending on the class. Aircraft such as airliners may have a better glide ratio than a hang glider, but would rarely be able to thermal because of their much higher forward speed and their much higher sink rate. (Note that the Boeing 767 in the Gimli Glider incident achieved a glide ratio of only 12:1.)

During landing, a high lift/drag ratio is desirable. Some aircraft therefore employ flaps, to increase their performance at lower speeds. Experiments with lifting bodies show that a lift/drag ratio below about 2 makes landing very difficult because of the high rate of descent. Many birds also adapt the contour of their wings before landing.

The loss of height can be measured at several speeds and plotted on a "polar curve" to calculate the best speed to fly in various conditions, such as when flying into wind or when in sinking air. Other polar curves can be measured after loading the glider with water ballast. As mass increases, the best glide ratio is achieved at higher speeds. (The glide ratio is not increased.)

Soaring

Soaring animals and aircraft may alternate glides with periods of soaring in rising air. Five principal types of lift are used: thermals, ridge lift, lee waves, convergences and dynamic soaring. Dynamic soaring is used predominately by birds, and some model aircraft, though it has also been achieved on rare occasions by piloted aircraft.

Examples of soaring flight by birds are the use of:

- Thermals and convergences by raptors such as vultures
- Ridge lift by gulls near cliffs

- Wave lift by migrating birds
- Dynamic effects near the surface of the sea by albatrosses

For humans, soaring is the basis for three air sports: gliding, hang gliding and paragliding.

Insect Wing

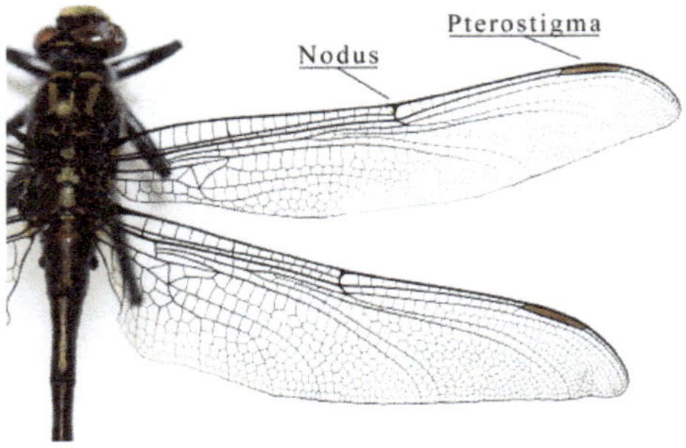

Original veins and wing posture of a dragonfly

Insect wings are adult outgrowths of the insect exoskeleton that enable insects to fly. They are found on the second and third thoracic segments (the mesothorax and metathorax), and the two pairs are often referred to as the forewings and hindwings, respectively, though a few insects lack hindwings, even rudiments. The wings are strengthened by a number of longitudinal veins, which often have cross-connections that form closed "cells" in the membrane (extreme examples include the dragonflies and lacewings). The patterns resulting from the fusion and cross-connection of the wing veins are often diagnostic for different evolutionary lineages and can be used for identification to the family or even genus level in many orders of insects.

Hoverflies hovering to mate

Physically, some insects move their flight muscles directly, others indirectly. In insects with direct flight, the wing muscles directly attached to the wing base, so that a small downward movement of the wing base lifts the wing itself upward. Those insects with indirect flight have muscles that attach to and deform the thorax, causing the wings to move as well.

The wings are present in only one sex (often the male) in some groups such as velvet ants and Strepsiptera, or are selectively lost in "workers" of social insects such as ants and termites. Rarely, the female is winged but the male not, as in fig wasps. In some cases, wings are produced only at particular times in the life cycle, such as in the dispersal phase of aphids. Wing structure and colouration often vary with morphs, such as in the aphids, migratory phases of locusts and polymorphic butterflies. At rest, the wings may be held flat, or folded a number of times along specific patterns; most typically, it is the hindwings which are folded, but in a few groups such as the vespid wasps, it is the forewings.

A cockchafer's hardened forewings raised, hindwings unfolding

How and why insect wings evolved is not well understood. Three main theories on the origins of insect flight are that wings developed from paranotal lobes, extensions of the thoracic terga; that they are modifications of movable abdominal gills as found on aquatic naiads of mayflies; and that insect wings arose from the fusion of pre-existing endite and exite structures each with pre-existing articulation and tracheation.

Morphology

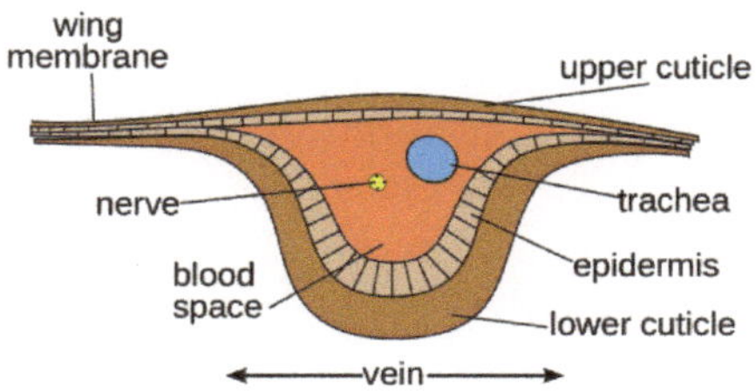

Internal

Each of the wings consists of a thin membrane supported by a system of veins. The membrane is formed by two layers of integument closely apposed, while the veins are formed where the two layers remain separate; sometimes the lower cuticle is thicker and more heavily sclerotized under a vein. Within each of the major veins there is a nerve and a trachea, and, since the cavities of the veins are connected with the hemocoel, hemolymph can flow into the wings.

As the wing develops, the dorsal and ventral integumental layers become closely apposed over most of their area forming the wing membrane. The remaining areas form channels, the future veins, in which the nerves and tracheae may occur. The cuticle surrounding the veins becomes thickened and more heavily sclerotized to provide strength and rigidity to the wing. Two types of hair may occur on the wings: microtrichia, which are small and irregularly scattered, and macrotrichia, which are larger, socketed, and may be restricted to veins. The scales of Lepidoptera and Trichoptera are highly modified macrotrichia.

Venation

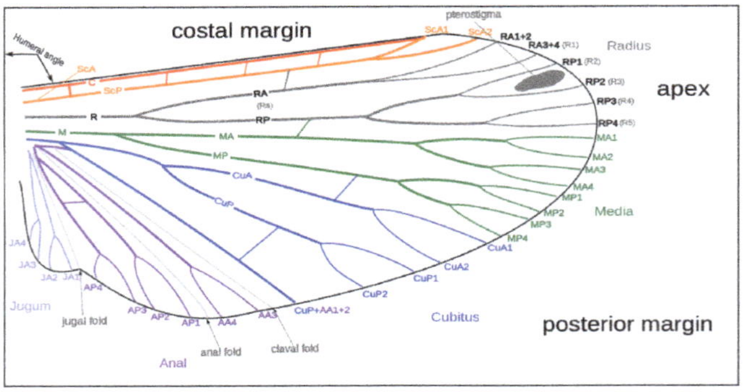

Venation of insect wings, based on the Comstock–Needham system

In some very small insects, the venation may be greatly reduced. In Chalcidoidea (Chalcid wasps), for instance, only the subcosta and part of the radius are present. Conversely, an increase in venation may occur by the branching of existing veins to produce accessory veins or by the development of additional, intercalary veins between the original ones, as in the wings of Orthoptera (grasshoppers and crickets). Large numbers of cross-veins are present in some insects, and they may form a reticulum as in the wings of Odonata (dragonflies and damselflies) and at the base of the forewings of Tettigonioidea and Acridoidea (katydids and grasshoppers respectively).

The archedictyon is the name given to a hypothetical scheme of wing venation proposed for the very first winged insect. It is based on a combination of speculation and fossil data. Since all winged insects are believed to have evolved from a common ancestor, the archediction represents the "template" that has been modified (and streamlined) by natural selection for 200 million years. According to current dogma, the archedictyon contained 6–8 longitudinal veins. These veins (and their branches) are named according to a system devised by John Comstock and George Needham—the Comstock–Needham system:

Costa (C) – the leading edge of the wing

Subcosta (Sc) – second longitudinal vein (behind the costa), typically unbranched

Radius (R) – third longitudinal vein, one to five branches reach the wing margin

Media (M) – fourth longitudinal vein, one to four branches reach the wing margin

Cubitus (Cu) – fifth longitudinal vein, one to three branches reach the wing margin

Anal veins (A1, A2, A3) – unbranched veins behind the cubitus

The costa (C) is the leading marginal vein on most insects, although sometimes there is a small vein above the costa called the precosta, although in almost all extant insects, the precosta is fused with the costa; The costa rarely ever branches because it is at the leading edge, which is associated at its base with the humeral plate. The trachea of the costal vein is perhaps a branch of the subcostal trachea. Located after the costa is the third vein, the subcosta, which branches into two separate veins: the anterior and posterior. The base of the subcosta is associated with the distal end of the neck of the first axillary. The fourth vein is the radius (R), which is branched into five separate veins. The radius is generally the strongest vein of the wing. Toward the middle of the wing, it forks into a first undivided branch (R1) and a second branch, called the radial sector (Ra), which subdivides dichotomously into four distal branches (R2, R3, R4, R5). Basally, the radius is flexibly united with the anterior end of the second axillary (2Ax).

The fifth vein of the wing is the media. In the archetype pattern (A), the media forks into two main branches: a media anterior (MA), which divides into two distal branches (MA1, MA2), and a median sector, or media posterior (MP), which has four terminal branches (M1, M2, M3, M4). In most modern insects the media anterior has been lost, and the usual "media" is the four-branched media posterior with the common basal stem. In the Ephemerida, according to present interpretations of the wing venation, both branches of the media are retained, while in Odonata the persisting media is the primitive anterior branch. The stem of the media is often united with the radius, but when it occurs as a distinct vein its base is associated with the distal median plate (m') or is continuously sclerotized with the latter. The cubitus, the sixth vein of the wing, is primarily two branched. The primary forking of the takes place near the base of the wing, forming the two principal branches (Cu1, Cu2). The anterior branch may break up into a number of secondary branches, but commonly it forks into two distal branches. The second branch of the cubitus (Cu2) in Hymenoptera, Trichoptera, and Lepidoptera was mistaken by Comstock and Needham for the first anal. Proximally the main stem of the cubitus is associated with the distal median plate (m') of the wing base.

Postcubitus (Pcu) is the first anal of the Comstock–Needham system. The postcubitus, however, has the status of an independent wing vein and should be recognized as such. In nymphal wings, its trachea arises between the cubital trachea and the group of vannal tracheae. In the mature wings of more generalized insect the Postcubitus is always associated proximally with the cubitus and is never intimately connected with the flexor sclerite (3Ax) of the wing base. In Neuroptera, Mecoptera, and Trichoptera the postcubitus may be more closely associated with the vannal veins, but its base is always free from the latter. The postcubitus is usually unbranched; it is primitively two branched. The vannal veins (lV to nV) are the anal veins that are immediately associated with the third axillary, and which are directly affected by the movement of this sclerite that brings about the flexion of the wings. In number the vannal veins vary. from 1 to 12, according to the expansion of the vannal area of the wing. The vannal tracheae usually arise from a common tracheal stem in nymphal insects, and the veins are regarded as branches of a single anal vein. Distally the vannal veins are either simple or branched. Jugal Veins (J) of the jugal lobe of the wing is often occupied by a network of irregular veins, or it may be entirely membranous; but sometimes it contains one or two distinct small veins, the first jugal vein, or vena arcuata, and the second jugal vein, or vena cardinalis (2J).

C-Sc cross-veins – run between the costa and subcosta

R cross-veins – run between adjacent branches of the radius

R-M cross-veins – run between the radius and media

M-Cu cross-veins – run between the media and cubitus

All the veins of the wing are subject to secondary forking and to union by cross-veins. In some orders of insects the cross-veins are so numerous that the whole venational pattern becomes a close network of branching veins and cross-veins. Ordinarily, however, there is a definite number of cross-veins having specific locations. The more constant cross-veins are the humeral cross-vein

(h) between costa and subcosta, the radial cross-vein (r) between R and the first fork of Rs, the sectorial cross-vein (s) between the two forks of R8, the median cross-vein (m–m) between M2 and M3, and the mediocubital cross-vein (m-cu) between media and cubitus.

The veins of insect wings are characterized by a convex-concave placement, such as those seen in mayflies (i.e., concave is "down" and convex is "up") which alternate regularly and by its triadic type of branching; whenever a vein forks there is always an interpolated vein of the opposite position between the two branches. A concave vein will fork into two concave veins (with the interpolated vein being convex) and the regular alteration of the veins is preserved. The veins of the wing appear to fall into an undulating pattern according to whether they have a tendency to fold up or down when the wing is relaxed. The basal shafts of the veins are convex, but each vein forks distally into an anterior convex branch and a posterior concave branch. Thus the costa and subcosta are regarded as convex and concave branches of a primary first vein, Rs is the concave branch of the radius, posterior media the concave branch of the media, Cu1 and Cu2 are respectively convex and concave, while the primitive Postcubitus and the first vannal have each an anterior convex branch and a posterior concave branch. The convex or concave nature of the veins has been used as evidence in determining the identities of the persisting distal branches of the veins of modern insects, but it has not been demonstrated to be consistent for all wings.

Fields

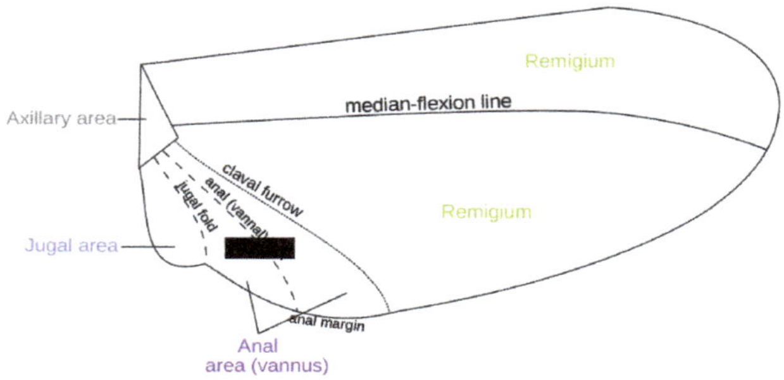

Wing areas are delimited and subdivided by fold-lines along which the wing can fold, and flexion-lines along which the wing can flex during flight. The fundamental distinction between the flexion-lines and the fold-lines is often blurred, as fold-lines may permit some flexibility or vice versa. Two constants that are found in nearly all insect wings are the claval (a flexion-line) and jugal folds (or fold line); forming variable and unsatisfactory boundaries. Wing foldings can very complicated, with transverse folding occurs in the hindwings of Dermaptera and Coleoptera, and in some insects the anal area can be folded like a fan. There are about four different fields found on the insect wings:

Remigium

Anal area (vannus)

Jugal area

Axillary area

Alula

Most veins and crossveins occur in the anterior area of the remigium, which is responsible for most of the flight, powered by the thoracic muscles. The posterior portion of the remigium is sometimes called the clavus; the two other posterior fields are the anal and jugal ares. When the vannal fold has the usual position anterior to the group of anal veins, the remigium contains the costal, subcostal, radial, medial, cubital, and postcubital veins. In the flexed wing the remigiumturns posteriorly on the flexible basal connection of the radius with the second axillary, and the base of the mediocubital field is folded medially on the axillary region along the plica basalis (bf) between the median plates (m, m') of the wing base.

The vannus is bordered by the vannal fold, which typically occurs between the postcubitus and the first vannal vein. In Orthoptera it usually has this position. In the forewing of Blattidae, however, the only fold in this part of the wing lies immediately before the postcubitus. In Plecoptera the vannal fold is posterior to the postcubitus, but proximally it crosses the base of the first vannal vein. In the cicada the vannal fold lies immediately behind the first vannal vein (lV). These small variations in the actual position of the vannal fold, however, do not affect the unity of action of the vannal veins, controlled by the flexor sclerite (3Ax), in the flexion of the wing. In the hindwings of most Orthoptera a secondary vena dividens forms a rib in the vannal fold. The vannus is usually triangular in shape, and its veins typically spread out from the third axillary like the ribs of a fan. Some of the vannal veins may be branched, and secondary veins may alternate with the primary veins. The vannal region is usually best developed in the hindwing, in which it may be enlarged to form a sustaining surface, as in Plecoptera and Orthoptera. The great fanlike expansions of the hindwings of Acrididae are clearly the vannal regions, since their veins are all supported on the third axillary sclerites on the wing bases, though Martynov (1925) ascribes most of the fan areas in Acrididae to the jugal regions of the wings. The true jugum of the acridid wing is represented only by the small membrane (Ju) mesad of the last vannal vein. The jugum is more highly developed in some other Polyneoptera, as in the Mantidae. In most of the higher insects with narrow wings the vannus becomes reduced, and the vannal fold is lost, but even in such cases the flexed wing may bend along a line between the postcubitus and the first vannal vein.

The Jugal Region, or Neala, is a region of the wing that is usually a small membranous area proximal to the base of the vannus strengthened by a few small, irregular veinlike thickenings; but when well developed it is a distinct section of the wing and may contain one or two jugal veins. When the jugal area of the forewing is developed as a free lobe, it projects beneath the humeral angle of the hindwing and thus serves to yoke the two wings together. In the Jugatae group of Lepidoptera it bears a long finger-like lobe. The jugal region was termed the neala ("new wing") because it is evidently a secondary and recently developed part of the wing.

The axillary region is region containing the axillary sclerites has in general the form of a scalene triangle. The base of the triangle (a-b) is the hinge of the wing with the body; the apex (c) is the distal end of the third axillary sclerite; the longer side is anterior to the apex. The point d on the anterior side of the triangle marks the articulation of the radial vein with the second axillary sclerite. The line between d and c is the plica basalis (bf), or fold of the wing at the base of the mediocubital field.

At the posterior angle of the wing base in some Diptera there is a pair of membranous lobes (squamae, or calypteres) known as the alula. The alula is well developed in the house fly. The outer squama (c) arises from the wing base behind the third axillary sclerite (3Ax) and evidently represents the jugal lobe of other insects (A, D); the larger inner squama (d) arises from the posterior scutellar margin of the tergum of the wing-bearing segment and forms a protective, hoodlike canopy over the halter. In the flexed wing the outer squama of the alula is turned upside down above the inner squama, the latter not being affected by the movement of the wing. In many Diptera a deep incision of the anal area of the wing membrane behind the single vannal vein sets off a proximal alar lobe distal to the outer squama of the alula.

Joints

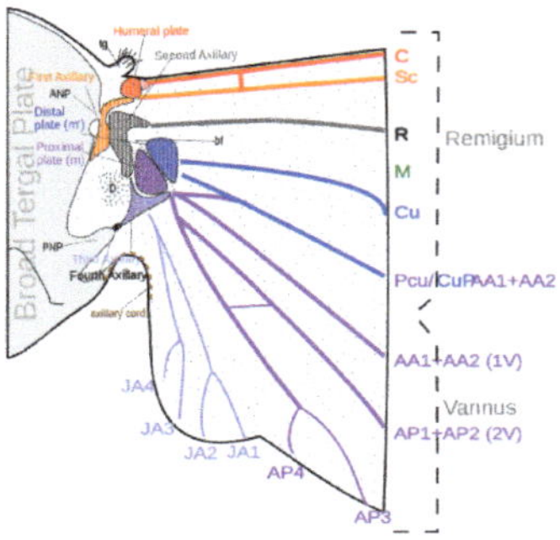

The various movements of the wings, especially in insects that flex the wings horizontally over the back when at rest, demand a more complicated articular structure at the wing base than a mere hinge of the wing with the body. Each wing is attached to the body by a membranous basal area, but the articular membrane contains a number of small articular sclerites, collectively known as the pteralia. The pteralia include an anterior humeral plate at the base of the costal vein, a group of axillaries (Ax) associated with the subcostal, radial, and vannal veins, and two less definite median plates (m, m') at the base of the mediocubital area. The axillaries are specifically developed only in the wing-flexing insects, where they constitute the flexor mechanism of the wing operated by the flexor muscle arising on the pleuron. Characteristic of the wing base is also a small lobe on the anterior margin of the articular area proximal to the humeral plate, which, in the forewing of some insects, is developed into a large, flat, scale-like flap, the tegula, overlapping the base of the wing. Posteriorly the articular membrane often forms an ample lobe between the wing and the body, and its margin is generally thickened and corrugated, giving the appearance of a ligament, the so-called axillary cord, continuous mesally with the posterior marginal scutellar fold of the tergal plate bearing the wing.

The articular sclerites, or pteralia, of the wing base of the wing-flexing insects and their relations to the body and the wing veins, shown diagrammatically, are as follows:

Humeral plates

First Axillary

Second Axillary

Third Axillary

Fourth Axillary

Median plates (m, m')

The humeral plate is usually a small sclerite on the anterior margin of the wing base, movable and articulated with the base of the costal vein. Odonata have their humeral plate greatly enlarged, with two muscles arising from the episternum inserted into the Humeral plates and two from the edge of the epimeron inserted into the axillary plate.

The first axillary sclerite (lAx) is the anterior hinge plate of the wing base. Its anterior part is supported on the anterior notal wing process of the tergum (ANP); its posterior part articulates with the tergal margin. The anterior end of the sclerite is generally produced as a slender arm, the apex of which (e) is always associated with the base of the subcostal vein (Sc), though it is not united with the latter. The body of the sclerite articulates laterally with the second axillary. The second axillary sclerite (2Ax) is more variable in form than the first axillary, but its mechanical relations are no less definite. It is obliquely hinged to the outer margin of the body of the first axillary, and the radial vein (R) is always flexibly attached to its anterior end (d). The second axillary presents both a dorsal and a ventral sclerotization in the wing base; its ventral surface rests upon the fulcral wing process of the pleuron. The second axillary, therefore, is the pivotal sclerite of the wing base, and it specifically manipulates the radial vein.

The third axillary sclerite (3Ax) lies in the posterior part of the articular region of the wing. Its form is highly variable and often irregular, but the third axillary is the sclerite on which is inserted the flexor muscle of the wing (D). Mesally it articulates anteriorly (f) with the posterior end of the second axillary, and posteriorly (b) with the posterior wing process of the tergum (PNP), or with a small fourth axillary when the latter is present. Distally the third axillary is prolonged in a process which is always associated with the bases of the group of veins in the anal region of the wing here termed the vannal veins (V). The third axillary, therefore, is usually the posterior hinge plate of the wing base and is the active sclerite of the flexor mechanism, which directly manipulates the vannal veins. The contraction of the flexor muscle (D) revolves the third axillary on its mesal articulations (b, f) and thereby lifts its distal arm; this movement produces the flexion of the wing. The Fourth Axillary sclerite is not a constant element of the wing base. When present it is usually a small plate intervening between the third axillary and the posterior notal wing process and is probably a detached piece of the latter.

The median plates (m, m') are also sclerites that are not so definitely differentiated as specific plates as are the three principal axillaries, but nevertheless they are important elements of the flexor apparatus. They lie in the median area of the wing base distal to the second and third axillaries and are separated from each other by an oblique line (bf) which forms a prominent convex fold during flexion of the wing. The proximal plate (m) is usually attached to the distal arm of the third axillary and perhaps should be regarded as a part of the latter. The distal plate (m') is less constantly present as a distinct sclerite and may be represented by a general sclerotization of the base of the mediocubital field of the wing. When the veins of this region are distinct at their bases, they are associated with the outer median plate.

Muscles

The diamond-shaped alary muscles (green) of the mosquito *Anopheles gambiae* and their structural relationship to the tube-like heart (also in green). Red depicts pericardial cells, blue cell nuclei.

The muscles that control flight in insects can take up to 10% to 30% of the total body mass. The muscles that control flight vary with the two types of flight found in insects: indirect and direct. Insects that use first, indirect, have the muscles attach to the tergum instead of the wings, as the name suggests. As the muscles contract, the thoracic box becomes distorted, transferring the energy to the wing. There are two "bundles" of muscles, those that span parallel to the tergum, the dorsolongitudinals, and those that are attached to the tegum and extend to the sternum, the dorsoventrals. In direct muscle, the connection is directly from the pleuron (thoracic wall) to individual sclerites located at the base of the wing. The subalar and basalar muscles have ligament attachments to the subalar and basalar sclerites. Here resilin, a highly elastic material, forms the ligaments connecting flight muscles to the wing apparatus.

In more derived orders of insects, such as Diptera (flies) and Hymenoptera (wasp), the indirect muscles occupy the greatest volume of the pterothorax and function as the primary source of power for the wingstroke. Contraction of the dorsolongitudinal muscles causes the severe arching of the notum which depresses the wing while contraction of the dorsoventral muscles causes opposite motion of notum. Other more primitive insects, such as Orthoptera (locusts), Coleoptera (beetles), and Odonata (dragonflies) use direct muscles that are responsible for developing the needed power for the up and down strokes.

Insect wing muscle is a strictly aerobic tissue. Per unit protein it consumes fuel and oxygen at rates taking place in a very concentrated and highly organized tissue so that the steady-state rates per unit volume represent an absolute record in biology. The fuel and oxygen rich blood is carried to the muscles through diffusion occurring in large amounts, in order to maintain the high level of energy used during flight. Many wing muscles are large and may be as large as 10 mm in length and 2 mm in width. Moreover, in some Diptera the fibres are of giant dimensions. For instance, in the very active *Rutilia*, the cross-section is 1800 μm long and more than 500 μm wide. The transport of fuel and oxygen from the surroundings to the sites of consumption and the reverse transport of carbon dioxide therefore represent a challenge to the biologist both in relation to transport in the liquid phase and in the intricate system of air tubes, i.e. in the tracheal system.

Coupling, Folding, and Other Features

In many insect species, the forewing and hindwing are coupled together, which improves the aerodynamic efficiency of flight. The most common coupling mechanism (e.g., Hymenoptera and Trichoptera) is a row of small hooks on the forward margin of the hindwing, or "hamuli", which lock onto the forewing, keeping them held together (hamulate coupling). In some other insect species (e.g., Mecoptera, Lepidoptera, and some Trichoptera) the jugal lobe of the forewing covers a portion of the hindwing (jugal coupling), or the margins of the forewing and hindwing overlap broadly (amplexiform coupling), or the hindwing bristles, or frenulum, hook under the retaining structure or retinaculum on the forewing.

When at rest, the wings are held over the back in most insects, which may involve longitudinal folding of the wing membrane and sometimes also transverse folding. Folding may sometimes occur along the flexion lines. Though fold lines may be transverse, as in the hindwings of beetles and earwigs, they are normally radial to the base of the wing, allowing adjacent sections of a wing to be folded over or under each other. The commonest fold line is the jugal fold, situated just behind the third anal vein, although, most Neoptera have a jugal fold just behind vein 3A on the forewings. It is sometimes also present on the hindwings. Where the anal area of the hindwing is large, as in Orthoptera and Blattodea, the whole of this part may be folded under the anterior part of the wing along a vannal fold a little posterior to the claval furrow. In addition, in Orthoptera and Blattodea, the anal area is folded like a fan along the veins, the anal veins being convex, at the crests of the folds, and the accessory veins concave. Whereas the claval furrow and jugal fold are probably homologous in different species, the vannal fold varies in position in different taxa. Folding is produced by a muscle arising on the pleuron and inserted into the third axillary sclerite in such a way that, when it contracts, the sclerite pivots about its points of articulation with the posterior notal process and the second axillary sclerite.

As a result, the distal arm of the third axillary sclerite rotates upwards and inwards, so that finally its position is completely reversed. The anal veins are articulated with this sclerite in such a way that when it moves they are carried with it and become flexed over the back of the insect. Activity of the same muscle in flight affects the power output of the wing and so it is also important in flight control. In orthopteroid insects, the elasticity of the cuticle causes the vannal area of the wing to fold along the veins. Consequently, energy is expended in unfolding this region when the wings are moved to the flight position. In general, wing extension probably results from the contraction of muscles attached to the basalar sclerite or, in some insects, to the subalar sclerite.

Flight

Australian Emperor in flight; it uses the direct flight mechanism.

Flight Mechanisms

Two groups of relatively large insects, the Ephemeroptera (mayflies) and the Odonata (dragonflies and damselflies) have the flight muscles attached directly to their wings; the wings can beat no faster than the rate at which nerves can send impulses to command the muscles to beat. All other living winged insects fly using a different mechanism, involving indirect flight muscles which cause the thorax to vibrate; the wings can beat faster than the rate at which the muscles receive nerve impulses. This mechanism evolved once, and is the defining feature (synapomorphy) for the infraclass Neoptera.

Aerodynamics

There are two basic aerodynamic models of insect flight. Most insects use a method that creates a spiralling leading edge vortex. Some very small insects use the fling and clap or Weis-Fogh mechanism in which the wings clap together above the insect's body and then fling apart. As they fling open, the air gets sucked in and creates a vortex over each wing. This bound vortex then moves across the wing and, in the clap, acts as the starting vortex for the other wing. Circulation and lift are increased, at the price of wear and tear on the wings.

Many insects can hover by beating their wings rapidly, requiring sideways stabilization as well as lift.

A few insects use gliding flight, without the use of thrust. It is found in some species of arboreal ants, known as gliding ants.

Evolution

Sometime in the Carboniferous Period, some 350 million years ago, when there were only two major land masses, insects began flying. How and why insect wings developed, however, is not well understood, largely due to the scarcity of appropriate fossils from the period of their development in the Lower Carboniferous. Three main theories on the origins of insect flight are that wings developed from paranotal lobes, extensions of the thoracic terga; that they are modifications of movable abdominal gills as found on aquatic naiads of mayflies; or that they developed from thoracic protrusions used as radiators.

Fossils

Holotype wing of the extinct Cimbrophlebia brooksi. 49.5 Million Years old; "Boot Hill", Klondike Mountain Formation, Washington, USA.

Fossils from the Devonian (400 million years ago) are all wingless, but by the Carboniferous (320 million years ago), more than 10 different genera of insects had fully functional wings. There is little preservation of transitional forms between the two periods. The earliest winged insects are from this time period (Pterygota), including the Blattoptera, Caloneurodea, primitive stem-group Ephemeropterans, Orthoptera and Palaeodictyopteroidea. Very early Blattopterans (during the Carboniferous) had a very large discoid pronotum and coriaceous forewings with a distinct CuP vein (an unbranched wing vein, lying near the claval fold and reaching the wing posterior margin). Even though the oldest definitive insect fossil is the Devonian *Rhyniognatha hirsti*, estimated at 396–407 million years old, it possessed dicondylic mandibles, a feature associated with winged insects.

During the Permian, the dragonflies *Odonata* were the dominant aerial predator and probably dominated terrestrial insect predation as well. True Odonata appeared in the Permian and all are amphibian. Their prototypes are the oldest winged fossils, go back to the Devonian, and are different from other wings in every way. Their prototypes may have had the beginnings of many modern attributes even by late Carboniferous and it is possible that they even captured small vertebrates, for some species had a wing span of 71 cm. The earliest beetle-like species during the Permian had pointed, leather like forewings with cells and pits. Hemiptera, or true bugs had appeared in the form of Arctiniscytina and Paraknightia having forewings with unusual venation, possibly diverging from Blattoptera.

A single large wing from a species of Diptera in the Triassic (10 mm instead of usual 2–6 mm) was found in Australia (Mt. Crosby).This family Tilliardipteridae, despite of the numerous 'tipuloid' features, should be included in Psychodomorpha sensu Hennig on account of loss of the convex distal 1A reaching wing margin and formation of the anal loop.

Hypotheses

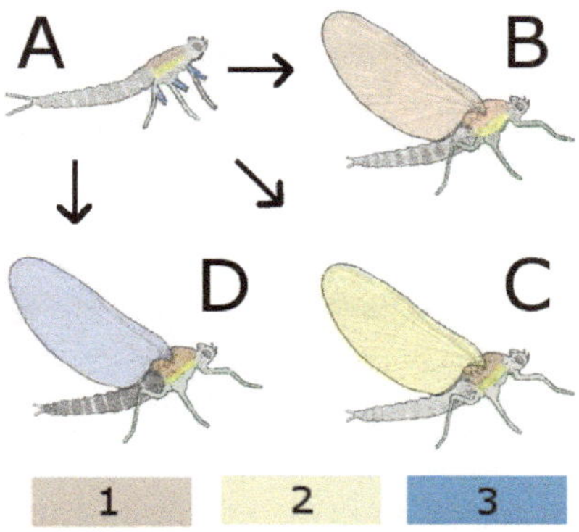

Diagram of different theoriesA Hypothetical wingless ancestorB Paranotal theory:Hypothetical insect with wings from the back (Notum)C Hypothetical insect with wings from page (Pleurum)D Epicoxal theoryHypothetical insect with wings from annex of the legs1 Notum (back)2 Pleurum (page)3 Exit (outer attachments of the legs)

- *Paranotal hypothesis*: This hypothesis suggests that the insect's wings developed from paranotal lobes, a preadaptation found in insect fossils that is believed to have assisted stabilization while hopping or falling. In favor of this hypothesis is the tendency of most insects, when startled while climbing on branches, to escape by dropping to the ground. Such lobes would have served as parachutes and enable the insect to land more softly. The theory suggests that these lobes gradually grew larger and in a later stage developed a joint with the thorax. Even later would appear the muscles to move these crude wings. This model implies a progressive increase in the effectiveness of the wings, starting with parachuting, then gliding and finally active flight. Still, lack of substantial fossil evidence of the development of the wing joints and muscles poses a major difficulty to the theory, as does the seemingly spontaneous development of articulation and venation, and it has been largely rejected by experts in the field.
- *Epicoxal hypothesis*:This theory suggested that a possible origin for insect wings might have been the movable abdominal gills found in many aquatic insects, such as on naiads of mayflies. According to this theory these tracheal gills, which started their way as exits of the respiratory system and over time were modified into locomotive purposes, eventually developed into wings. The tracheal gills are equipped with little winglets that perpetually vibrate and have their own tiny straight muscles.
- *Endite-exite hypothesis*:The hypothesis with perhaps the strongest evidence is that which stems from the adaptation of endites and exites, appendages on the respective inner and

outer aspects of the primitive arthropod limb. This was advanced by Trueman based on a study by Goldschmidt in 1945 on *Drosophila melanogaster*, in which a *pod* variation displayed a mutation transforming normal wings to what was interpreted as a triple-jointed leg arrangement with some additional appendages but lacking the tarsus, where the wing's costal surface normally would be. This mutation was reinterpreted as strong evidence for a dorsal exite and endite fusion, rather than a leg, with the appendages fitting in much better with this hypothesis. The innervation, articulation and musculature required for the evolution of wings are already present in podomeres.

- *Paranota plus leg gene recruitment hypothesis*:The fossil larvae of Coxoplectoptera provided important new clues to the disputed question of the evolutionary origin of insect wings. Before the larvae fossil discovery the paranotal-hypothesis and the leg-exite-hypothesis have been considered as incompatible alternative explanations, which have both been supported by a set of evidences from the fossil record, comparative morphology, developmental biology and genetics. The expression of leg genes in the ontogeny of the insect wing has been universally considered as conclusive evidence in favour of the leg-exite-hypothesis, which proposes that insect wings are derived from mobile leg appendages (exites). However, the larvae of Coxoplectoptera show that the abdominal gills of mayflies and their ancestors, which are generally considered as corresponding structures to insect wings, articulated within the dorsal tergite plates. This cannot be seen in modern mayfly larvae, because their abdominal tergites and sternites are fused to rings, without any traces left even in embryonic development. If larval gills and wings are corresponding ("serial homologous") structures and thus share the same evolutionary origin, the new results from Coxoplectoptera demonstrate that also wings are of tergal origin, as proposed by the classical paranotal-hypothesis. Staniczek, Bechly & Godunko (2011) therefore suggested a new hypothesis that could reconcile the apparently conflicting evidence from paleontology and developmental genetics: wings first originated as stiff outgrowths of tergal plates (paranota), and only later in evolution became mobile, articulated appendages through secondary recruiting of leg genes.

Suggestions have been made that wings may have evolved initially for sailing on the surface of water as seen in some stoneflies. An alternative idea is that it drives from directed aerial gliding descent—a preflight phenomena found in some apterygote, a wingless sister taxa to the winged insects. The earliest fliers were similar to dragonflies with two sets of wings, direct flight muscles, and no ability to fold their wings over their abdomens. Most insects today, which evolved from those first fliers, have simplified to either one pair of wings or two pairs functioning as a single pair and using a system of indirect flight muscles.

Natural selection has played an enormous role in refining the wings, control and sensory systems, and anything else that affects aerodynamics or kinematics. One noteworthy trait is wing twist. Most insect wings are twisted, as are helicopter blades, with a higher angle of attack at the base. The twist generally is between 10 and 20 degrees. In addition to this twist, the wing surfaces are not necessarily flat or featureless; most larger insects have wing membranes distorted and angled between the veins in such a way that the cross-section of the wings approximates an airfoil. Thus, the wing's basic shape already is capable of generating a small amount of lift at zero angle of attack. Most insects control their wings by adjusting tilt, stiffness, and flapping frequency of the wings with tiny muscles in the thorax (below). Some insects evolved other wing features that are not advantageous for flight, but play a role in something else, such as mating or protection.

<table>
<tr><td colspan="3">Evolution of the ways the wings at rest to the body to create</td></tr>
<tr><td colspan="2" rowspan="2">wings do not fold back (recent Archaeoptera)

over the back against one another (damselflies, mayflies)</td><td>spread laterally (large bubbles)</td></tr>
<tr><td></td></tr>
<tr><td rowspan="5">Folding (Neoptera)</td><td></td><td></td></tr>
<tr><td colspan="2">wings not foldable (e.g., stoneflies)</td></tr>
<tr><td rowspan="3">Folding</td><td>fan-fold (e.g., front wings of wasps)</td></tr>
<tr><td>Cross fold (such as the rear wing of the beetle)</td></tr>
<tr><td>Subjects folding (such as the rear wing of the earwigs)</td></tr>
</table>

Some insects, occupying the biological niches that they do, need to be incredibly maneuverable. They must find their food in tight spaces and be capable of escaping larger predators – or they may themselves be predators, and need to capture prey. Their maneuverability, from an aerodynamic viewpoint, is provided by high lift and thrust forces. Typical insect fliers can attain lift forces up to three times their weight and horizontal thrust forces up to five times their weight. There are two substantially different insect flight mechanisms, and each has its own advantages and disadvantages – just because odonates have a more primitive flight mechanism does not mean they are less able fliers; they are, in certain ways, more agile than anything that has evolved afterward.

Morphogenesis

While the development of wings in insects is clearly defined in those who are members of Endopterygota, which undergo complete metamorphosis; in these species, the wing develops while in the pupal stage of the insects life cycle. However, insects that undergo incomplete metamorphosis do not have a pupal stage, therefore they must have a different wing morphogenesis. Insects such as those that are hemimetabolic have wings that start out as buds, which are found underneath the exoskeleton, and do not become exposed until the last instar of the nymph.

The first indication of the wing buds is of a thickening of the hypodermis, which can be observed in insect species as early the embryo, and in the earliest stages of the life cycle. During the development of morphological features while in the embryo, or embryogenesis, a cluster of cells grow underneath the ectoderm which later in development, after the lateral ectoderm has grown dorsally to form wind imaginal disc. An example of wing bud development in the larvae, can be seen in those of White butterflies (*Pieris*). In the second instar the histoblast become more prominent, which now form a pocket-like structure. As of the third and fourth instars, the histoblast become more elongated. This greatly extended and evaginated, or protruding, part is what becomes the wing. By the close of the last instar, or fifth, the wing is pushed out of the wing-pocket, although continues to lie under the old larval cuticle while in its prepupal stage. It is not until the butterfly is in its pupal stage that the wing-bud becomes exposed, and shortly after eclosion, the wing begins to expand and form its definitive shape.

The development of tracheation of the wings begin before the wing histoblast form, as it is important to note that they develop near a large trachea. During the fourth instar, cells from the epithelium of this trachea become greatly enlarged extend into the cavity of the wing bud, with each cell having developed a closely coiled tracheole. Each trachcole is of unicellular origin, and is at first intracellular in position; while tracheae are of multicellular origin and the lumen of each is intercellular in position. The development of tracheoles, each coiled within a single cell of the epithelium of a trachea, and the subsequent opening of communication between the tracheoles and the lumen of the trachea, and the uncoiling and stretching out of the tracheoles, so that they reach all parts of the wing.

In the earlier stages of its development, the wing-bud is not provided with special organs of respiration such as tracheation, as it resembles in this respect the other portions of the hypodermis of which it is still a part. It should be noted, however, that the histoblast is developed near a large trachea, a cross-section of which is shown in, which represents sections of these parts of the first, second, third and fourth instars respectively. At the same time the tracheoles uncoil, and extend in bundles in the forming vein-cavities of the wing-bud. At the molt that marks the beginning of the pupal stadium stage, they become functional. At the same time, the larval tracheoles degenerate; their function having been replaced by the wing tracheae.

Nomenclature

Most of the nomenclature of insect orders is based on the Ancient Greek word for wing.

Scientific Name	**linguistic root**	**Translation of the Scientific name**	**English Name**
Anisoptera	(*aniso-*)	Unequal wings	Dragonfly
Aptera	(*a-*), not	Wingless	Apterygotans, now obsolete
Apterygota	(*pterygion* small wing) (*a-*), not	Wingless	Apterygotans
Coleoptera	(*koleos*, sheath)	Hardened wings	Beetles
Dermaptera	(*derma*, skin, leather)	Leather wings	Earwigs
Diaphanopterodea	(*diaphanes*, transparent or translucent)	With transparent wings	diaphanopteroideans
Dictyoptera	(*diktyon*, net-work)	Wings with netted venation	Cockroaches, mantises and termites
Diptera	(dyo-, *two*)	Two wings	Flies
Embioptera	(*en*, inside; βίος *bios*, life)	Interior living winged insects	Webspinners
Endopterygota	(*entos*, inside; small wing)	Inside wings	Holometabolous insects
Ephemeroptera	(*ephemeros* about one day long)	Short lived winged insects	Mayflies
Exopterygota	(*exo*, external)	Outdoor flying insects	Insects that undergo incomplete metamorphosis
Hemiptera	(*hemi-*, half)	Halfwinged insects	Hemiptera (true bugs, leafhoppers, aphids, etc.)

Heteroptera	(*hetero-*, differ-ent)	Different winged	True bugs
Homoptera	(*homo-*, similar)	Same winged	now obsolete
Hymenoptera	(*hymenion*, membrane)	Insects with wings of thin membranes	bees, wasps, ants, etc.
Isoptera	(*ison*, equal)	Same winged	Termites
Lepidoptera	(*lepis*, scale)	Scaled wings	Butterflies & Moths
Lonchopteridae	(*lonchi*, lance)	Lance wings	Lance flies
Mecoptera	(*mekos*, length)	Long wings	Snake flies, etc.
Megaloptera	(*megalo-*, large)	Large wings	Dobsonflies, fishflies
Neuroptera	(*neuron*, vein)	Veined wing	Lacewings, owlflies, antlions, etc.
Neoptera	(*neos*, new, young)	New wings	Includes all currently living orders of flying insects except mayflies and dragonflies
Oligoneoptera	(*oligon-*, few) (*neos* or new)	New with little veins	Division of the Neoptera
Orthoptera	(*ortho-*, straight)	Straight wings	Grasshoppers, katydids, and crickets
Palaeodictyoptera	(*palaios-*, old) δίκτυον (*diktyon* mean-ing network)	Old veined wings	Primitive palaeozoic paleopterous insects
Palaeoptera	(*Palaios*, old)	Old wings	Mayflies, dragonflies, and several fossil orders
Paraneoptera	(*Para-*) (*neos*, new)	Part of Neoptera, mostly with piercing mouthparts	True bugs, lice, barklice, thrips
Phthiraptera	(*phtheir*, lice) ἀ, *a-*, not	Lice without wings	Animal lice
Plecoptera	(*plekein*, fold)	Folded wings	Stoneflies
Polyneoptera	(*polys*, many)	Many veined wings	Neoptera with hemimetabolous devel-opment
Psocoptera	(*psocho*, to rub)	Rubbing wings	Barklice, booklice
Pterygota	(*pterygion*, wing)	Winged insects	In class, unlike Apterygota, includ-ing winged and wingless secondary systems
Raphidioptera	(*rhaphis*, needle)	Needle wings	Snakeflies
Siphonaptera	(*siphon*, tube)ἀ- or without	Wingless siphon	Fleas
Strepsiptera	(*strepsis*, to turn around)	Rotating or twisted wings	twisted-winged parasites
Thysanoptera	(*thysanoi*, fringes)	Fringe winged	Thrips
Trichoptera	(*trichoma*, hair)	Haired wings	Caddisflies
Zoraptera	(*zōros* meaning strong)	Strong wings	Zorapterans
Zygoptera	(*zeugos* meaning pair)	Paired wings	Damselflies

Adaptations

Variation

Insect wings are fundamental in identifying and classifying species as there is no other set of structures in studying insects more significant. Each order and insect family has distinctive wing shapes and features. In many cases, even species may be distinguished from each other by differences of color and pattern. For example, just by position one can identify species, albeit to a much lesser extent. Though most insects fold their wings when at rest, dragonflies and some damselflies rest with their wings spread out horizontally, while groups such as the caddisflies, stoneflies, alderflies, and lacewings hold their wings sloped roof-like over their backs. A few moths wrap their wings around their bodies, while many flies and most butterflies close their wings together straight upward over the back.

Many times the shape of the wings correlates with the type of insect flight. The best-flying insects tend to have long, slender wings. In many species of Sphingidae (sphinx moths), the forewings are large and sharply pointed, forming with the small hindwings a triangle that is suggestive of the wings of fast, modern airplanes. Another, possibly more important correlation, is that of the size and power of the muscles to the speed and power of flight. In the powerfully flying insects, the wings are most adapted for the stresses and aerodynamics of flight. The veins are thicker, stronger, and closer together toward the front edge (or "leading edge") and thinner yet flexible toward the rear edge (or "trailing edge"). This makes the insect wing an excellently constructed airfoil, capable of exerting both propulsion and lift while minimizing drag.

Variation of the wing beat may also occur, not just amongst different species, but even among individuals at different times. In general, the frequency is dependent upon the ratio between the power of the wing muscles and the resistance of the load. Large-winged, light-bodied butterflies may have a wing beat frequency of 4–20 per second whereas small-winged, heavy-bodied flies and bees beat their wings more than 100 times a second and mosquitoes can beat up to 988–1046 times a second. The same goes for flight; though it is generally difficult to estimate the speed of insects in flight, most insects can probably fly faster in nature than they do in controlled experiments.

Coleoptera

In species of Coleoptera (beetles), the only functional wings are the hindwings. The hindwings are longer than the elytra, folded longitudinally and transversely under the elytra. The wing is rotated forwards on its base into flight position. This action spread the wing and unfolded longitudinally and transversely. There is the spring mechanism in the wing structure, sometimes with the help of abdomen movement, to keep the wing in folded position. The beetle wing venation is reduced and modified due to the folding structure, which include:

Cross folding in the wings of beetles	
	The hindwing, spread: by folding lines, it is divided into five fields that are completed each to the rear.

	The same wing, half folded: The two joints of the cross-folding form an obtuse angle. The right is already in the wings folded in three layers. With greater resolution, the third arch of the wing margin in the first and second is visible. To the left of the fifth arch appears in the fourth.
	The same wing, folded completely: The five fields are aligned (The elytra have been removed).

- Costa (C), Subcosta posterior (ScP) – at the leading wing marginal, fused for most of the length.
- Radius anterior (RA) – divided into two branches beyond the middle of the wing.
- Radius posterior (RP) – basal connection is lost.
- Media posterior (MP) – branches, long and strong vein.
- Cubitus anterior (CuA)
- Anal veins (AA, AP) – veins behind the cubitus, separated by anal fold.

In most species of beetles, the front pair of wings are modified and sclerotised (hardened) to form elytra and they protect the delicate hindwings which are folded beneath. The elytra are connected to the pterathorax; being called as such because it is where the wings are connected. The elytra are not used for flight, but tend to cover the hind part of the body and protect the second pair of wings (*alae*). The elytra must be raised in order to move the hind flight wings. A beetle's flight wings are crossed with veins and are folded after landing, often along these veins, and are stored below the elytra. In some beetles, the ability to fly has been lost. These include some ground beetles (family Carabidae) and some "true weevils" (family Curculionidae), but also some desert and cave-dwelling species of other families. Many of these species have the two elytra fused together, forming a solid shield over the abdomen. In a few families, both the ability to fly and the elytra have been lost, with the best known example being the glow-worms of the family Phengodidae, in which the females are larviform throughout their lives.

Lepidoptera

The two pairs of wings are found on the middle and third segment, or mesothorax and metathorax respectively. In the more recent genera, the wings of the second segment are much more pronounced, however some more primitive forms have similarly sized wings of both segments. The wings are covered in scales arranged like shingles, forming the extraordinary variety seen in color. The mesothorax is evolved to have more powerful muscles to propel moth or butterfly through the air, with the wing of said segment having a stronger vein structure. The largest superfamily, Noctuidae, has the wings modified to act as Tympanal or hearing organs Modifications in the wing's venation include:

- Costa (C) – not found in Butterflies.
- Subcosta (Sc) + Radius 1 (Sc+R1) – at the leading wing marginal, fused or very close for most of the length, in hindwing fused and well developed in the humeral area, subcosta never branches in butterfly.
- Radius (R2-R5) – radius divides into branches beyond the middle of the wing up to five branches in Papilionidae. On forewing, the last R is stalked in all butterflies except Hesperiidae is separated.
- Radius sector (Rs) – in hindwing.
- Media (M1-M3) – the basal section has been lost.
- Cubitus anterior (CuA1-CuA2) – CuP section has been lost.
- Anal veins (A, 1A+2A, 3A) – either one vein A, or two veins 1A+2A, 3A.
- Humeral vein – The hindwing of most butterflies has the humeral vein, except Lycaenidae There is the enlargement of the humeral area of the hindwing which is overlapped with the forewing. The humeral vein strengthened the hindwing overlapped area so that the two wings coupling better.

The wings, head parts of thorax and abdomen of Lepidoptera are covered with minute scales, from which feature the order 'Lepidoptera' derives its names, the word "lepteron" in Ancient Greek meaning 'scale'. Most scales are lamellar, or blade-like and attached with a pedicel, while other forms may be hair-like or specialized as secondary sexual characteristics. The lumen or surface of the lamella, has a complex structure. It gives color either due to the pigmentary colors contained within or due to its three-dimensional structure. Scales provide a number of functions, which include insulation, thermoregulation, aiding gliding flight, amongst others, the most important of which is the large diversity of vivid or indistinct patterns they provide which help the organism protect itself by camouflage, mimicry, and to seek mates.

Odonata

Species of Odonata (Damselflies and dragonflies) both have two pairs of wings which are about equal in size and shape and are clear in color. There are five, if the R+M is counted as 1, main vein stems on dragonfly and damselfly wings, and wing veins are fused at their bases and the wings cannot be folded over the body at rest, which also include:

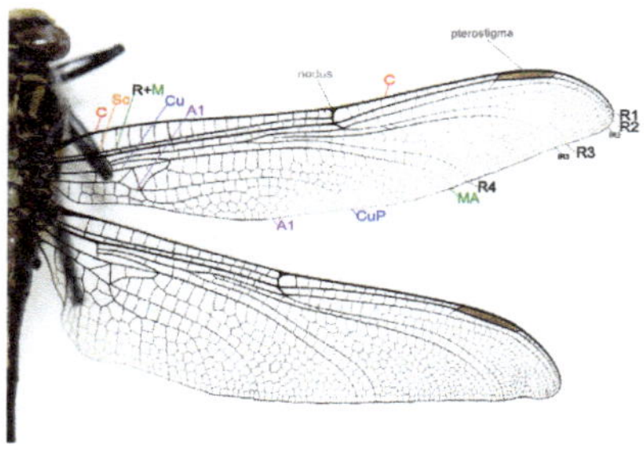

Dragonfly

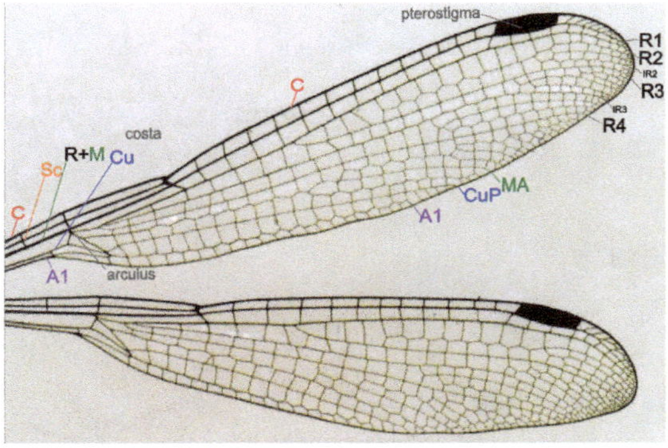

Damselfly

- Costa (C) – at the leading edge of the wing, strong and marginal, extends to the apex of the wing.
- Subcosta (Sc) – second longitudinal vein, it is unbranched, joins C at nodus.
- Radius and Media (R+M) – third and fourth longitudinal vein, the strongest vein on the wing, with branches, R1-R4, reach the wing margin, the media anterior (MA) are also reach the wing margin. IR2 and IR3 are intercalary veins behind R2 and R3 respectively.
- Cubitus (Cu) – fifth longitudinal vein, cubitus posterior (CuP) is unbranched and reach the wing margin.
- Anal veins (A1) – unbranched veins behind the cubitus.
- A nodus is formed where the second main vein meets the leading edge of the wing. The black pterostigma is carried near the wing tip.

The main veins and the crossveins form the wing venation pattern. The venation patterns are different in different species. There may be very numerous crossveins or rather few. The Australian Flatwing Damselfly's wings are one of the few veins patterns. The venation pattern is useful for species identification. Almost all Anisoptera settle with the wings held out sideways or slightly downward, however most Zygoptera settle with the wings held together, dorsal surfaces apposed. The thorax of Zygoptera is so oblique that when held in this way the wings fit neatly along the top of the abdomen. They do not appear to be held straight up as in butterflies or mayflies. In a few zygopteran families the wings are held horizontally at rest, and in one anisopteran genus (e.g. Cordulephya, Corduliidae) the wings are held in the typical damselfly resting position. Adult species possess two pairs of equal or subequal wings. There appear to be only five main vein stems. A nodus is formed where the second main vein (subcosta) meets the leading edge of the wing. In most families a conspicuous pterostigma is carried near the wing tip. Identification as Odonata can be based on the venation. The only likely confusion is with some lacewings (order Neuroptera) which have many crossveins in the wings. Until the early years of the 20th century Odonata were often regarded as being related to lacewings and were given the ordinal name Paraneuroptera, but any resemblance between these two orders is entirely superficial. In Anisoptera the hindwing is broader than the forewing and in both wings a crossvein divides the discoidal cell into a Triangle and Supertriangle.

Orthoptera

Species of Orthoptera (grasshoppers and crickets) have forewings that are tough opaque tegmina, narrow which are normally covering the hindwings and abdomen at rest. The hindwings are board membranous and folded in fan-like manner, which include the following venation:

- Costa (C) – at the leading marginal of the forewing and hindwing, unbranched.
- Subcosta (Sc) – second longitudinal vein, unbranched.
- Radius (R) – third longitudinal vein, branched to Rs in forewing and hindwing.
- Media anterior (MA) – fourth longitudinal vein, branched in basal part as Media posterior (MP).
- Cubitus (Cu) – fifth longitudinal vein, on forewing and hindwing dividing near the wing base into branched CuA, and unbranched CuP.
- Anal veins (A) – veins behind the cubitus, unbranched, two in forewing, many in hindwing.

Phasmatodea

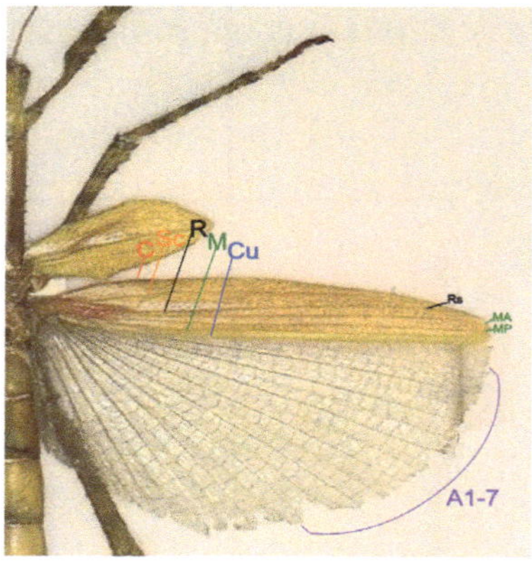

- Costa (C) – at the leading marginal of the hindwing, unbranched, absent in forewing.
- Subcosta (Sc) – second longitudinal vein, unbranched.
- Radius (R) – third longitudinal vein, branched to Rs in hindwing, unbranched in forewing.

- Media anterior (MA) – fourth longitudinal vein, branched in basal part as Media posterior (MP).
- Cubitus (Cu) – fifth longitudinal vein, unbranched.
- Anal veins (A) – veins behind the cubitus, unbranched, two in forewing, many in hindwing 1A-7A in one group and the rest in another group.

Stick insect have forewings that are tough, opaque tegmina, short and covering only the base part of the hindwings at rest. Hindwings from costa to Cubitus are tough and opaque like the forewings. The large anal area are membranous and folded in fan-like manner. There are no or very few branching in stick insect wing veins.

Dermaptera

Unfolding of earwig wing	
	The front and rear wings at rest: The front wing covers most of the hindwing, with only the joint projects in the form of a quarter circle forward with a central white spot under the forewing. On the right hand side of the forewing is opened to the right (blue arrow), which from this perspective appears narrower than it is with the rear wing still folded completely. .
	The front wing is open to the left (blue arrow) with the right side of the forewing removed; the hindwing is half open. With greater resolution, the multiple folding is shown, resembling a fan which is parallel to the lines b and c. The arrow points to the e point where the fan is closed again, having been folded by 180°..

Other orders such as the Dermaptera (earwigs), Orthoptera (grasshoppers, crickets), Mantodea (praying mantis) and Blattodea (cockroaches) have rigid leathery forewings that aren't flapped while flying, sometimes called *tegmen* (pl. *tegmina*), *elytra*, or *pseudoelytron*.

Hemiptera

In Hemiptera (true bugs), the forewings may be hardened, though to a lesser extent than in the beetles. For example, the anterior part of the front wings of stink bugs is hardened, while the posterior part is membranous. They are called *hemelytron* (pl. *hemelytra*). They are only found in the suborder Heteroptera; the wings of the Homoptera, such as the cicada, are typically entirely membranous. Both forewings and hindwings of Cicada are membranous, most species are glass-like although some are opaque. Cicadas are not good fliers and most fly only a few seconds. When flying, forewing and hindwing are hooked together by a grooved coupling along the hindwing costa and forewing margin. Most species have a basic venation as shown in the following picture.

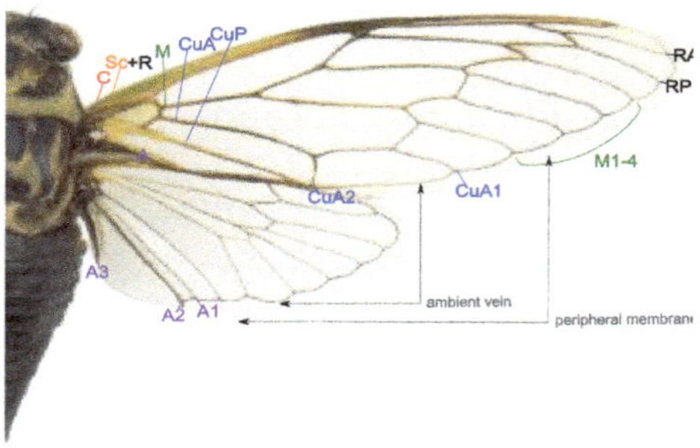

- Costa (C) – at the leading wing marginal, in forewing extends to the node and lies close to Sc+R.
- Subcosta + Radius (Sc+R) – in forewing Sc and R fused together to the node. Radial sector (Rs) arises near the node and unbranches.
- Radius anterior (RA)
- Radius posterior (RP)
- Media (M) – branches to M1 to M4.
- Cubitus anterior (CuA) – branches to CuA1 and CuA2.
- Cubitus posterior (CuP) – unbranches.
- Anal veins (A) – veins behind the cubitus, 1A and 2A fused in the forewing, CuP and 2A are folded.

Also notice there are the ambient veins and peripheral membranes on the margin of both wings.

Diptera

In the Diptera (true flies), there is only one pair of functional wings, with the posterior pair of wings are reduced to halteres, which help the fly to sense its orientation and movement, as well as to improve balance by acting similar to gyroscopes. In Calyptratae, the very hindmost portion of the wings are modified into somewhat thickened flaps called calypters which cover the halteres.

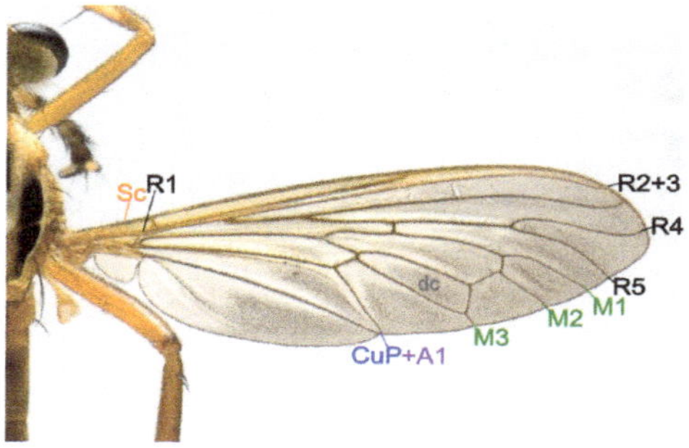

- Costa (C) – not found in Diptera.
- Subcosta (Sc) – became the leading wing vein, unbranched.
- Radius (R) – branched to R1-R5.
- Media (M) – branched to M1-M4.
- Cubitus anterior(CuA)- unbranched, CuP is reduced in Diptera. Some species CuA and 1A are separated, some species meets when reaching the wing margin, some species fused.
- Anal veins (A) – only two anal veins 1A and 2A are present, 2A is not distinctive in some species.
- Discal Cell (dc) – well defined in most species.

Blattodea

Species of Blattodea (cockroaches) have a forewing, are also known as tegmen, that is more or less sclerotized. It is used in flight as well as a form of protection of the membranous hindwings. The veins of hindwing are about the same as front wing but with large anal lobe folded at rest between CuP and 1A. The anal lobe usually folded in a fan-like manner.

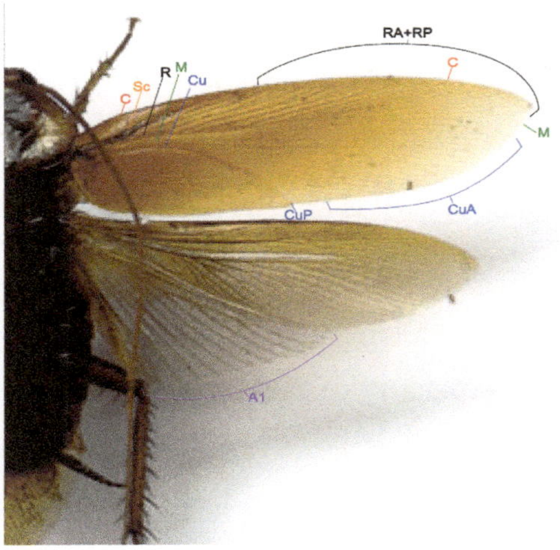

- Costa (C) – at the leading edge of the wing.
- Subcosta (Sc) – second longitudinal vein, it is relatively short.
- Radius (R) – third longitudinal vein, with many pectinate branches.
- Media (M) – fourth longitudinal vein, reach the wing margin.
- Cubitus anterior (CuA) – fifth longitudinal vein, with dichotomous branches occupy large part of tegmen.
- Cubitus posterior (CuP) – is unbranched, curved and reach the wing margin.
- Anal veins (A) – veins behind the cubitus.

Hymenoptera

An example of Longitudinal folding in wasps (Vespidae)	
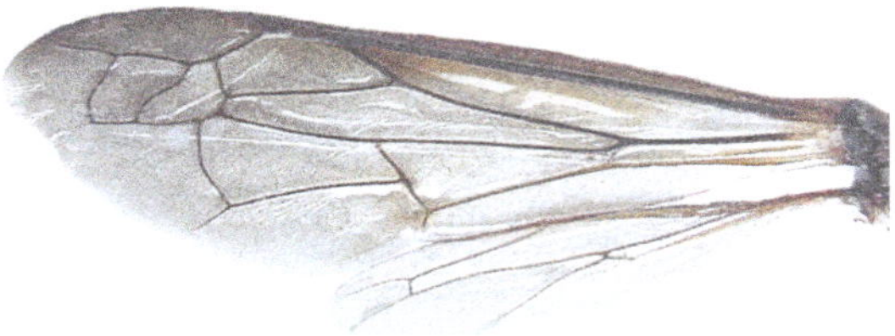	The main fold line of the forewing seen half way up as a bright horizontal line. The wing part that is behind this line is turned back down. The narrow strip at the front edge of the wing is in front of the first strong wire folded forward and down.
	So in rest position, the outer lining forms the tough outer edge of the wing, which protects the sides of the abdomen as a shock absorber. The rear wing is covered largely by the forewing.

The Hymenoptera adults, include sawflies, wasps, bees and non-working ants, all of which have two pairs of membranous wings.

- Costa (C) – not found in Hymenoptera.
- Subcosta (Sc) – unbranched.
- Radius (R) – branched to R1-R5.
- Media (M) – M is unbranched, in forewing M is fused with Rs for part of its length.
- Cubitus (CuA) – unbranched, CuP is absent in Hymenoptera.
- Anal veins (A) – only two anal veins 1A and 2A are present, 2A is not distinctive in some species.
- Wing-coupling – Row of hooks on the leading edge of hindwing engage the hind margin of the forewing, strongly couple the wings in flight.
- Line of wing folding – Some species, including Vespidae, the forewing are longitudinally folded along the 'line of wing folding' at rest.
- Pterostigma – is present for some species.

The forward margin of the hindwing bears a number of hooked bristles, or "hamuli", which lock onto the forewing, keeping them held together. The smaller species may have only two or three hamuli on each side, but the largest wasps may have a considerable number, keeping the wings gripped together especially tightly. Hymenopteran wings have relatively few veins compared with many other insects, especially in the smaller species.

Other Families

Termites are relatively poor fliers and are readily blown downwind in wind speeds of less than 2 km/h, shedding their wings soon after landing at an acceptable site, where they mate and attempt to form a nest in damp timber or earth. Wings of most termites have three heavy veins along the basal part of the front edge of the forewing and the crossveins near the wing tip are angled, making trapezoidal cells. Although subterranean termite wings have just two major veins along the front edge of the forewing and the cross veins towards the wingtip are perpendicular to these veins, making square and rectangular cells.

Species of Thysanoptera (thrips) have slender front and hindwings with long fringes of hair, called fringed wings. While species of Trichoptera (caddisfly) have hairy wings with the front and hindwings clothed with setae.

Bat Wing Development

The order Chiroptera, comprising all bats, has evolved the unique mammalian adaptation of flight. Bat wings are modified tetrapod forelimbs. Because bats are mammals, the skeletal structures in their wings are morphologically homologous to the skeletal components found in other tetrapod forelimbs. Through adaptive evolution these structures in bats have undergone dramatic morphological changes. This evolutionary innovation involved many morphological changes such as webbed digits, elongation of the forelimb, and reduction in bone thickness. Recently, there have been comparative studies of mouse and bat forelimb development to understand the genetic basis of morphological evolution. Consequently, the bat wing is a valuable evo-devo model for studying the evolution of vertebrate limb diversity.

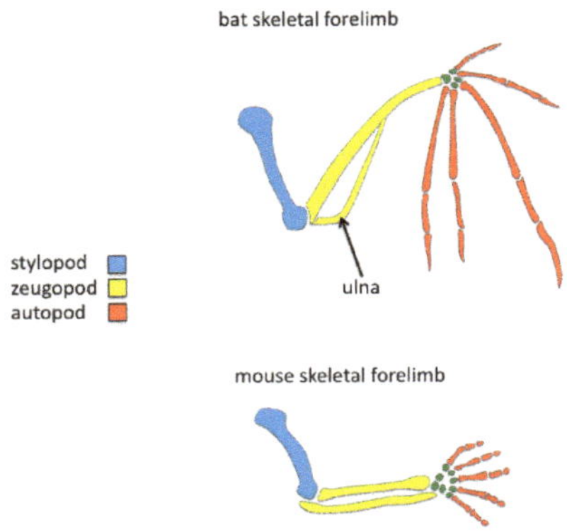

diagram showing homologous skeletal structures found in bat and mouse.

Comparisons to Mouse Limb Development

Tetrapod limb development involves many signaling molecules such as FGF, BMP, SHH and WNT. The apical ectodermal ridge is a structure found at the distal most tip which becomes a key signaling center for the developing limb. Surprisingly many of the same signaling pathways known to play a role in tetrapod limb development have been found to play a role in bat forelimb development but the timing, intensity, and spatial gene expression of some orthologous genes have changed. Since mice are also mammals, it is convenient to compare morphology and development of forelimbs between mice and bats; these comparisons may elucidate the genetic basis of adaptive bat wing development.

Although many of the molecular mechanisms involved in limb development are conserved between mouse and bat, there are a number of differences primarily seen in gene expression patterns. Surprisingly, the coding regions of many of these genes with different expression domains are highly conserved between mouse and bat. Thus, it is likely that this major morphological transition was a consequence of cis-regulatory changes. Researchers can study the genetic basis of bat wing development by using comparative in situ hybridization to examine gene expression domains and using experimental embryology in mice and bats.

Presence of Webbed Digits

Formation of the bat wing membrane (the patagium) allowed a greater surface area of the wing necessary for flight. All vertebrate limb formation initially has tissue between the digits after which apoptosis occurs to separate the digits. BMP signals are most likely responsible for the interdigit apoptosis as they are expressed in the interdigit tissue and blocking BMP signaling will prevent interdigital apoptosis. However, in bats, BMP genes are still expressed in the interdigits and yet interdigit apotosis is repressed. FGF signaling has been associated with blocking cell death. Interestingly, *fgf8* is expressed in bat interdigit tissue during a time when apoptosis occurs which does not occur in mice. Thus, FGFs may play a role in blocking the apoptotic effects of BMPs in the bat wing interdigit. Finally, applying ectopic BMPs and FGF antagonists to developing bat wings results in apoptosis of the patagium.

Elongation of Forelimb

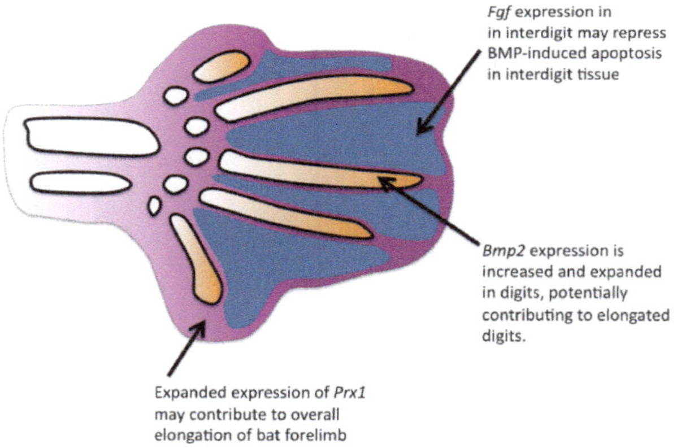

Simplified diagram showing expanded gene expression domains in developing bat forelimb potentially contributing to the morphological changes resulting in the bat wing.

One major difference in bat forearms is that their skeletal limb structures are elongated. This elongation of the forelimb skeleton is required to support the wing membrane. Comparative in situ hybridization studies have revealed that the expression domain of *fgf8* in bat forelimb AER are expanded in comparison to the mouse forelimb, suggesting that expanded expression of *fgf8* may contribute to the larger size of the bat forelimb. Because the mouse and bat orthologs are conserved, there is likely to be a regulatory change in *fgf8*. In mice, one gene known to regulate limb growth is *prx1*, which encodes a transcription factor. The expression patterns of *prx1* in bats differs from mice in that *prx1* has an expanded expression domain and is upregulated. Researchers found that the coding region of *prx1* in bats is nearly identical to mice but found a bat-specific *prx1* enhancer. When they replaced the bat *prx1* enhancer with the endogenous enhancer found in mice, these transgenic mice had slightly increased forlimbs. Comparative studies have established that bat digits undergo a more rapid rate of chondrocyte proliferation. In addition to interdigit apoptosis, BMPs have been shown to affect chrondrocyte proliferation and digit length in mice. *Bmp-2* shows increased expression in the digits of bats compared with mice, suggesting that a change in the BMP pathway has occurred to give rise to longer bat digits.

Reduction in Bone Thickness

Another major difference in bat forelimbs is in the density of their skeletal limbs. The bones found in their forelimbs are reduced to achieve a light body weight required for flight. In particular, their ulna is reduced in width and fused to the other zeugopod element, the radius. One of the possible molecular pathways involved in reduction of bat skeletal forelimb thickness is differences in SHH expression. Mice with *shh* null mutations lose their ulna structure. Another good candidate for bat bone reduction is *Hox-d13*, a gene belonging to the Hox gene family. In situ hybridization studies have found that the *Hoxd13* expression domain in bat limbs has been shifted posteriorly in comparison to mouse. This observed difference in the expression pattern of *Hoxd13* may also explain reduced size and density of the ulna found in bats. Overall, these studies suggest that the molecular changes responsible for the evolution of wings in bats is due to genetic regulatory changes.

Flight Feather

Red kite (Milvus milvus) in flight, showing remiges and rectrices

Flight feathers (P*ennae volatus*) are the long, stiff, asymmetrically shaped, but symmetrically paired pennaceous feathers on the wings or tail of a bird; those on the wings are called remiges (singular remex) while those on the tail are called rectrices (singular rectrix). The primary function of the flight feathers is to aid in the generation of both thrust and lift, thereby enabling flight. The flight feathers of some birds have evolved to perform additional functions, generally associated with territorial displays, courtship rituals or feeding methods. In some species, these feathers have developed into long showy plumes used in visual courtship displays, while in others they create a sound during display flights. Tiny serrations on the leading edge of their remiges help owls to fly silently (and therefore hunt more successfully), while the extra-stiff rectrices of woodpeckers help them to brace against tree trunks as they hammer. Even flightless birds still retain flight feathers, though sometimes in radically modified forms.

The moult of their flight feathers can cause serious problems for birds, as it can impair their ability to fly. Different species have evolved different strategies for coping with this, ranging from dropping all their flight feathers at once (and thus becoming flightless for some relatively short period of time) to extending the moult over a period of several years.

Remiges

Remiges are located on the posterior side of the wing. Ligaments attach the long *calami* (quills) firmly to the wing bones, and a thick, strong band of tendinous tissue known as the *postpatagium* helps to hold and support the remiges in place. Corresponding remiges on individual birds are symmetrical between the two wings, matching to a large extent in size and shape (except in the case of mutation or damage), though not necessarily in pattern. They are given different names depending on their position along the wing.

Primaries

Primaries are connected to the manus (the bird's "hand", composed of carpometacarpus and phalanges); these are the longest and narrowest of the remiges (particularly those attached to the phalanges), and they can be individually rotated. These feathers are especially important for flapping flight, as they are the principal source of thrust, moving the bird forward through the air. Most thrust is generated on the downstroke of flapping flight. However, on the upstroke (when the bird often draws its wing in close to its body), the primaries are separated and rotated, reducing air resistance while still helping to provide some thrust. The flexibility of the remiges on the wingtips of large soaring birds also allows for the spreading of those feathers, which helps to reduce the creation of wingtip vortices, thereby reducing drag. The barbules on these feathers, friction barbules, are specialized with large lobular barbicels that help grip and prevent slippage of overlying feathers and are present in most of the flying birds.

Bald eagle (Haliaeetus leucocephalus) in flight with primaries spread to decrease drag and improve lift

Species vary somewhat in the number of primaries they possess. The number in non-passerines generally varies between 9 and 11, but grebes, storks and flamingos have 12, and ostriches have 16. While most modern passerines have ten primaries, some have only nine. Those with nine are missing the most distal primary (sometimes called the remicle) which is typically very small and sometimes rudimentary in passerines.

The outermost primaries—those connected to the phalanges—are sometimes known as pinions.

Secondaries

Secondaries are connected to the ulna. In some species, the ligaments that bind these remiges to the bone connect to small, rounded projections, known as quill knobs, on the ulna; in other species, no such knobs exist. Secondary feathers remain close together in flight (they cannot be individually separated like the primaries can) and help to provide lift by creating the airfoil shape of the bird's wing. Secondaries tend to be shorter and broader than primaries, with blunter ends. They vary in number from 6 in hummingbirds to as many as 40 in some species of albatross. In general, larger and longer-winged species have a larger number of secondaries. Birds in more than 40 non-passerine families seem to be missing the fifth secondary feather on each wing, a state known as diastataxis. In these birds, the fifth set of secondary covert feathers does not cover any remex, possibly due to a twisting of the feather papillae during embryonic development. Loons, grebes, pelicans, hawks and eagles, cranes, sandpipers, gulls, parrots, and owls are among the families missing this feather.

Primary (left) and secondary (right) feathers of the common buzzard (Buteo buteo); note the asymmetrical orientation of the shafts

Tertials

Tertials arise in the brachial region and are not considered true remiges as they are not supported by attachment to the corresponding bone, in this case the humerus. These elongated "true" tertials act as a protective cover for all or part of the folded primaries and secondaries, and do not qualify as flight feathers as such. However, many authorities use the term tertials to refer to the shorter, more symmetrical innermost secondaries of passerines (arising from the olecranon and performing the same function as true tertials) in an effort to distinguish them from other secondaries. The term humeral is sometimes used for birds such as the albatrosses and pelicans that have a long humerus.

Tectrices

The calami of the flight feathers are protected by a layer of non-flight feathers called *covert* feathers or *tectrices* (singluar *tectrix*), at least one layer of them both above and beneath the flight feathers of the wings as well as above and below the rectrices of the tail. These feathers may vary widely in size — in fact, the upper tail tectrices of the male peafowl, rather than its rectrices, are what constitute its elaborate and colorful "train".

Emargination

The outermost primaries of large soaring birds, particularly raptors, often show a pronounced narrowing at some variable distance along the feather edges. These narrowings are called either notches or emarginations depending on the degree of their slope. An emargination is a gradual change, and can be found on either side of the feather. A notch is an abrupt change, and is only found on the wider trailing edge of the remige. (Both are visible on the primary in the photo showing the feathers; they can be found about halfway along both sides of the left hand feather—a shallow notch on the left, and a gradual emargination on the right.) The presence of notches and emarginations creates gaps at the wingtip; air is forced through these gaps, increasing the generation of lift.

Alula

Male mallard (Anas platyrhynchos) landing, showing outspread alulae

Feathers on the alula or bastard wing are not generally considered to be flight feathers in the strict sense; though they are asymmetrical, they lack the length and stiffness of most true flight feathers. However, alula feathers are definitely an aid to slow flight. These feathers—which are attached to the bird's "thumb" and normally lie flush against the anterior edge of the wing—function in the same way as the slats on an airplane wing, allowing the wing to achieve a higher than normal angle of attack – and thus lift – without resulting in a stall. By manipulating its thumb to create a gap between the alula and the rest of the wing, a bird can avoid stalling when flying at low speeds or landing.

Delayed Development in Hoatzins

The development of the remiges (and alulae) of nestling hoatzins is much delayed compared to the development of these feathers in other young birds, presumably because young hoatzins are equipped with claws on their first two digits. They use these small rounded hooks to grasp branches when clambering about in trees, and feathering on these digits would presumably interfere with that functionality. Most youngsters shed their claws sometime between their 70th and 100th day of life, but some retain them— though callused-over and unusable— into adulthood.

Rectrices

Rectrices, which help the bird to brake and steer in flight, lie in a single horizontal row on the rear margin of the anatomic tail. Only the central pair are attached (via ligaments) to the tail bones; the remaining rectrices are embedded into the rectricial bulbs, complex structures of fat and muscle that surround those bones. Rectrices are always paired, with a vast majority of species having six pairs. They are absent in grebes and some ratites, and greatly reduced in size in penguins. Many grouse species have more than 12 rectrices. Some species (including ruffed grouse, hazel grouse and common snipe) have a number that varies among individuals. Domestic pigeons have a highly variable number, due to centuries of selective breeding.

Numbering Conventions

In order to make the discussion of such topics as moult processes or body structure easier, ornithologists assign a number to each flight feather. By convention, the numbers assigned to primary feathers always start with the letter P (*P1, P2, P3, etc.)*, those of secondaries with the letter S, those of tertials with T and those of rectrices with R.

Most authorities number the primaries descendantly, starting from the innermost primary (the one closest to the secondaries) and working outwards; others number them ascendantly, from the most distal primary inwards. There are some advantages to each method. Descendant numbering follows the normal sequence of most birds' primary moult. In addition, in the event that a species is missing the small distal 10th primary, as some passerines are, its lack does not impact the numbering of the remaining primaries. Ascendant numbering, on the other hand, allows for uniformity in the numbering of non-passerine primaries, as they almost invariably have four attached to the manus regardless of how many primaries they have overall. This method is particularly useful for indicating wing formulae, as the outermost primary is the one with which the measurements begin.

Secondaries are always numbered ascendantly, starting with the outermost secondary (the one closest to the primaries) and working inwards. Tertials are also numbered ascendantly, but in this case, the numbers continue on consecutively from that given to the last secondary (e.g. ... S5, S6, T7, T8, ... etc.).

Rectrices are always numbered from the centermost pair outwards in both directions.

Specialized Flight Feathers

Male pin-tailed whydah (Vidua macroura) showing modified rectrices

The flight feathers of some species have undergone evolutionary changes which allow them to provide additional functionality.

In some species, for example, either remiges or rectrices make a sound during flight. These sounds are most often associated with courtship or territorial displays. The outer primaries of male broad-tailed hummingbirds produce a distinctive high-pitched trill, both in direct flight and in power-dives during courtship displays; this trill is diminished when the outer primaries are worn, and absent when those feathers have been moulted. During the northern lapwing's zigzagging display flight, the bird's outer primaries produce a humming sound. The outer primaries of the male American woodcock are shorter and slightly narrower than those of the female, and are likely the source of the whistling and twittering sounds made during his courtship display flights. Male club-winged manakins use modified secondaries to make a clear trilling courtship call. A curve-tipped secondary on each wing is dragged against an adjacent ridged secondary at high speeds (as many as 110 times per second—slightly faster than a hummingbird's wingbeat) to create a stridulation much like that produced by some insects. Both Wilson's and common snipe have modified outer tail feathers which make noise when they are spread during the birds' roller coaster display flights; as the bird dives, wind flows through the modified feathers and creates a series of rising and falling notes, which is known as "winnowing". Differences between the sounds produced by these two former conspecific subspecies—and the fact that the outer two pairs of rectrices in Wilson's snipe are modified, while only the single outermost pair are modified in common snipe—were among the characteristics used to justify their splitting into two distinct and separate species.

Flight feathers are also used by some species in visual displays. Male standard-winged and pennant-winged nightjars have modified P2 primaries (using the descendant numbering scheme explained above) which are displayed during their courtship rituals. In the standard-winged nightjar, this modified primary consists of an extremely long shaft with a small "pennant" (actually a large web of barbules) at the tip. In the pennant-winged nightjar, the P2 primary is an extremely long (but otherwise normal) feather, while P3, P4 and P5 are successively shorter; the overall effect is a broadly forked wingtip with a very long plume beyond the lower half of the fork.

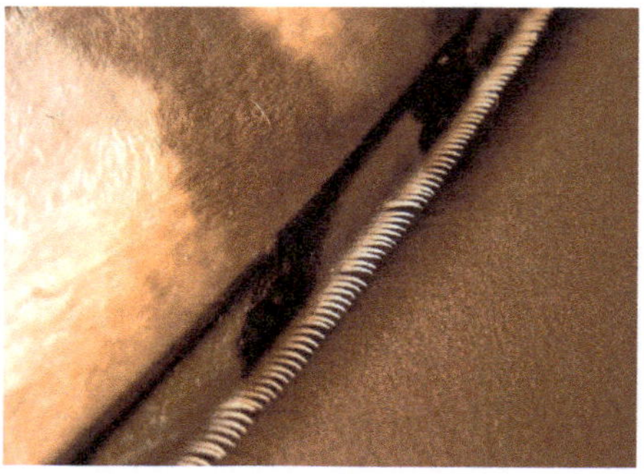

Leading edge of an owl feather, showing serrations

Males of many species, ranging from the widely introduced ring-necked pheasant to Africa's many whydahs, have one or more elongated pairs of rectrices, which play an often-critical role in their courtship rituals. The outermost pair of rectrices in male lyrebirds are extremely long and strongly curved at the ends. These plumes are raised up over the bird's head (along with a fine spray of modified uppertail coverts) during his extraordinary display. Rectrix modification reaches its pinnacle among the birds of paradise, which display an assortment of often bizarrely modified feathers, ranging from the extremely long plumes of the ribbon-tailed astrapia (nearly three times the length of the bird itself) to the dramatically coiled twin plumes of the magnificent bird-of-paradise.

Owls have remiges which are serrated rather than smooth on the leading edge. This adaptation disrupts the flow of air over the wings, eliminating the noise that airflow over a smooth surface normally creates, and allowing the birds to fly and hunt silently.

The rectrices of woodpeckers are proportionately short and very stiff, allowing them to better brace themselves against tree trunks while feeding. This adaptation is also found, though to a lesser extent, in some other species that feed along tree trunks, including woodcreepers and treecreepers.

Scientists have not yet determined the function of all flight feather modifications. For instance, male swallows in the genera *Psalidoprocne* and *Stelgidopteryx* have tiny recurved hooks on the leading edges of their outer primaries, but the function of these hooks is not yet known; some authorities suggest they may produce a sound during territorial or courtship displays.

Vestigiality in Flightless Birds

Double-wattled cassowary, (Casuarius casuarius) showing modified remiges

Over time, a small number of bird species have lost their ability to fly. Some of these, such as the steamer ducks, show no appreciable changes in their flight feathers. Some, such as the Titicaca grebe and a number of the flightless rails, have a reduced number of primaries.

The remiges of ratites are soft and downy; they lack the interlocking hooks and barbules that help to stiffen the flight feathers of other birds. In addition, the emu's remiges are proportionately much reduced in size, while those of the cassowaries are reduced both in number and structure, consisting merely of 5–6 bare quills. Most ratites have completely lost their rectrices; only the ostrich still has them.

Penguins have lost their differentiated flight feathers. As adults, their wings and tail are covered with the same small, stiff, slightly curved feathers as are found on the rest of their bodies.

The ground-dwelling kakapo, which is the world's only flightless parrot, has remiges which are shorter, rounder and more symmetrically vaned than those of parrots capable of flight; these flight feathers also contain fewer interlocking barbules near their tips.

Moult

Eurasian jackdaw (Corvus monedula), showing moult of central rectrices

Once they have finished growing, feathers are essentially dead structures. Over time, they become worn and abraded, and need to be replaced. This replacement process is known as moult (molt in the United States). The loss of wing and tail feathers can affect a bird's ability to fly (sometimes dramatically) and in certain families can impair the ability to feed or perform courtship displays. The timing and progression of flight feather moult therefore varies among families.

For most birds, moult begins at a certain specific point, called a focus (plural foci), on the wing or tail and proceeds in a sequential manner in one or both directions from there. For example, most passerines have a focus between the innermost primary (P1, using the numbering scheme explained above) and outermost secondary (S1), and a focus point in the middle of the center pair of rectrices. As passerine moult begins, the two feathers closest to the focus are the first to drop. When replacement feathers reach roughly half of their eventual length, the next feathers in line (P2 and S2 on the wing, and both R2s on the tail) are dropped. This pattern of drop and replacement continues until moult reaches either end of the wing or tail. The speed of the moult can vary somewhat within a species. Some passerines that breed in the Arctic, for example, drop many more flight feathers at once (sometimes becoming briefly flightless) in order to complete their entire wing moult prior to migrating south, while those same species breeding at lower latitudes undergo a more protracted moult.

Young white-bellied sea eagle (Haliaeetus leucogaster) in flight, showing moult waves in wings

In many species, there is more than one focus along the wing. Here, moult begins at all foci simultaneously, but generally proceeds only in one direction. Most grouse, for example, have two wing foci: one at the wingtip, the other between feathers P1 and S1. In this case, moult proceeds descendantly from both foci. Many large, long-winged birds have multiple wing foci.

Birds that are heavily "wing-loaded"—that is, heavy-bodied birds with relatively short wings—have great difficulty flying with the loss of even a few flight feathers. A protracted moult like the one described above would leave them vulnerable to predators for a sizeable portion of the year. Instead, these birds lose all their flight feathers at once. This leaves them completely flightless for a period of three to four weeks, but means their overall period of vulnerability is significantly shorter than it would otherwise be. Eleven families of birds, including loons, grebes and most waterfowl, have this moult strategy.

The cuckoos show what is called saltatory or transilient wing moults. In simple forms this involves the moulting and replacement of odd-numbered primaries and then the even-numbered primaries. There are however complex variations with differences based on life history.

Arboreal woodpeckers, which depend on their tails—particularly the strong central pair of rectrices—for support while they feed, have a unique tail moult. Rather than moulting their central tail feathers first, as most birds do, they retain these feathers until last. Instead the second pair of rectrices (both R2 feathers) are the first to drop. (In some species in the genera *Celeus* and *Dendropicos*, the third pair is the first dropped.) The pattern of feather drop and replacement proceeds as described for passerines (above) until all other rectrices have been replaced; only then are the central tail rectrices moulted. This provides some protection to the growing feathers, since they're always covered by at least one existing feather, and also ensures that the bird's newly strengthened tail is best able to cope with the loss of the crucial central rectrices. Ground-feeding woodpeckers, such as the wrynecks, do not have this modified moult strategy; in fact, wrynecks moult their outer tail feathers first, with moult proceeding proximally from there.

Age Differences in Flight Feathers

Western gull chick about 3 weeks old flapping its developing wings

There are often substantial differences between the remiges and rectrices of adults and juveniles of the same species. Because all juvenile feathers are grown at once—a tremendous energy burden to the developing bird—they are softer and of poorer quality than the equivalent feathers of adults, which are moulted over a longer period of time (as long as several years in some cases). As a result, they wear more quickly.

As feathers grow at variable rates, these variations lead to visible dark and light bands in the fully formed feather. These so-called *growth bars* and their widths have been used to determine the daily nutritional status of birds. Each light and dark bar correspond to around 24 hours and the use of this technique has been called *ptilochronology* (analogous to dendrochronology).

In general, juveniles have feathers which are narrower and more sharply pointed at the tip. This can be particularly visible when the bird is in flight, especially in the case of raptors. The trailing edge of the wing of a juvenile bird can appear almost serrated, due to the feathers' sharp tips, while that of an older bird will be straighter-edged. The flight feathers of a juvenile bird will also be uniform in length, since they all grew at the same time. Those of adults will be of various lengths and levels of wear, since each is moulted at a different time.

The flight feathers of adults and juveniles can differ considerably in length, particularly among the raptors. Overall, juveniles tend to have slightly longer rectrices, and shorter, broader wings (with shorter outer primaries, and longer inner primaries and secondaries) than do adults of the same species. However, there are many exceptions. In longer-tailed species, such as swallow-tailed kite, secretary bird and European honey buzzard, for example, juveniles have shorter rectrices than adults do. Juveniles of some *Buteo* buzzards have narrower wings than adults do, while those of large juvenile falcons are longer. It is theorized that the differences help young birds compensate for their inexperience, weaker flight muscles and poorer flying ability.

Wing Formula

Measuring primary lengths, one of the steps in determining a bird's wing formula

A wing formula describes the shape of distal end of a bird's wing in a mathematical way. It can be used to help distinguish between species with similar plumages, and thus is particularly useful for those who ring (band) birds.

To determine a bird's wing formula, the distance between the tip of the most distal primary and the tip of its greater covert (the longest of the feathers that cover and protect the shaft of that primary) is measured in millimeters. In some cases, this results in a positive number (e.g., the primary extends beyond its greater covert), while in other cases it is a negative number (e.g. the primary is completely covered by the greater covert, as happens in some passerine species). Next, the longest primary feather is identified, and the differences between the length of that primary and that of all remaining primaries and of the longest secondary are also measured, again in millimeters. If any primary shows a notch or emargination, this is noted, and the distance between the feather's tip and any notch is measured, as is the depth of the notch. All distance measurements are made with the bird's wing closed, so as to maintain the relative positions of the feathers.

While there can be considerable variation across members of a species—and while the results are obviously impacted by the effects of moult and feather regeneration—even very closely related species show clear differences in their wing formulas.

Primary Extension

Comparison of primary extensions: chiffchaff (left) and willow warbler

Marsh warbler (Acrocephalus palustris) has a long primary projection

The distance that a bird's longest primaries extend beyond its longest secondaries (or tertials) when its wings are folded is referred to as the primary extension or primary projection. As with wing formulae, this measurement is useful for distinguishing between similarly plumaged birds; however, unlike wing formulae, it is not necessary to have the bird in-hand to make the measurement. Rather, this is a useful *relative* measurement—some species have long primary extensions, while others have shorter ones. Among the *Empidonax* flycatchers of the Americas, for example, the dusky flycatcher has a much shorter primary extension than does the very similarly plumaged Hammond's flycatcher. Europe's common skylark has a long primary projection, while that of the near-lookalike Oriental skylark is very short.

As a general rule, species which are long distance migrants will have longer primary projection than similar species which do not migrate or migrate shorter distances.

Flying and Gliding Animals

Greylag geese (Anser anser). Birds are one of only four taxonomic groups to have evolved powered flight.

A number of animals have evolved aerial locomotion, either by powered flight or by gliding. Flying and gliding animals (*volant* animals) have evolved separately many times, without any single ancestor. Flight has evolved at least four times, in the insects, pterosaurs, birds, and bats. Gliding has evolved on many more occasions. Usually the development is to aid canopy animals in getting from tree to tree, although there are other possibilities. Gliding, in particular, has evolved among rainforest animals, especially in the rainforests in Asia (most especially Borneo) where the trees are tall and widely spaced. Several species of aquatic animals, and a few amphibians have also evolved to acquire this gliding flight ability, typically as a means of evading predators.

Types

Animal aerial locomotion can be divided into two categories—powered and unpowered. In unpowered modes of locomotion, the animal uses on aerodynamics forces exerted on the body due to wind or falling through the air. In powered flight, the animal uses muscular power to generate aerodynamic forces. Animals using unpowered aerial locomotion cannot maintain altitude and speed due to unopposed drag, while animals using powered flight can maintain steady, level flight as long as their muscles are capable of doing so.

Unpowered

These modes of locomotion typically require an animal start from a raised location, converting that potential energy into kinetic energy and using aerodynamic forces to control trajectory and angle of descent. Energy is continually lost to drag without being replaced, thus these methods of locomotion have limited range and duration.

- Falling: decreasing altitude under the force of gravity, using no adaptations to increase drag or provide lift.
- Parachuting: falling at an angle greater than 45° from the horizontal with adaptations to increase drag forces. Very small animals may be carried up by the wind. Some gliding animals may use their gliding membranes for drag rather than lift, to safely descend.
- Gliding flight: falling at an angle less than 45° from the horizontal with lift from adapted aerofoil membranes. This allows slowly falling directed horizontal movement, with streamlining to decrease drag forces for aerofoil efficiency and often with some maneuverability in air. Gliding animals have a lower aspect ratio (wing length/breadth) than true flyers.

Powered Flight

Powered flight has evolved only four times (in birds, bats, pterosaurs, and insects). It uses muscular power to generate aerodynamic forces and to replace energy lost to drag.

- Flapping: moving wings to produce lift and thrust. May ascend without the aid of the wind, as opposed to gliders and parachuters.

Externally Powered

Ballooning and soaring are not powered by muscle, but rather by external aerodynamic sources of energy: the wind and rising thermals, respectively. Both can continue as long as the source of external power is present. Soaring is typically only seen in species capable of powered flight, as it requires extremely large wings.

- Ballooning: being carried up into the air from the aerodynamic effect on long strands of silk in the wind. Certain silk-producing arthropods, mostly small or young spiders, secrete a special light-weight gossamer silk for ballooning, sometimes traveling great distances at high altitude.
- Soaring: gliding in rising or otherwise moving air that requires specific physiological and morphological adaptations that can sustain the animal aloft without flapping its wings. The rising air is due to thermals, ridge lift or other meteorological features. Under the right conditions, soaring creates a gain of altitude without expending energy. Large wingspans are needed for efficient soaring.

Many species will use multiple of these modes at various times; a hawk will use powered flight to rise, then soar on thermals, then descend via free-fall to catch its prey.

Evolution and Ecology

Gliding and Parachuting

While gliding occurs independently from powered flight, it has some ecological advantages of its own. Gliding is a very energy-efficient way of travelling from tree to tree. An argument made is that many gliding animals eat low energy foods such as leaves and are restricted to gliding because of this, whereas flying animals eat more high energy foods such as fruits, nectar, and insects. In contrast to flight, gliding has evolved independently many times (more than a dozen times among extant vertebrates), however these groups have not radiated nearly as much as have groups of flying animals.

Worldwide, the distribution of gliding animals is uneven as most inhabit rain forests in Southeast Asia. (Despite seemingly suitable rain forest habitats, few gliders are found in India or New Guinea and none in Madagascar.) Additionally, a variety of gliding vertebrates are found in Africa, a family of hylids (flying frogs) lives in South America and several species of gliding squirrels are found in the forests of northern Asia and North America. Various factors produce these disparities. In the forests of Southeast Asia, the dominant canopy trees (usually dipterocarps) are taller than the canopy trees of the other forests. A higher start provides a competitive advantage of further glides and farther travel. Gliding predators may more efficiently search for prey. The lower abundance of insect and small vertebrate prey for carnivorous animals (such as lizards) in Asian forests (In Australia, many mammals (and all mammalian gliders) possess, to some extent, prehensile tails.

An abundance of lianas (woody vines) may obstruct gliders but aid climbers with prehensile tails. The differing populations may relate to the prevalence in South America (compared to Africa or Southeast Asia) of animals with prehensile tails. Arguably, gliding animals prosper in Southeast Asia because the forests are more open (with more room to glide) than those in South America. In dense forests, a prehensile tail enables better tree to tree movement. Also, South American rainforests tend to have more as there are fewer large animals to eat them compared to Africa and Asia; these lianas

Because small animals necessarily have higher surface to volume ratios than larger species of similar form, aerodynamic forces have a greater effect on them, resulting in much lower terminal velocity in free fall and amplifying the effects of even small alterations to body surface area. These small changes provide incremental benefits towards further development of gliding.

Powered Flight

Powered flight has evolved unambiguously only four times—birds, bats, pterosaurs, and insects. In contrast to gliding, which has evolved more frequently but typically gives rise to only a handful of species, all three extant groups of powered flyers have a huge number of species, suggesting that flight is a very successful strategy once evolved. Bats, after rodents, have the most species of any mammalian order, about 20% of all mammalian species. Birds have the most species of any class of terrestrial vertebrates. Finally, insects (most of which fly at some point in their life cycle) have more species than all other animal groups combined.

The evolution of flight is one of the most striking and demanding in animal evolution, and has attracted the attention of many prominent scientists and generated many theories. Additionally, because flying animals tend to be small and have a low mass (both of which increase the surface area to mass ratio), they tend to fossilize infrequently and poorly compared to the larger, heavier-boned terrestrial species they share habitat with. Fossils of flying animals tend to be confined to exceptional fossil deposits formed under highly specific circumstances, resulting in a generally poor fossil record, and a particular lack of transitional forms. Furthermore, as fossils do not preserve behavior or muscle, it can be difficult to discriminate between a poor flyer and a good glider.

Insects were the first to evolve flight, approximately 350 million years ago. The developmental origin of the insect wing remains in dispute, as does the purpose prior to true flight. One suggestion is that wings initially were used to catch the wind for small insects that live on the surface of the water, while another is that they functioned in parachuting, then gliding, then flight for originally arboreal insects.

Pterosaurs were the next to evolve flight, approximately 200 million years ago. These reptiles were close relatives of the dinosaurs (and sometimes mistakenly considered dinosaurs by laymen), and reached enormous sizes, with some of the last forms being the largest flying animals ever to inhabit the Earth, having wingspans of over 9.1 m (30 ft). However, they spanned a large range of sizes, down to a 250 mm (10 in) wingspan in *Nemicolopterus*.

Birds have an extensive fossil record, along with many forms documenting both their evolution from small theropod dinosaurs and the numerous bird-like forms of theropod which did not survive the mass extinction at the end of the Cretaceous. Indeed, *Archaeopteryx* is arguably the most famous transitional fossil in the world, both due to its mix of reptilian and avian anatomy and the luck of being discovered only two years after Darwin's publication of *On the Origin of Species*. However, the ecology and this transition is considerably more contentious, with various scientists supporting either a "trees down" origin (in which an arboreal ancestor evolved gliding, then flight) or a "ground up" origin (in which a fast-running terrestrial ancestor used wings for a speed boost and to help catch prey).

Bats are the most recent to evolve (about 60 million years ago), most likely from a gliding ancestor, though their poor fossil record has hindered more detailed study.

Only a few animals are known to have specialised in soaring: the larger of the extinct pterosaurs, and some large birds. Powered flight is very energetically expensive for large animals, but for soaring their size is an advantage, as it allows them a low wing loading, that is a large wing areas relative to their weight, which maximizes lift. Soaring is very energetically efficient.

Biomechanics

Gliding and Parachuting

During a free-fall with no aerodynamic forces, the object accelerates due to gravity, resulting in increasing velocity as the object descends. During parachuting, animals use the aerodynamic forces on their body to counteract the force or gravity. Any object moving through air experiences a drag force that is proportion to surface area and to velocity squared, and this force will partially counter the force of gravity, slowing the animal's descent to a safer speed. If this drag is oriented at an angle to the vertical, the animal's trajectory will gradually become more horizontal, and it will cover horizontal as well as vertical distance. Smaller adjustments can allow turning or other maneuvers. This can allow a parachuting animal to move from a high location on one tree to a lower location on another tree nearby.

During gliding, lift plays an increased role. Like drag, lift is proportional to velocity squared. Gliding animals will typically leap or drop from high locations such as trees, just as in parachuting, and as gravitational acceleration increases their speed, the aerodynamic forces also increase. Because the animal can utilize lift and drag to generate greater aerodynamic force, it can glide at a shallower angle than parachuting animals, allowing it to cover greater horizontal distance in the same loss of altitude, and reach trees further away.

Powered Flight

Unlike most air vehicles, in which the objects that generate lift (wings) and thrust (engine/propeller) are separate and the wings remained fixed, flying animals use their wings to generate both lift and thrust by moving them relative to the body. This has made the flight of organisms considerably harder to understand than that of vehicles, as it involves varying speeds, angles, orientations, areas, and flow patterns over the wings.

A bird or bat flying through the air at a constant speed moves its wings up and down (usually with some fore-aft movement as well). Because the animal is in motion, there is some airflow relative to its body which, combined with the velocity of with wings, generates a faster airflow moving over the wing. This will generate lift force vector pointing forwards and upwards, and a drag force vector pointing rearwards and upwards. The upwards components of these counteract gravity, keeping the body in the air, while the forward component provides thrust to counteract both the drag from the wing and from the body as a whole. Pterosaur flight likely worked in a similar manner, though no living pterosaurs remain for study.

Insect flight is considerably different, due to their small size, rigid wings, and other anatomical differences. Turbulence and vortices play a much larger role in insect flight, making it even more complex and difficult to study than the flight of vertebrates. There are two basic aerodynamic models of insect flight. Most insects use a method that creates a spiralling leading edge vortex. Some very small insects use the fling-and-clap or Weis-Fogh mechanism in which the wings clap together above the insect's body and then fling apart. As they fling open, the air gets sucked in and creates a vortex over each wing. This bound vortex then moves across the wing and, in the clap, acts as the starting vortex for the other wing. Circulation and lift are increased, at the price of wear and tear on the wings.

Limits and Extremes

Flying/Soaring

- Largest. The largest known flying animal was formerly thought to be *Pteranodon*, a pterosaur with a wingspan of up to 7.5 metres (25 ft). However, the more recently discovered azhdarchid pterosaur *Quetzalcoatlus* is much larger, with estimates of the wingspan ranging from 9 to 12 metres (30 to 39 ft). Some other recently discovered azhdarchid pterosaur species, such as *Hatzegopteryx*, may have also wingspans of a similar size or even slightly larger. Although it is widely thought that *Quetzalcoatlus* reached the size limit of a flying animal, it should be noted that the same was once said of *Pteranodon*. The heaviest living flying animals are the kori bustard and the great bustard with males reaching 21 kilograms (46 lb). The wandering albatross has the greatest wingspan of any living flying animal at 3.63 metres (11.9 ft). Among living animals which fly over land, the Andean condor and the marabou stork have the largest wingspan at 3.2 metres (10 ft). Studies have shown that it is physically possible for flying animals to reach 18-metre (59 ft) wingspans, but there is no firm evidence that any flying animal, not even the azhdarchid pterosaurs, got that large.
- Smallest. There is no real minimum size for getting airborne. Indeed, there are many bacteria floating in the atmosphere that constitute part of the aeroplankton. However, to move about under one's own power and not be overly affected by the wind requires a certain amount of size. The smallest flying vertebrates are the bee hummingbird and the bumblebee bat, both of which may weigh less than 2 grams (0.071 oz). They are thought to represent the lower size limit for endotherm flight.
- Fastest. The fastest of all known flying animals is the peregrine falcon, which when diving travels at 300 kilometres per hour (190 mph) or faster. The fastest animal in flapping horizontal flight may be the Mexican free-tailed bat, said to attain about 160 kilometres per hour (99 mph) based on ground speed by an aircraft tracking device; that measurement does not separate any contribution from wind speed, so the observations could be caused by strong tailwinds.
- Slowest. Most flying animals need to travel forward to stay aloft. However, some creatures can stay in the same spot, known as hovering, either by rapidly flapping the wings, as do hummingbirds, hoverflies, dragonflies, and some others, or carefully using thermals, as do some birds of prey. The slowest flying non-hovering bird recorded is the American woodcock, at 8 kilometres per hour (5.0 mph).
- Highest flying. There are records of a Rüppell's vulture *Gyps rueppelli*, a large vulture, being sucked into a jet engine 11,550 metres (37,890 ft) above Côte d'Ivoire in West Africa. The animal that flies highest most regularly is the bar-headed goose *Anser indicus*, which migrates directly over the Himalayas between its nesting grounds in Tibet and its winter quarters in India. They are sometimes seen flying well above the peak of Mount Everest at 8,848 metres (29,029 ft).

Gliding/Parachuting

- Most efficient glider. This can be taken as the animal that moves most horizontal distance per metre fallen. Flying squirrels are known to glide up to 200 metres (660 ft), but have measured glide ratio of about 2. Flying fish have been observed to glide for hundreds of metres on the drafts on the edge of waves with only their initial leap from the water to provide height, but may be obtaining additional lift from wave motion. On the other hand, Albatrosses have measured lift/drag ratios of 20, and thus fall just 1 meter (foot) for every 20 in still air.

- Most maneuverable glider. Many gliding animals have some ability to turn, but which is the most maneuverable is difficult to assess. Even Paradise tree snakes, Chinese gliding frogs, and gliding ants have been observed as having considerable capacity to turn in the air.

Extant Flying and Gliding Animals

Invertebrates

Arthropods

A bee in flight.

- Insects (flying). The first of all animals to evolve flight, insects are also the only invertebrates that have evolved flight. The species are too numerous to list here. Insect flight is an active research field.
    - Gliding bristletails (gliding). Directed aerial gliding descent is found in some tropical arboreal bristletails, an ancestrally wingless sister taxa to the winged insects. The bristletails median caudal filament is important for the glide ratio and gliding control
    - Gliding ants (gliding). The flightless workers of these insects have secondarily gained some capacity to move through the air. Gliding has evolved independently in a number of arboreal ant species from the groups Cephalotini, Pseudomyrmecinae, and Formicinae (mostly *Camponotus*). All arboreal dolichoderines and non-cephalotine myrmicines except *Daceton armigerum* do not glide. Living in the rainforest canopy like many other gliders, gliding ants use their gliding to return to the trunk of the tree they live on should they fall or be knocked off a branch. Gliding was first discovered for *Cephalotes atreus* in the Peruvian rainforest. *Cephalotes atreus* can make 180 degree turns, and locate the trunk using visual cues, succeeding in landing 80% of the time. Unique among gliding animals, Cephalotini and Pseudomyrmecinae ants glide abdomen first, the Forminicae however glide in the more conventional head first manner. The following page has some good videos of gliding ants.

- Gliding immature insects. The wingless immature stages of some insect species that have wings as adults may also show a capacity to glide. These include some species of cockroach, mantid, katydid, stick insect and true bug.

- Spiders. Although typically flightless some may engage in aerial locomotion as described below.
    - Ballooning spiders (parachuting). The young of some species of spiders travel through the air by using silk draglines to catch the wind, as may some smaller species of adult spider, such the money spider family. This behavior is commonly known as "ballooning". Ballooning spiders make up part of the aeroplankton.
    - Gliding spiders (gliding). Some species of arboreal spider of the genus *Selenops* can glide back to the trunk of a tree should they fall.

Molluscs

- Flying squid (gliding). Several oceanic squids, such as the Pacific flying squid, will leap out of the water to escape predators, an adaptation similar to that of flying fish. Smaller squids will fly in shoals, and have been observed to cover distances as long as 50 metres (160 ft). Small fins towards the back of the mantle do not produce much lift, but do help stabilize the motion of flight. They exit the water by expelling water out of their funnel, indeed some squid have been observed to continue jetting water while airborne providing thrust even after leaving the water. This may make flying squid the only animals with jet-propelled aerial locomotion. The neon flying squid has been observed to glide for distances over 30 metres (100 ft), at speeds of up to 11.2 metres per second (37 ft/s) .

Vertebrates

Fish

Band-winged flying fish, with enlarged pectoral fins

- Flying fish (gliding). There are over 50 species of flying fish belonging to the family Exocoetidae. They are mostly marine fishes of small to medium size. The largest flying fish can reach lengths of 45 centimetres (18 in) but most species measure less than 30 cm (12 in) in length. They can be divided into two-winged varieties and four-winged varieties. Before the fish leaves the water it increases its speed to around 30 body lengths per second and as it breaks the surface and is freed from the drag of the water it can be traveling at around 60 kilometres per hour (37 mph). The glides are usually up to 30–50 metres (100–160 ft) in length, but some have been observed soaring for hundreds of metres using the updraft on the leading edges of waves. The fish can also make a series of glides, each time dipping the tail into the water to produce forward thrust. The longest recorded series of glides, with the fish only periodically dipping its tail in the water, was for 45 seconds (Video here). It has been suggested that the genus *Exocoetus* is on an evolutionary borderline between flight and gliding. It flaps its enlarged pectoral fins when airborne, but still seems only to glide, as there is no hint of a power stroke. It has been found that some flying fish can glide as effectively as some flying birds.

- Halfbeaks (gliding). A group related to the Exocoetidae, one or two hemirhamphid species possess enlarged pectoral fins and show true gliding flight rather than simple leaps. Marshall (1965) reports that *Euleptorhamphus viridis* can cover 50 metres (160 ft) in two separate hops.
- Freshwater butterflyfish (possibly gliding). *Pantodon buchholzi* has the ability to jump and possibly glide a short distance. It can move through the air several times the length of its body. While it does this, the fish flaps its large pectoral fins, giving it its common name. However, it is debated whether the freshwater butterfly fish can truly glide, Saidel et al. (2004) argue that it cannot.
- Freshwater hatchetfish (possibly flying). There are 9 species of freshwater hatchetfish split among 3 genera. Freshwater hatchetfish have an extremely large sternal region that is fitted with a large amount of muscle that allows it to flap its pectoral fins. They can move in a straight line over a few metres to escape predators.

Amphibians

Gliding has evolved independently in two families of tree frogs, the Old World Rhacophoridae and the New World Hylidae. Within each lineage there are a range of gliding abilities from non-gliding, to parachuting, to full gliding.

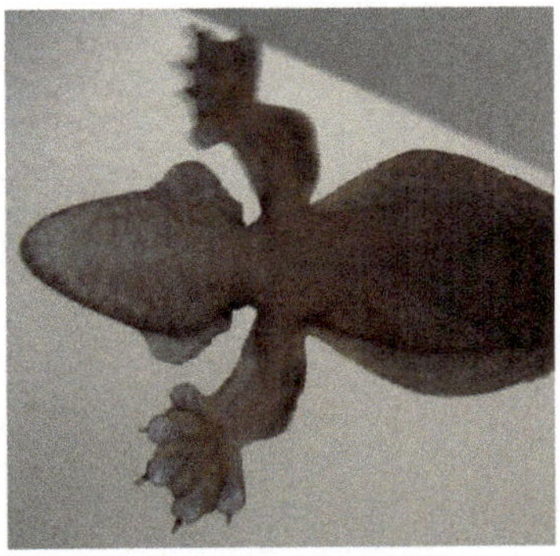

The underside of Kuhl's flying gecko Ptychozoon kuhli. Note the gliding adaptations: flaps of skin on the legs, feet, sides of the body, and on the sides of the head.

- Rhacophoridae flying frogs (gliding). A number of the Rhacophoridae, such as Wallace's flying frog (*Rhacophorus nigropalmatus*), have adaptations for gliding, the main feature being enlarged toe membranes. For example, the Malayan flying frog glides using the membranes between the toes of its limbs, and small membranes located at the heel, the base of the leg, and the forearm. Some of the frogs are quite accomplished gliders, for example, the Chinese flying frog *Rhacophorus dennysi* can maneuver in the air, making two kinds of turn, either rolling into the turn (a banked turn) or yawing into the turn (a crabbed turn).

- Hylidae flying frogs (gliding). The other frog family that contains gliders.

Reptiles

- *Draco* lizards (gliding). There are 28 species of lizard of the genus *Draco*, found in Sri Lanka, India, and Southeast Asia. They live in trees, feeding on tree ants, but nest on the forest floor. They can glide for up to 60 metres (200 ft) and over this distance they lose only 10 metres (30 ft) in height. Unusually, their patagium (gliding membrane) is supported on elongated ribs rather than the more common situation among gliding vertebrates of having the patagium attached to the limbs. When extended, the ribs form a semicircle on either side the lizard's body and can be folded to the body like a folding fan.
- Gliding Lacertids (gliding). There are two species of gliding lacertid, of the genus *Holaspis*. Found in Africa. They have fringed toes and tail sides and can flatten their bodies for gliding/parachuting.
- *Ptychozoon* flying geckos (gliding). There are six species of gliding gecko, of the genus *Ptychozoon*, from Southeast Asia. These lizards have small flaps of skin along their limbs, torso, tail, and head that catch the air and enable them to glide.
- *Lupersaurus* flying geckos (gliding). A possible sister-taxon to Ptychozoon which has similar flaps and folds and also glides.
- *Thecadactylus* flying geckos (gliding). At least some species of Thecadactylus, such as T. rapicauda, are known to glide.
- *Cosymbotus* flying gecko (gliding). Similar adaptations to Ptychozoon are found in the two species of the gecko genus *Cosymbotus*.
- *Chrysopelea* snakes (gliding). Five species of snake from Southeast Asia, Melanesia, and India. The paradise tree snake of southern Thailand, Malaysia, Borneo, Philippines, and Sulawesi is the most capable glider of those snakes studied. It glides by stretching out its body sideways and opening its ribs so the belly is concave, and by making lateral slithering movements. It can remarkably glide up to 100 metres (330 ft) and make 90 degree turns.

Birds

Birds are a successful group of flying vertebrate.

- Birds (flying, soaring) — Most of the approximately 10,000 living species can fly, (flightless birds are the exception). Bird flight is one of the most studied forms of aerial locomotion in animals. See List of soaring birds for birds that can soar as well as fly.

Mammals

- Flying phalangers or wrist-winged gliders (subfamily Petaurinae) (gliding). Possums found in Australia, and New Guinea. The gliding membranes are hardly noticeable until they jump. On jumping, the animal extends all four legs and stretches the loose folds of skin. The subfamily contains seven species. Of the six species in the genus *Petaurus*, the sugar glider and the Biak glider are the most common species. The lone species in the genus *Gymnobelideus*, Leadbeater's possum has only a vestigial gliding membrane.

- Greater glider (*Petauroides volans*) (gliding). The only species of the genus *Petauroidae* of the family Pseudocheiridae. This marsupial is found in Australia, and was originally classed with the flying phalangers, but is now recognised as separate. Its flying membrane only extends to the elbow, rather than to the wrist as in Petaurinae.

- Feather-tailed possums (family Acrobatidae) (gliding). This family of marsupials contains two genera, each with one species. The feathertail glider (*Acrobates pygmaeus*), found in Australia is the size of a very small mouse and is the smallest mammalian glider. The feathertail possum (*Distoechurus pennatus*) is found in New Guinea, but does not glide. Both species have a stiff-haired feather-like tail.

Townsends's big-eared bat, (Corynorhinus townsendii) displaying the "hand wing"

- Bats (flying). There are approximately 1,240 bat species, representing about 20% of all classified mammal species.

- Flying squirrels (subfamily Petauristinae) (gliding). There are 43 species divided between 14 genera of flying squirrel. Flying squirrels are found almost worldwide in tropical (Southeast Asia, India, and Sri Lanka), temperate, and even Arctic environments. They tend to be nocturnal. When a flying squirrel wishes to cross to a tree that is further away than the distance possible by jumping, it extends the cartilage spur on its elbow or wrist. This opens out the flap of furry skin (the patagium) that stretches from its wrist to its ankle. It glides spread-eagle and with its tail fluffed out like a parachute, and grips the tree with its claws when it lands. Flying squirrels have been reported to glide over 200 metres (660 ft).

- Anomalure or scaly-tailed flying squirrels (family Anomaluridae) (gliding). These brightly coloured African rodents are not squirrels but have evolved to a resemble flying squirrels by convergent evolution. There are seven species, divided in three genera. All but one species have gliding membranes between their front and hind legs. The genus *Idiurus* contains two particularly small species known as flying mice, but similarly they are not true mice.

- Colugos or "flying lemurs" (order Dermoptera) (gliding). There are two species of colugo. Despite their common name, colugos are not lemurs; true lemurs are primates. Molecular evidence suggests that colugos are a sister group to primates; however, some mammalogists suggest they are a sister group to bats. Found in Southeast Asia, the colugo is probably the mammal most adapted for gliding, with a patagium that is as large as geometrically possible. They can glide as far as 70 metres (230 ft) with minimal loss of height.
- Sifaka and possibly some other primates (possible limited gliding/parachuting). A number of primates have been suggested to have adaptations that allow limited gliding and/or parachuting: sifakas, indris, galagos and saki monkeys. Most notably, the sifaka, a type of lemur, has thick hairs on its forearms that have been argued to provide drag, and a small membrane under its arms that has been suggested to provide lift by having aerofoil properties.
- Cats and maybe others. (very limited parachuting). If they fall cats spread their bodies to maximize drag, a very limited form of parachuting. Cats have an innate 'righting reflex' that allows them to rotate their bodies so they fall feet first. Some other animals may show similar very limited parachuting. There are also anecdotal accounts of less limited parachuting, or even semi-gliding, in palm civets.

Extinct Flying and Gliding Animals

Pterosaurs included the largest known flying animals

Reptiles

- Extinct reptiles similar to *Draco* (gliding). There are a number of unrelated extinct lizard-like reptiles with similar "wings" to the *Draco* lizards. *Icarosaurus, Coelurosauravus, Weigeltisaurus, Mecistotrachelos,* and *Kuehneosaurus*. The largest of these, *Kuehneosaurus*, has a wingspan of 30 centimetres (12 in), and was estimated to be able to glide about 30 metres (100 ft).
- Sharovipterygidae (gliding). These strange reptiles from the Upper Triassic of Kyrgyzstan and Poland unusually had a membrane on their elongated hind limbs, extending their otherwise normal, flying-squirrel-like patagia significantly. The forelimbs are in contrast much smaller.

- *Longisquama insignis* (possibly gliding/parachuting). This small reptile may have had long paired feather-like scales on its back, however it has been more recently argued that the scales form just a single dorsal frill. If paired, they may have been used for parachuting. "Everything you can make out is consistent with it being a small, tree-living, gliding animal, which is precisely the thing you'd expect birds to evolve out of," says Larry Martin, senior curator at the Natural History Museum at the University of Kansas.
- Pterosaurs (flying). Pterosaurs were the first flying vertebrates, and are generally agreed to have been sophisticated flyers. They had large wings formed by a patagium stretching from the torso to a dramatically lengthened fourth finger. There were hundreds of species, most of which are thought to have been intermittent flappers, and many soarers. The largest known flying animals are pterosaurs.
- *Hypuronector* (gliding). This bizarre drepanosaur displays limb proportions, particularly the elongated forelimbs, that are consistent with a flying or gliding animal with patagia.

Non-avian dinosaurs

- Theropods (gliding/flying). There were several species of theropod dinosaur thought to be capable of gliding or flying, that are not classified as birds (though they are closely related). Some species (*Microraptor gui*, *Microraptor zhaoianus*, *Cryptovolans pauli*, and *Changyuraptor*) have been found that were fully feathered on all four limbs, giving them four 'wings' that they are believed to have used for gliding or flying. One species, *Deinonychus antirrhopus*, may display partial volancy, with the young being capable of flight while the adults are flightless, a characteristic also seen in some modern birds like the Horned coot and the Flying steamer duck.
- *Yi* is unique among gliding dinosaurs for the development of membranous wings, unlike the feathered airfoils of other theropods. Much like modern anomalures it developed a bony rod to help support the wing, albeit on the wrist and not the elbow.

Fish

- Thoracopteridae (gliding) is a lineage of Triassic flying fish-like Perleidiformes, having converted their pectoral and pelvic fins into broad wings very similar to those of their modern counterparts. The Ladinian genus *Potanichthys* is the oldest member of this clade, as well as the earliest aerial vertebrate known, suggesting that these fish began exploring aerial niches soon after the Permian-Triassic extinction event.

Mammals

Volaticotherids predate bats as mammalian aeronauts by at least 110 million years

- *Volaticotherium antiquum* (gliding). The earliest known flying or gliding mammal. This squirrel-sized animal belonged to a now extinct ancestral line and was not related to modern day flying or gliding mammals, such as bats or gliding marsupials. It lived around 164 million years ago and used a fur-covered skin membrane to glide through the air. The closely related *Argentoconodon* is also thought to have been able to glide, based on postcranial similarities; it lived around 165 million years ago.
- Several species of extinct bat have been found, like *Icaronycteris*, *Palaeochiropteryx*, and *Onychonycteris*.
- A gliding metatherian (possibly a marsupial) is known from the Paleocene of Itaboraí, Brazil.

References

- Alexander, David E. Nature's Flyers: Birds, Insects, and the Biomechanics of Flight. 2002(hardcover) and 2004(paperback). Baltimore: The Johns Hopkins University Press. ISBN 0-8018-6756-8.
- Brooke, Michael and Tim Birkhead (editors). The Cambridge Encyclopedia of Ornithology. 1991. Cambridge: Cambridge University Press. ISBN 0-521-36205-9.
- Campbell, Bruce, and Elizabeth Lack (editors). A Dictionary of Birds. 1985. Calton: T&A D Poyse. ISBN 0-85661-039-9.
- Strahan, the Australian Museum (1983). Ronald, ed. Complete Book of Australian Mammals: The National Photographic Index of Australian Wildlife (1 ed.). Sydney: Angus & Robertson. ISBN 0207144540.
- Welch, John (1999). Van Sickle's Modern Airmanship. City: McGraw-Hill Professional. pp. 856–858. ISBN 0-07-069633-0. There are four main kinds of lift which the soaring pilot may use....
- Triplehorn, Charles A.; Johnson Norman F. (2005). Borror and DeLong's introduction to the study of insects (7th ed.). Thomson Brooks/Cole. ISBN 0-03-096835-6
- del Hoyo, Josep; Elliott, Andrew; Sargatal, Jordi, eds. (1992), Handbook of the Birds of the World, Volume 1: Ostrich to Ducks, Barcelona: Lynx Edicions, ISBN 84-87334-10-5
- del Hoyo, Josep; Elliott, Andrew; Sargatal, Jordi, eds. (1997), Handbook of the Birds of the World, Volume 4: Sandgrouse to Cuckoos, Barcelona: Lynx Edicions, ISBN 84-87334-22-9
- del Hoyo, Josep; Elliott, Andrew; Christie, David, eds. (2004), Handbook of the Birds of the World, Volume 9: Cotingas to Pipits and Wagtails, Barcelona: Lynx Edicions, ISBN 84-87334-69-5
- Ehrlich, Paul R.; Dobkin, Darryl A.; Wheye, Darryl; Pimm, Stuart L. (1994), The Birdwatcher's Handbook, Oxford University Press, ISBN 0-19-858407-5
- Jenni, Lukas; Winkler, Raffael (1994), Moult and Ageing of European Passerines, London: Academic Press, ISBN 0-12-384150-X
- Corlett, Richard T.; Primack, Richard B. (2011). Tropical rain forests : an ecological and biogeographical comparison (2nd ed.). Chichester: Wiley-Blackwell. pp. 197, 200. ISBN 978-1444332551.
- van der Beld, John — "Nature of Australia — A portrait of the island continent", co-published by William Collins Pty. Ltd. and ABC Enterprises for the Australian Broadcasting Corporation, Sydney, 1988 (revised edition 1992), ISBN 0-7333-0241-6
- Russell, Rupert — "Spotlight on Possums", published by University of Queensland Press, St. Lucia, Queensland, 1980, ISBN 0-7022-1478-7
- Troughton, Ellis — "Furred Animals of Australia", published by Angus and Robertson (Publishers) Pty. Ltd, Sydney, in 1941 (revised edition 1973), ISBN 0-207-12256-3

- Morcombe, Michael & Irene — “Mammals of Australia”, published by Australian Universities Press Pty. Ltd, Sydney, 1974, ISBN 0-7249-0017-9
- Ride, W. D. L. — “A Guide to the Native Mammals of Australia”, published by Oxford University Press, Melbourne, 1970, ISBN 0 19 550252 3
- Serventy, Vincent — “Wildlife of Australia”, published by Thomas Nelson (Australia) Ltd., Melbourne, 1968 (revised edition 1977), ISBN 0-17-005168-4

Permissions

All chapters in this book are published with permission under the Creative Commons Attribution Share Alike License or equivalent. Every chapter published in this book has been scrutinized by our experts. Their significance has been extensively debated. The topics covered herein carry significant information for a comprehensive understanding. They may even be implemented as practical applications or may be referred to as a beginning point for further studies.

We would like to thank the editorial team for lending their expertise to make the book truly unique. They have played a crucial role in the development of this book. Without their invaluable contributions this book wouldn't have been possible. They have made vital efforts to compile up to date information on the varied aspects of this subject to make this book a valuable addition to the collection of many professionals and students.

This book was conceptualized with the vision of imparting up-to-date and integrated information in this field. To ensure the same, a matchless editorial board was set up. Every individual on the board went through rigorous rounds of assessment to prove their worth. After which they invested a large part of their time researching and compiling the most relevant data for our readers.

The editorial board has been involved in producing this book since its inception. They have spent rigorous hours researching and exploring the diverse topics which have resulted in the successful publishing of this book. They have passed on their knowledge of decades through this book. To expedite this challenging task, the publisher supported the team at every step. A small team of assistant editors was also appointed to further simplify the editing procedure and attain best results for the readers.

Apart from the editorial board, the designing team has also invested a significant amount of their time in understanding the subject and creating the most relevant covers. They scrutinized every image to scout for the most suitable representation of the subject and create an appropriate cover for the book.

The publishing team has been an ardent support to the editorial, designing and production team. Their endless efforts to recruit the best for this project, has resulted in the accomplishment of this book. They are a veteran in the field of academics and their pool of knowledge is as vast as their experience in printing. Their expertise and guidance has proved useful at every step. Their uncompromising quality standards have made this book an exceptional effort. Their encouragement from time to time has been an inspiration for everyone.

The publisher and the editorial board hope that this book will prove to be a valuable piece of knowledge for students, practitioners and scholars across the globe.

Index